LEARNING WITH LIGHT AND SHADOWS

MEDIA PERFORMANCE HISTORIES

The series MEDIA PERFORMANCE HISTORIES is part of the TECHNE collection directed by Dániel Margócsy and Koen Vermeir:

TECHNE
KNOWLEDGE, TECHNIQUE, AND MATERIAL CULTURE

8

Learning with Light and Shadows

Educational Lantern and Film Projection, 1860–1990

Edited by

NELLEKE TEUGHELS AND KAAT WILS

BREPOLS

This book is part of a series of publications within the framework of 'B-magic. The Magic Lantern and its Cultural Impact as Visual Mass Medium in Belgium (1830-1940)', a project funded by Fonds Wetenschappelijk Onderzoek, Vlaanderen – FWO and Fonds de la Recherche Scientifique – FNRS under the Excellence of Science (EOS) project number 30802346.

D/2022/0095/188
ISBN 978-2-503-59904-5
eISBN 978-2-503-59905-2
DOI 10.1484/M.TECHNE-EB.5.127928

ISSN 2736-7452
eISSN 2736-7460

Printed in the EU on acid-free paper.

Contents

List of Illustrations 7

List of Contributors 13

Acknowledgements 15

Introduction 17
Nelleke Teughels Kaat Wils

Introducing Light Projection in Education

The Emergence of the Projected Image as a Teaching Tool in Higher Education (1860–1914) 27
Frank Kessler Sabine Lenk

Taking the University to the People. The Role of Lantern Lectures in Extramural Adult Education in Early Twentieth-Century Brussels and Antwerp 51
Margo Buelens-Terryn

The Photographic Turn in Visual Teaching Aids: Films and Slides for Schools in the Netherlands, 1911–1926 77
Jamilla Notebaard Nico de Klerk

Agents of Change

'Deep and Lasting Traces'. How and Why Belgian Teachers Integrated the Optical Lantern in their Teaching (1895–1940) 101
Wouter Egelmeers

Progressive Education and Early Uses of Film in Swiss Schools 123
Audrey Hostettler

Objects and Spaces of Change

Teachers' Agency and the Introduction of New Materialities of Schooling: The Projection Lantern and Classroom Transformations in Antwerp Municipal Schools, *c.* 1900–1940 145
Nelleke Teughels

Casting Long Shadows on the Teaching of Experimental Physics: The Projection Techniques of Physicist Robert Wichard Pohl (1884–1976) 169
Michael Markert

DIY versus Ditmar 1006: The Economics, Institutional Politics, and Media Ecology of Classroom Projectors Made in 1950s Austria 197
Joachim Schätz

The Political Made Visual

Lantern Slides in Geography Lessons: Imperial Visual Education for Children in the British Colonial-Era 219
Sabrina Meneghini

Complex Associations: On the Emotional Impact of Educational Film in the German Democratic Republic (1950–1990) 245
Kerrin von Engelhardt

List of Illustrations

Fig. 1.1. 'Hippopotamus', possibly Carpenter & Westley, *c.* 1850. Robert Vrielynck Collection, courtesy of MuHKA. 32

Fig. 1.2. 'Le Coquelicot', Léon Roup, *c.* 1925–1926. Fonds Léon Roup, courtesy of Archives, patrimoine et réserve précieuse, Université libre de Bruxelles. 32

Fig. 1.3. Scan of a positive for a lecture in physiology, probably by Paul Héger, *c.* 1895–1899. Massart Collection, courtesy of Archives, patrimoine et réserve précieuse, Université libre de Bruxelles. 42

Fig. 1.4. Three busts, self-made slides in the collection of Alfred A. Schmidt, professor of art history in Fribourg, possibly 1960s. Courtesy of Diathèque, Section d'histoire de l'art, Université de Lausanne. 47

Fig. 2.1. Example of an announcement of lantern lectures by Brussels popular universities. *Le Peuple*, 19 November 1904, p. 3. 56

Table. 2.1. The number of unique lantern lectures in extramural education mentioned in the newspapers that took place in Brussels and Antwerp (1902–1904 and 1922–1924). Author's database. 56

Table 2.2. The number of unique extramural education lantern lectures mentioned in the newspapers that took place in Brussels and its communes (1902–1904 and 1922–1924), and the official population numbers (according to general censuses, 1900 and 1920). The municipalities in which lantern activity grew between the first and the second sample are shown in italics. Author's database; Vrielinck, Sven, *De territoriale indeling van België (1795–1963). Bestuursgeografisch en statistisch repertorium van de gemeenten en de supracommunale eenheden (administratief en gerechtelijk). Met de officiële uitslagen van de volkstellingen*, 3 vols (Leuven: Universitaire Pers Leuven, 2000), III, pp. 1667–1777. 62

Fig. 2.2. Locations of the subsections of the Catholic Flemish University held in Brussels and Antwerp (1922–1924). Author's database. 63

Table 2.3. The number of unique extramural education lantern lectures mentioned in the newspapers that took place in Antwerp and its suburban municipalities (1902–1904 and 1922–1924), and the official population numbers (according to general censuses, 1900 and 1920). The municipalities in which lantern activity grew between the first and the second sample are shown in italics. Author's database; Vrielinck, *De territoriale indeling van België (1795–1963)*, pp. 1669–1735. 64

Table 2.4. Themes discussed during extramural adult education lantern lectures held in Brussels and Antwerp (1902–1904). Author's database. 72

Table 2.5. Themes discussed during extramural adult education lantern lectures held in Brussels and Antwerp (1922–1924). Author's database. 73

Fig. 3.1. Logo of the Lantern Slide Association on a newsletter circulated in 1911. Internationaal Instituut voor Sociale Geschiedenis. Documentatiecollectie Cultuur Nederland. Doos 12.8. Cultuur (V-Z) folder 8. Vereeniging tot het houden van voordrachten met lichtbeelden. 80

Fig. 3.2. Titlepage of David van Staveren's brochure on cinema and education (1919). David van Staveren, *De Bioscoop en het Onderwijs* (Leiden: A. W. Sijthoff's Uitgeversmaatschappij, 1919.) 89

Fig. 3.3. Royal Institute for the Tropics in Amsterdam (Mauritskade), 1936. Royal Institute for the Tropics: <https://www.kit.nl/nl/over-ons/geschiedenis/#media-0-6017>. 91

Fig. 3.4. Amsterdam Lyceum, site of the didactic experiments of Révész and Hazewinkel Image via:<https://archief.amsterdam/beeldbank/detail/abce87bb-8603-df67–64a8–54b9b91e50b0/media/703c42c4–94b9–9bf3-ffa3-c557ca946a5c?mode= detail&view= horizontal&q = Amsterdams%20Lyceum&rows= 1&page= 4>. 96

Fig. 4.1. Six brightly coloured slides from the series 'Relief'. Archive of Sint-Ursula Instituut, Onze-Lieve-Vrouw-Waver. 108

Fig. 4.2. 'Convertisseurs Bessemer et marteau pilon', slide 36 of the slide series 'Metaalnijverheid'. Archive of Heilig Graf secondary school, Turnhout. 109

Fig. 4.3. 'Slide 2'. Like the other drawn slides in this series, Wagner drew this slide of a 'moorland in the land of the Eburones' himself. He based it on a picture taken in the Belgian province of Limburg that was published in Massart, J., *Pour la Protection de la nature en Belgique*. The slide was reproduced in Wagner, R., *Nieuwe methode voor de behandeling der Belgische geschiedenis. 1ste tijdperk: Oud België. Platenalbum voor de leerlingen* (Antwerp: Krick, 1924), n.p. 113

Fig. 4.4. Three slides from the series 'Relief' highlighting specific geological layers in pink ink. Archive of Sint-Ursula Instituut, Onze-Lieve-Vrouw-Waver. 118

Fig. 4.5. 'Four à bassin', slide 8 of a series on the production of glass published by Mazo. Slide 13 of the slide series 'Glasnijverheid'. Archive of Heilig Graf secondary school, Turnhout. 120

Fig. 5.1. On the set of the SAFU film *Schleuse* (*Lock*, 1931). ETH-Bibliothek Zürich, Bildarchiv / Photographisches Institut der ETH Zürich / PI_31-A-0002 / CC BY-SA 4.0. 137

Fig. 5.2. Poster made by André Ehrler about the film *The Yellow Cruise*, 1939. AEG Archives privées 95.2.5.18. 138

Fig. 5.3. Poster made by André Ehrler about the film *The Yellow Cruise*, 1939. AEG Archives privées 95.2.5.18. 139

Fig. 5.4. Still from the SAFU film *Die Lachmöwe*. Lichtspiel Kinemathek Bern. 141

Fig. 6.1. Blueprint for the construction of a projection screen, a projector stand and a column for a bust for the boys' school in Napelsstraat, Antwerp. Design by the Antwerp city buildings service, 1927. Felixarchief Antwerpen (FA) 697#5829. 147

Fig. 6.2. Ground plan of the school in Boulevard de Hainaut, Brussels, considered a model example of an urban school by Felix Narjoux. Narjoux, Félix, *Les écoles publiques. Construction et installation en Belgique et Hollande. Documents officiels services intérieurs et extérieurs — Bâtiments scolaires — Mobiliers scolaires — Services Annexes* (Paris: Ve A. Morel et Cie, 1878), p. 128. 156

Fig. 6.3. Drawing of a model classroom by Felix Narjoux, indicating the ideal organisation of classroom space and the ideal dimensions of windows, furniture and the spaces in between. Narjoux, Félix, *Les écoles publiques. Construction et installation en Belgique et Hollande. Documents officiels services intérieurs et extérieurs — Bâtiments scolaires — Mobiliers scolaires — Services Annexes* (Paris: Ve A. Morel et Cie, 1878), p. 69. 158

Fig. 6.4. Plans of the new rear annex, Institute for Higher Education for Girls, Lange Gasthuisstraat, Antwerp, including a projection room on the first floor (1911). FA 595#311 161

Fig. 6.5. The film and slide projectors exhibited at the Antwerp Exposition of modern teaching aids in 1922. FA FOTO-FO-#2316. 162

Fig. 7.1. Robert Wichard Pohl during his lecture in 1952, film still. (Fritz Lüty, *[Bunsenstraße 9]*, 16 mm film (Göttingen, 1952). Family archive Pohl, Göttingen. 171

Fig. 7.2. Pohl's 'streamline apparatus (Stromlinienapparat)' in a catalogue of company Spindler & Hoyer. Spindler & Hoyer, Liste 52: Zwei Apparate zur Hydro- und Aerodynamik (Göttingen, n. d.), p. 4 <http://vlp.mpiwg-berlin.mpg.de/references?id= lit18183> [accessed 5 August 2020]. 173

Fig. 7.3. Portrait of the four new professors in 1923. From left to right M. Reich, M. Born, J. Franck, and R. W. Pohl. Family archive Pohl, Göttingen. 175

Fig. 7.4. Snapshot of Pohl and his assistant in the pre-1926 Göttingen lecture hall adjusting one of the demonstration experiments (source unknown; the picture is mentioned as an anonymous gift in a letter of Pohl to his wife in 1931). Letter of Robert Wichard Pohl to his wife Tussa, February 17 1931. Family archive Pohl, Göttingen. 176

Fig. 7.5. Sperber (left) and Pohl (right) during an action-equals-reaction demonstration in Pohl's textbook. Robert Wichard Pohl, *Einführung in die Mechanik, Akustik und Wärmelehre*, 3. und 4., umgearb. und erg. Aufl. (Berlin [u.a.]: Springer, 1941), p. 20. 178

Fig. 7.6. A 'German Lantern'. Wright, Lewis, Optical Projection: A Treatise on the Use of the Lantern in Exhibition and Scientific Demonstration, ed. by Russel S. Wright, 4th edition (London: Longmans, Green, 1906), p. 154. 179

Fig. 7.7. A projecting device for demonstrations in lectures on chemistry. Stock, Alfred, 'Ein Projektionsapparat für die Chemievorlesung', Zeitschrift für Elektrochemie und Angewandte Physikalische Chemie, 17.23 (1911), p. 997 (with kind permission of Wiley-VCH and the Deutsche Bunsen-Gesellschaft für physikalische Chemie e.V.). 180

Fig. 7.8. Projecting device at Pohl's lecture hall around 1930. Family archive Pohl, Göttingen. 182

Fig. 7.9. Physics lecture hall at Göttingen as opened in 1905. Göttinger Vereinigung zur Förderung der angewandten Physik und Mathematik, Die Physikalischen Institute der Universität Göttingen: Festschrift im Anschlusse an die Einweihung der Neubauten am 9. Dezember 1905 (Leipzig: Teubner, 1906), p. 61. 183

Fig. 7.10. The lecture hall after rebuilding in 1926. Family archive Pohl, Göttingen. 184

Fig. 7.11. Pohl's Experimental Table as sold by Spindler & Hoyer. Spindler & Hoyer, *Liste 50*, p. 1. 185

Fig. 7.12. Picture of the complete apparatus taken by an earlier mechanic for his lecture preparation. Teaching collection of the I. Physikalisches Institut, Göttingen University, unknown photographer. 187

Fig. 7.13. Pohl's electrostatic generator with translucent isolators particularly suited for shadow projection. Teaching collection of the I. Physikalisches Institut, Göttingen University, photographers: Marie-Luise Ahlig & Lara Siegers. 188

Fig. 7.14. Professor Ansgar Reiners and lecture-mechanic Michael Hillmann performing the demonstration from Fig. 5 during the introductory lecture on experimental physics in winter semester 2020/2021. Still from the streamed and recorded lecture, 18.11.2020. 189

Fig. 7.15. The new physics lecture hall of the Technische Universität Berlin in 1930. P.-L. Flouquet, 'Le Mobilier Moderne à l'École', BATIR, No. 16, 1934, 598–600 (p. 600). The University's name is neither mentioned in the caption nor the article itself. However, a comparison of that image with a photograph from another source (F 8106, TU Architekturmuseum) indicates the location. 191

Fig. 7.16. Pohl presenting stationary waves using an animated rubber band. Filmstill, Lüty 1952, copy from the Family archive Pohl, Göttingen. 193

Fig. 8.1. The display of the voltmeter built into the Ditmar 1006 projector, with the attached series resistor on the left. — Josef Sikora, *Der 16mm-Schmalfilmprojektor 'Ditmar 1006'* (Vienna: Bundesstaatliche Hauptstelle für Lichtbild und Bildungsfilm, 1950), slide 10 of 15. Collection of Pädagogische Hochschule Steiermark. 205

Fig. 8.2. The Ditmar 1006 projector together with a slide projector on a projection table, as recommended by SHB-Film's pedagogues. — Josef Sikora, *Der 16mm-Schmalfilmprojektor 'Ditmar 1006'* (Vienna: Bundesstaatliche Hauptstelle für Lichtbild und Bildungsfilm, 1950), slide 15 of 15. Collection of Pädagogische Hochschule Steiermark. 207

Fig. 8.3. How to carefully rewind a reel by hand. — Josef Sikora, *Der 16mm-Schmalfilmprojektor 'Ditmar 1006'* (Vienna: Bundesstaatliche Hauptstelle für Lichtbild und Bildungsfilm, 1950), slide 14 of 15. Collection of Pädagogische Hochschule Steiermark. 215

Fig. 9.1. Alfred Hugh Fisher, The Hangi — greeting between Maggie Papekura and a Maori woman, 1910, New Zealand, photographic print. GBR/0115/RCS/Fisher 27/7016. Reproduced by kind permission of the Syndics of Cambridge University Library. 229

Fig. 9.2. Alfred Hugh Fisher, Chief Tikitere, and Maoris, 1910, New Zealand, photographic print. GBR/0115/RCS/Fisher 26/6997. Reproduced by kind permission of the Syndics of Cambridge University Library. 231

Fig. 9.3. Alfred Hugh Fisher, Chief Tikitere, and Maoris [Chief Tikitere], 1910, New Zealand, photographic print. GBR/0115/RCS/Fisher 26/6996. Reproduced by kind permission of the Syndics of Cambridge University Library. 232

Fig. 9.4. Alfred Hugh Fisher, Hilarion: the Belvedere, 1908, Cyprus, photographic print. GBR/0115/RCS/Fisher 8/1179. Reproduced by kind permission of the Syndics of Cambridge University Library. 234

Fig. 9.5. Alfred Hugh Fisher, Jack Burman (a typical Burman), 1907–12–1908–01, Myanmar, oil on board. GBR/0115/RCS/RCMS 10/5/359. Reproduced by kind permission of the Syndics of Cambridge University Library. 239

Fig. 9.6. Alfred Hugh Fisher, Cutting Black Concord Grapes [Cutting "Black Concord" Grapes: R. Thompson's Farm, Grantham, Ontario], 1908–09, Canada, photographic print. GBR/0115/RCS/Fisher 11/2421. Reproduced by kind permission of the Syndics of Cambridge University Library. 240

Fig. 9.7. Alfred Hugh Fisher, Cutting Black Concord Grapes, 1908–09, Canada, photographic print. GBR/0115/RCS/Fisher 11/2422. Reproduced by kind permission of the Syndics of Cambridge University Library. 241

Fig. 10.1. Film still from *Lernen für die Zukunft* (GDR, 1980): Students in class. 260

Fig. 10.2. Film still from *Lernen für die Zukunft* (GDR, 1980): Young adults dancing. 260

Fig. 10.3. Film still from *Lernen für die Zukunft* (GDR, 1980): At the production site. 260

Fig. 10.4. Film still from *Lernen für die Zukunft* (GDR, 1980): Teacher and students in class. 261

Fig. 10.5. Film still from *Lernen für die Zukunft* (GDR, 1980): Young adults at leisure. 261

Fig. 10.6. Film still from *Lernen für die Zukunft* (GDR, 1980): Labourer programming a machine. 261

Fig. 10.7. Film still from *Lernen für die Zukunft* (GDR, 1980): In the control centre. 262

Fig. 10.8. Film still from *Lernen für die Zukunft* (GDR, 1980): Automated production. 262

Fig. 10.9. Film still from *Lernen für die Zukunft* (GDR, 1980): Cargo transport wagon. 262

Fig. 10.10. Film still from *Lernen für die Zukunft* (GDR, 1980): Young woman dancing. 265

Fig. 10.11. Film still from *Lernen für die Zukunft* (GDR, 1980): Young woman soldering. 265

Fig. 10.12. Film still from *Lernen für die Zukunft* (GDR, 1980): Hands at punched paper tape. 265

(All Fig. 10: Copyright: German Federal Archive).

List of Contributors

Audrey Hostettler is a doctoral candidate at the University of Lausanne (Switzerland). Her dissertation investigates the early uses of film in Swiss classrooms.

Frank Kessler is a professor of Media History at Utrecht University. His research interests lie in the field of early cinema and nineteenth-century media and visual culture.

Jamilla Notebaard is a historian of science who works as a PhD-student in the project 'Projecting Knowledge' at Utrecht University, where she studies the affordances of the lantern in academia.

Joachim Schätz works as university assistant at the University of Vienna's Department of Theater, Film and Media Studies. He leads the FWF-funded research project 'Educational film practice in Austria'.

Kaat Wils is professor of modern European cultural history at KU Leuven. Her research is situated in the fields of the history of education, gender history, the history of the human sciences and the history of healing and medicine.

Kerrin von Engelhardt is a Research Assistant at the Department of History of Education, Humboldt University Berlin and leads the case study 'The Myth of Scientific Neutrality. The School Educational Film in the Cold War'.

Margo Buelens-Terryn (she/her) is a PhD candidate in the B-magic project at the University of Antwerp. Her research focuses on the lecture circuit with the projection lantern in Antwerp and Brussels *c.* 1900-*c.* 1920.

Michael Markert is a historian of science focused on the development and use of university collections in the natural sciences. He currently works at the Thüringer Universitäts- und Landesbibliothek.

Nelleke Teughels is a historian at Geheugen Collectief, a project agency specialized in public history. Her research ranges from nineteenth- and twentieth-century shopping culture and consumption history over the history of world exhibitions to the use of film and lantern projection in Belgian education.

Nico de Klerk is a film historian and archivist and a postdoc researcher in the project 'Projecting knowledge: the magic lantern as a tool for mediated science communication in the Netherlands, 1880–1940', at Utrecht University.

Sabine Lenk, (University of Antwerp, Université libre de Bruxelles) is a film and media scholar. She has worked for film archives in Belgium, France, Germany, Luxembourg, UK, and the Netherlands. As one of the co-authors of the B-magic project, she conducts research on the educative role of the lantern in religious communities and spiritual circles.

Sabrina Meneghini is a researcher and curator and holds a PhD in Visual Culture from De Montfort University's Photographic History Research Centre (UK). Based at Cambridge University Library's Royal Commonwealth Society Department, she explores imperial British visual education.

Wouter Egelmeers is a PhD candidate in the B-magic project at KU Leuven. His research focuses on the impact of the optical lantern on Belgian education between 1880 and 1940.

Acknowledgements

This volume is a product of the research project 'B-magic. The Magic Lantern and Its Cultural Impact as Visual Mass Medium in Belgium (1830–1940)', funded by the Research Foundation Flanders (FWO) and the Fonds de la Recherche Scientifique (FNRS) — Excellence of Science (EOS). The book's foundations were laid in 2020 during the symposium 'Teaching Science with Light Projection: Regimes of Vision in the Classroom, 1880–1940' at the biannual conference of the European Association for the History of Science. Additional inspiration was generated during the conference 'Sound and Vision: Exploring the Role of Audio and Visual Technologies in the History of Education', organized in April 2021 in cooperation with BENGOO, the Belgian-Dutch Association for the History of Education. We would like to express our warmest gratitude to co-organizers Wouter Egelmeers, Sarah van Ruyskensvelde and Pieter Verstraete. We are equally grateful to Leen Bokken, Elisa Seghers, the members of the B-magic team and its director Kurt Vanhoutte, the members of the Research Group Cultural History since 1750 at KU Leuven and Alexander Sterkens from Brepols for their support.

NELLEKE TEUGHELS
KAAT WILS

Introduction

In one scene of the 1969 film *The Prime of Miss Jean Brodie*, taking place in mid-1930s Edinburgh, we see Maggie Smith, as the titular character and unconventional teacher in an all girls' school, set up an optical lantern.[1] She positions the lantern stand in the middle of the classroom before carefully placing the projector on top of it, as her pupils start pouring in after recess. Once she has plugged in the apparatus, she asks two of the girls to close the curtains, and instructs another pupil to pull down a projection screen. 'The girls in the back may sit up on their desks,' she announces, before switching on the projector. As she enters the first slide in the lantern, the single beam of light conjures up an image on the screen. The pupils sit and watch, enthralled, as their teacher takes them on a virtual excursion to Rome and Egypt.

Although the projection device may have varied, for those who went to school sometime in the twentieth century, this will likely be a familiar scenario. Millions of people around the world have been instructed by way of light projection of still or moving images, captured on lantern slides, film, filmstrips or diapositive slides. They were used for teaching pupils of all ages and in various settings, from primary and secondary school children in classrooms, museums or projection halls to adults at workplaces, in convention halls, in

* This work was supported by the Research Foundation — Flanders (FWO) and the Fonds de la Recherche Scientifique (FNRS) under Grant of Excellence of Science (EOS), project number 30802346 (B-magic project 'The Magic Lantern and Its Cultural Impact as Visual Mass Medium in Belgium (1830–1940)').

1 The optical lantern, also known as the magic lantern, was an image projector that used a light source and one or more lenses to magnify and project painted, printed or photographic images from transparent glass plates (called lantern slides) onto a wall or a projection screen. The term 'magic lantern' is most often used in media history and media archaeology to refer to a light projector that was invented in the seventeenth century and for a long time was primarily used for entertainment. When, in the nineteenth century, photographic lantern slides were introduced and the mass-production of slides made them more affordable, the magic lantern became increasingly used by lecturers, scientists and teachers for instructional purposes. Many practitioners then rejected the term 'magic lantern' in favour of more 'scientific' sounding names such as 'optical lantern' or 'sciopticon'. Crompton, Dennis, Richard Franklin, and Stephen Herbert, 'Naming of Parts', in *Servants of Light. The Book of the Lantern*, ed. by Dennis Crompton, Richard Franklin, and Stephen Herbert (London: The Magic Lantern Society, 1997), pp. 4–7.

Nelleke Teughels • Geheugen Collectief, nelleke@geheugencollectief.be

Kaat Wils • KU Leuven, kaat.wils@kuleuven.be

Learning with Light and Shadows: Educational Lantern and Film Projection, 1860-1990, ed. by Nelleke Teughels and Kaat Wils, TECHNE-MPH, 8 (Turnhout, 2022), pp. 17-24
© BREPOLS PUBLISHERS 10.1484/M.TECHNE-MPH-EB.5.131492

auditoriums or cinema's. In the course of the century, new models of existing projectors and new projection technology entered those spaces of instruction, in conjunction with or as alternatives to teaching aids already in place.

Educational Light Projection: What's Under This Umbrella Term?

In the nineteenth-century, growing criticism was voiced about the exclusive textual mediation of knowledge through books and lectures. The abstract character of contemporary instruction and the passive role it assigned to the audience had to give way to a more sensorial approach to teaching that was learner-centred and based on the use of various visual teaching aids. This 'progressive education' became increasingly popular in Europe in the second half of the century and has persisted, albeit with slight modifications, throughout most of the twentieth century.[2] Didactic drawings, classroom wall charts, textbook illustrations and physical objects such as mounted animals were the first new media to be incorporated into instruction. When, in the second half of the century, approximately two hundred years after its invention, photographic slides were introduced and the mass-production of slides made them more affordable, the magic lantern became increasingly used by lecturers, scientists and teachers for instructional purposes. The first initiatives by policymakers to promote the use of light projection of still images in schools and teachers' training date from the late nineteenth century.[3] In the first decades of the twentieth century new forms of light projection, such as film, episcopes for the projection of opaque images, and filmstrips came on the market as alternatives to lantern slide projection. Each of these were promoted as holding great advantages for use in educational settings: film, of course, had the advantage of being able to show and clarify the movement of creatures, objects and phenomena. Episcopes allowed for the projection of a wide range of images that were often already available to the teacher (e.g. textbook images) or that were cheaper to acquire (such as postcards, hand drawn images, or newspaper clippings), and in any case less fragile than lantern slides. In the 1920s, filmstrips made their appearance, spooled rolls of 35 mm (sometimes also 30 mm) positive film on which twenty to fifty images were reproduced back to back in sequential order. As such, they were a lightweight, less bulky, and shatterproof substitute for lantern slides. Many of these technologies survived well into the second half of the twentieth century, which saw the introduction of new types of light projection into education. Two of those, the 35 mm slide projector and the overhead projector, can be considered direct descendants of the magic lantern. Throughout the entire period under consideration, however, instead of relying on commercial solutions,

---

2 Depaepe, Marc, *Order in Progress: Everyday Educational Practice in Primary Schools, Belgium, 1880–1970*, Studia Paedagogica, 29 (Leuven: University Press, 2000), 51; Depaepe, Marc, Frank Simon, and Angelo Van Gorp, 'The Canonization of Ovide Decroly as a "Saint" of the New Education', *History of Education Quarterly*, 43.2 (2003), 224–49.

3 Crompton, and others, 'Naming of Parts', pp. 4–7; Egelmeers, Wouter, and Nelleke Teughels, '"A Thousand Times More Interesting": Introducing the Optical Lantern into the Belgian Classroom, 1880–1920', *History of Education* (2021) <https//:doi.org/10.1080/0046760X.2021.1918271>; Ruchatz, Jens, *Licht und Wahrheit. Eine Mediumgeschichte der fotografische Projektion* (Munich: Wilhelm Fink, 2003).

educators and instructors increasingly made use of homemade or user-modified forms of instructional media to fit their everyday needs and individual pedagogical convictions.

Schools were not the only sites of instruction where these media became widely used. The market for educational light projection technology included, among others, universities, museums, organizations for adult education, and religious societies. By opting for the term 'educational light projection' we wish to include all instances in which light projection was used to inform, instruct, teach, or convince an audience about ideas, facts, skills, and morals, both inside and outside the classroom.

This collection of essays focuses on European educational light projection, from its first appearance at the end of the nineteenth century through the 1990s, when digital image projection started to gradually replace analogue film, slide and overhead projectors. The geographical limitation to Europe stems from our acknowledgment that educational light projection in itself already makes up an impossibly large field of study that needs to be narrowed down in order to add to our understanding of it. Its uses varied greatly around the world, depending on local needs, pedagogical convictions, and the political and economic situation. However, although no such thing as 'the' European use of light projection in education existed, within Europe, through the essays collected here, we can identify various transnational actors and trends.

Tracing the History of Projection Technology

Despite being viewed and remembered by millions of teachers and learners, educational projection technology constitutes an under-researched aspect of the history of education and media history. Over the last two decades, researchers in those fields have increasingly turned their attention to the introduction and proliferation of high-tech educational media in Western education, including film projectors, television, and computers.[4] However, most of the historical scholarship has focused on the priorities of and promotional efforts by the industry, or on governmental initiatives to introduce commercially produced media into the classroom.[5] As such, the dominant voices in these studies have been those of actors

4 See for example Bianchi, William, *Schools of the Air: A History of Instructional Programs on Radio in the United States* (Jefferson, N. C.: McFarland & Co, 2008); Cuban, Larry, *Teachers and Machines. The Use of Classroom Technology since 1920* (New York: Teachers College Press, 1986); Cunningham, Peter, 'Moving Images: Propaganda Film and British Education 1940–1945,' *Paedagogica Historica*, 36.1 (2000), 389–406; Orgeron, Devin, Marsha Orgeron, and Dan Streible, *Learning with the Lights Off: Educational Film in the United States* (New York: Oxford University Press, 2012); Peterson, Jennifer, '"The Five-Cent University": Educational Films and the Drive to Uplift the Cinema,' in *Education in the School of Dreams: Travelogues and Early Nonfiction Films* (Durham: Duke University Press, 2013), 101–36; Quillien, Anne, *Lumineuses projections! La projection fixe éducative* (Chasseneuil du Poitou: Réseau Canopé, 2016); and Taggart, Robert J.,'The Promise and Failure of Educational Television in a Statewide System: Delaware, 1964–1971,' *American Educational History Journal*, 34.1–2 (2007), 111–22.

5 See for example Alexander, Geoff, *Academic Films for the Classroom. A History* (Jefferson: McFarland, 2010); Bak, Meredith, 'Democracy and Discipline: Object Lessons and the Stereoscope in American Education, 1870–1920,' *Early Popular Visual Culture*, 10.2 (2012), 147–67; Fuchs, Ekhard, Anne Bruch, and Michael Annegarn-Gläss, 'Educational Films: A Historical Review of Media Innovation in Schools,' *Journal of Education Media*, 8.1 (2016), 1–13; Vignaux, Valérie, 'Le film fixe Pathéorama (1921) ou généalogie d'une invention,' *Trema*, 41 (2014), <https://journals.openedition.org/trema/3128> [accessed 3 January 2020]; Warmington, Paul,

outside of the school walls. How teachers developed their own stance towards projection technology, how their individual interests shaped their own unique pedagogical practices of media use and production, and how the materialities of schooling — school buildings, projection equipment, furniture, location — impacted the educational use of projection technology remain underexplored areas of study. Moreover, media historians and historians of education have favoured research into educational film, largely overlooking the impact of devices for still image projection in shifting norms of educational practice. This is at least partly due to the long-standing assumption that the advent of moving images would quickly erode lantern projection's popularity.[6]

However, as this collection will demonstrate, there is substantial historical evidence that still image projectors had a much greater and longer-lasting impact on educational practices than hitherto assumed: after the successful introduction of the magic lantern into the classroom during the first decades of the twentieth century, the projection of still images would stay common didactic practice up until today. In some schools, the lantern would stay in use until well into the 1950s, after which it was gradually replaced by the 35 mm slide projector.[7] Moreover, in the first decades of the twentieth century, other new educational media such as episcopes for projection of opaque images and filmstrips came on the market as alternatives to lantern slide and film projection. Each of these had its own advantages, resulting in the mixing, modifying, creating or simply ignoring of certain types of light projection technology, depending on schools' financial situation and teachers' pedagogical and didactic convictions, practical considerations and everyday needs.

This volume aims to challenge the prevalent top-down approach to the introduction of new visual technology and to question the dominant discourse that characterizes the relation of visual media technology to teachers as one of consumption. It wishes to further investigate the power relations and ideologies at play, with specific attention to the production by and participation of users.

There are, however, important scholarly precursors and precedents that inspired this volume. In 1998, the twentieth International Standing Conference of the History of Education (ISCHE) conference held in Kortrijk and Leuven (Belgium) addressed the importance of the visual in education. This resulted in a *Paedagogica Historica* Supplementary Series volume on 'The Challenge of the Visual in the History of Education', which contains valuable and innovative reflections on the significance of visual sources in the history of education.[8] Through an analysis of wall charts, school architecture, photographs, religious icons, textbook illustrations, cigarette cards, and educational film, the contributions explore

---

Angelo Van Gorp, and Ian Grosvenor, eds, 'Education in motion: producing methodologies for researching documentary film on education' *Paedagogica Historica*, special issue, 47.4 (2011); Willis, Artemis, 'Between Nonfiction Screen Practice and Nonfiction Peep Practice: The Keystone "600 set" and the Geographical Mode of Representation,' *Early Popular Visual Culture*, 13.4 (2015), 293–312.

6 Musser, Charles, *The emergence of cinema. The American screen to 1907* (Berkeley: University of California Press, 1990).

7 Teughels, Nelleke, 'Expectation versus reality: how visual media use in Belgian Catholic secondary schools was envisioned, encouraged and put into practice (c. 1900–1940),' *Paedagogica Historica* (2021) https://doi.org/10.1080/00309230.2020.185653.

8 Depaepe, Marc, and Bregt Henkens, eds, 'The Challenge of the Visual in the History of Education', *Paedagogica Historica*, supplementary series 6 (Ghent: CSHP, 2000).

theoretical and methodological approaches to visual sources and their role and meaning in the making of educational space.

Since then, many other articles, edited volumes and special issues focusing on the visual and its relevance in educational contexts or on methods of analysis of educational visual media have been published.[9] In 2005, Martin Lawn and Ian Grosvenor put the spotlight on the *Materialities of Schooling: Design, Technology, Objects, Routines*.[10] Although none of the contributions to their volume explicitly focuses on projection equipment, this work offers valuable insight into the trajectories of new artefacts and technology before and after they arrive in school, how they influence classroom routines, and how they create meaning.

Learning with the Lights Off, edited by Devin Orgeron, Marsha Orgeron, and Dan Streible, brings together essays by film scholars, historians, and archivists to demonstrate how educators, corporations, governmental actors, and independent filmmakers envisioned and used film as a medium to transform pedagogical practices.[11] It acknowledges the central place of the visual in educational practice of the current and previous century and offers historians of education as well as film historians a deeper understanding of how past and current pedagogical practices were shaped by film projection. Moreover, teachers' role in the uptake and use of this new medium runs like a thread through the entire volume.

The introduction and use of lantern slide and filmstrip projection in French schools have been the subject of an edited volume published in 2016 under supervision of Anne Quillien.[12] A year later, *Paedagogica Historica* published a special issue on 'the visual in histories of education', rightfully addressing new and more systematic ways of exploiting the potential for visual sources to add new insights into the history of education.[13]

However, in-depth explorations of the intermedial relationships between various types of light projection in the multimodal educational contexts in which they functioned, remain scarce. For example, there is clear evidence that film projection and lantern projection were sometimes used — literally — side by side in the classroom: film provided teachers with a way to analyse phenomena that were full of movement, like the eruption of a volcano, whereas lantern slides were more suitable to show shapes, colours and structures in more detail, such as stratigraphic layers or the specific physical characteristics of volcanic or magmatic rock. Nevertheless, few studies address how teachers mixed and matched projection equipment and images from various sources to meet everyday needs. Much work on how these intermedial relationships resulted from the materiality of projection equipment, and how this materiality was connected to changing views on educational purposes and school infrastructure remains to be done. Most of the existing scholarship

9 See for example, Grosvenor, Ian, Martin Lawn, and Kate Rousmaniere, eds, *Silences and Images: The Social History of the Classroom* (New York: Peter Lang, 1999); Meitzner, Ulrike, Kevin Myers, and Nick Peim, eds, *Visual History. Images of Education* (Bern: Peter Lang, 2005); Comàs Rubì, Francesca, ed., 'Photography and the History of Education,' special issue of *Educació i Història. Revista d'Historia de l'Educacío*, 15 (2010); Warmington, Paul, Angelo Van Gorp, and Ian Grosvenor, eds, 'Education in Motion: Uses of Documentary Film in Educational Research,' special issue of *Paedagogica Historica*, 47.4 (2011).

10 Lawn, Martin, and Ian Grosvenor, eds, *Materialities of schooling: design, technology, objects, routines* (Oxford: Symposium Books, 2005).

11 Orgeron and others, eds, *Learning with the Lights off*.

12 Quillien, ed., *Lumineuses projections!*

13 Dussel, Inés, and Karin Priem, 'Images and Films as Objects to Think With: A Reappraisal of Visual Studies in Histories of Education', special issue of *Paedagogica Historica*, 53.6 (2017), 641–49.

focuses on (attempts at) educational modernization, which obscures the intriguing relationship between didactic continuity and innovation in the use of projection technology. This volume wants to redress these imbalances. It offers a collection of essays that highlight how everyday demands and preferences transformed the 'ideal' instructional culture as put forward by policymakers, producers and specialized journals, into distinctive didactic practices that worked around or went beyond the pre-imposed ways of usage of visual media products. As such, it also exposes and challenges the determinist discourse on new visual technology's beneficial impact on educational practices that emphasizes the progress and modernization brought on by its implementation.

Structure of This Book

This volume is divided into four sections: 'Introducing Light Projection in Education', 'Agents of Change', 'Objects and Spaces of Change', and 'The Political Made Visual'. The first section shows how the introduction of the lantern in late nineteenth century education was a European-wide phenomenon, but it equally reveals how differently this process played out/developed in different geographical, educational and disciplinary contexts. As Frank Kessler and Sabine Lenk demonstrate in their contribution on the introduction of the lantern in higher education, shaking off the historical association of the lantern with entertainment constituted a challenge for academics, anxious as many of them were to protect the seriousness of their academic persona. Discipline-specific characteristics, such as the degree in which scientific research relied on the use of images, could however stimulate a swift introduction of the lantern into university auditoria, alongside infrastructural conditions and transnational scholarly networks. While resistance against the lantern was real within circles of higher education, this was much less the case in the field of adult education. In her contribution on early twentieth-century adult education in Belgium, Margo Buelens-Terryn demonstrates how the lantern was a much used devise in the search to reach a working class audience. High hopes were placed on this 'modern' tool, which was expected to catch the attention of those who were, after a long day of work, too tired to concentrate on a purely spoken lecture. Whether the optical lantern — and by extension visual education — effectively stimulated the process of learning remained however first and foremost a question of (dis)belief. As Jamilla Notebaard and Nico de Klerk discuss in their contribution on the position of lantern slides and films in Dutch education, the issue became a subject of scientific enquiry among scholars in pedagogy from the early 1920s onwards. In these first research experiments, the didactic use of lantern slides was more positively assessed than the use of film, while the crucial role of oral comments was stressed in both cases.

The second part of the book zooms in on exactly this crucial and often overlooked role of teachers in the use of light projection media in the classroom. In his contribution on Belgian teachers, Wouter Egelmeers analyses the different reasons that motivated teachers to include slide projection in their lessons, such as the capacity to offer an experience of immersion and the potential to stimulate students' observational skills. With these aims in mind, teachers proved to be highly creative in using and enhancing existing slides and combining them with self-made photographs or drawings. A very similar conclusion is drawn by Audrey Hostettler in her article on early uses of film in Swiss schools. Even though external pressures and constrains were higher in the more commercialized landscape of educational film in Switzerland, teachers managed to develop their own uses of film, in particular with a view to fostering their students' active engagement.

The third section demonstrates how teachers' agency and everyday classroom practice were in turn strongly shaped by teaching spaces and the materiality of educational technology. In her contribution, Nelleke Teughels demonstrates the various ways in which this required changes to the spaces of education, necessary transformations which were often overlooked by the producers, pedagogues and policymakers that were so eager to modernize instruction. Whereas teachers' role in classroom innovations has often been characterized as that of a consumer, teachers are revealed to have been the driving force behind the spread of lantern projection in Antwerp municipal schools, overcoming or working around the challenges posed by the inadequate nature of school infrastructure and budgetary constraints. Michael Markert's case study of the German physics professor Robert Wichard Pohl shows how Pohl built up a comprehensive physical environment in which his shadow projections fulfilled a pivotal role, connecting physical demonstration equipment, chalkboards, lantern slides, textbooks, lecturers and students. As such, Markert's contribution acknowledges how educational spaces are multimodal contexts and stresses the importance of investigating the intermedial relationships between each of these teaching aids. Of course, these relationships are also the result of medium-specific properties, which could provide an answer to different classroom needs. In his contribution on the Austrian classroom film projector Ditmar 1006, Joachim Schätz investigates how, in 1950s Austria, these needs were defined, by whom, and how the discussions between various stakeholders resulted in the specific materiality, maintenance and use of the Ditmar 1006.

These power relations between policymakers and educators, users and viewers, but also between viewers and the people, places and things projected onto the screen are the main focus of the final section of this book. Visual culture and the practices associated with it are shaped and influenced by the social, cultural, and political histories that define social reality at a given place and at a given time. Moreover, as Sabrina Meneghini demonstrates in her study on the Colonial Office Visual Instruction Committee's creation and use of lantern slides for instructing school children about the British colonies, they are inextricably bound up with processes of identity creation. Therefore, visual culture and visual practices can hold a lot of power, as they instruct viewers and users about social norms. Moreover, the Committee identified the projection lantern as the ideal teaching tool to construct and deliver a structured visual framework, which highlights the mediating role of projection technology and how visual regimes are shaped by the interplay of visual representation, technology and practices. Kerrin von Engelhardt's contribution focuses on this interplay as well, in an exploration of the use of emotions in GDR educational science films. GDR agents considered educational film as a valid substitute for actual experience, and linked their power of persuasion exactly to their so-called realism and objectivity. Technology was seen as a necessary medium to make the pedagogical process more rational and effective. However, in order to increase educational effectiveness, the films were soon being discussed in terms of their ability to generate emotional involvement. The emotional effect of educational film became increasingly connected to the use of audio, and music in particular.

By addressing the impact of educational contexts and methods, teachers' agency, the materiality of teaching with light projection and the power relations involved in the shaping of the instructional culture surrounding teaching with projected images, this volume wishes to move beyond the dominant view of instructional technology as a one-way route to modernization and teaching efficiency. Instead, it wishes to expose the highly personal and creative ways in which educators, in response to everyday demands and restrictions, have defied professional pressures and marketing discourses about instructional use of projection equipment. By laying bare the underlying power relations, interests and ideologies, it can also lend insight into the intertwinement between media, material culture and classroom practices.

Introducing Light Projection in Education

FRANK KESSLER
SABINE LENK

The Emergence of the Projected Image as a Teaching Tool in Higher Education (1860–1914)

Prologue

On 18 April 1900 Antonie Ewoud Jan Holwerda, who held the chair of classical archaeology at Leiden University, gave a keynote lecture at the Second Dutch Philology Convention. He announced he was not going to discuss a specific issue in his field of research, but rather talk more generally about his discipline saying, '[…] however tempting it might be to talk about a question concerning Greek art, it is virtually impossible to do so for a large audience if one is unable to produce a whole series of slides'.[1]

On 22 November of that same year, Willem Vogelsang gave his inaugural lecture as *Privaat-Docent* in Art History at the University of Amsterdam. Summarizing his lecture, he explained that he was obliged to give the audience a rather ragtag collection of art-historical remarks than he would have preferred. 'But without slides it would have been impractical to do so for a large audience […]'.[2]

* This chapter was written as part of the research project 'B-magic. The Magic Lantern and Its Cultural Impact as Visual Mass Medium in Belgium (1830–1940)'. 'B-magic' is an Excellence of Science project (EOS-contract 30802346, 2018–2023), supported by the Research Foundation Flanders (FWO) and the Fonds de la Recherche Scientifique (FNRS). It is also part of the research project 'Projecting Knowledge — The Magic Lantern as a Tool for Mediated Science Communication in the Netherlands, 1880–1940', project number VC.GW17.079 / 6214, financed by the Dutch Research Council (NWO). We would like to thank Nelleke Teughels and Kaat Wils for their very helpful suggestions. We are particularly grateful to Nathalie Blancardi for generously sharing primary source material on the Art and Archaeology department at *Université de Lausanne* with us, as well as a course listing she established on the basis of Lausanne course catalogues. Thanks are also due to Sjaak Boone for pointing us to sources on early Dutch lens makers.

1 Holwerda, Antonie Ewoud Jan, *De betekenis der archeologie voor de studie der oudheid* (Leiden: E. J. Brill, 1900), p. 7. 'Hoe verleidelijk het toch ook is over een onderwerp van Grieksche kunst te spreken, voor een grooter gehoor is dat, wanneer men niet in staat is eene gansche reeks van lichtbeelden te produceren zoo goed als ondoenlijk.' Unless stated otherwise, all translations are ours.

2 Vogelsang, Willem, *Kunstwetenschappelijke opmerkingen* (Amsterdam: Scheltema en Holkema's Boekhandel, 1900), p. 38 'Doch dit was voor een groot gehoor zonder lichtbeelden ondoenlijk geweest […].'

Frank Kessler • Utrecht University, F.E.Kessler@uu.nl

Sabine Lenk • University of Antwerp, S.Lenk@uu.nl

Learning with Light and Shadows: Educational Lantern and Film Projection, 1860-1990, ed. by Nelleke Teughels and Kaat Wils, TECHNE-MPH, 8 (Turnhout, 2022), pp. 27-49

10.1484/M.TECHNE-MPH-EB.5.131493

These statements dating from 1900 allow us to make three observations: first it would seem that the optical lantern had by that time become a relatively common device for lectures in at least the academic fields of archaeology and art history and perhaps one that was more or less taken for granted. Second, that the lantern was considered particularly suitable for presenting visual material when addressing a large audience. Third, that such illustrations were deemed more or less a necessity for some types of scholarly lectures and that it was thought 'impractical' to discuss certain topics before an auditorium full of people without the ability to project slides. In other words we may assume that by 1900 the lecture illustrated with lantern slides as a form of knowledge transmission was an established form of communication in some scholarly fields thanks to the ability, made possible by the lantern, to present visual information simultaneously to a large audience. In turn the use of the medium allowed the speaker to discuss issues that without the support of the projected image were better not addressed as one could do so only in an unsatisfactory and inadequate manner.

Using this as our point of departure, we will address the following questions in this chapter: who or what were the driving forces that sought to introduce the projection lantern into academic teaching; what obstacles had to be overcome to do so; were there disciplines that were more inclined than others to adopt this didactic tool and for what reasons; and finally we will briefly address the question of how academics using projection could procure slides and what infrastructure was available for supply.

Our study is based on primary sources and archival research in lantern slide collections.[3] We start with a brief historical overview to explain the problematic reputation lantern projections had acquired and then focus on the period between 1860 and 1914, when the developments we want to examine took place. As might be expected, pictures played an important role in disciplines such as geology, zoology, botany, medicine, and astronomy as well as in theology and of course art history.[4] We must therefore limit our observations to just a few cases. We want in particular to demonstrate the crucial role of transnational networks in the dissemination of projection in academic teaching. An important figure in this context is the French Abbé Moigno, generally considered a pioneer of lantern pedagogy. As we will argue, despite his immense contribution to popular science communication, his influence on academic teaching was nonetheless limited; a point largely neglected in earlier studies of Moigno.

---

3 As archival research has been seriously hampered by the current pandemic, we have not been able to systematically investigate the use of the projected image in university teaching. We had no access to materials such as course catalogues, lecture notes or personal records that make it possible to reconstruct in more detail the way in which projected images were adopted by academic teachers in their day-to-day practice. Access to such documents provided insights, for example, into Willem Vogelsang's use of the lantern in his art history lectures. Cf. Notebaard, Jamilla, 'De Kunst van het geprojecteerde beeld. De didactische waarde van de projectielantaarn voor de kunsthistorische lessen van Willem Vogelsang (1875–1954)', *De Moderne Tijd*, 4.1–2 (2020), pp. 88–107; cf. also de Klerk, Nico, 'Art Historian Willem Vogelsang', *Projecting Knowledge Working Papers*, 1 (2021, revised version) <https://projectingknowledge.sites.uu.nl/wp-content/uploads/sites/482/2019/11/Working-paper-1.Vogelsang.January-2021–1.pdf> [accessed 1 March 2021].

4 See for example for the natural sciences, Daston, Lorraine, and Peter Galison, *Objectivity* (New York: Zone Books, 2007); for theology, e.g., Saint-Martin, Isabelle, 'Du vitrail à la lanterne magique, le catéchisme en images', in *Lanternes magiques, tableaux transparents*, ed. by Ségolène Le Men (Paris: Réunion des Musées Nationaux, 1995), pp. 105–20. The lantern as teaching tool is conspicuously absent in, e.g. Dorsman, Leen, and Peter Jan Knegtman, eds, *Van Lectio tot PowerPoint. Over de geschiedenis van het onderwijs aan de Nederlandse universiteiten* (Hilversum: Verloren, 2011).

The Lantern's Reputation, or Obstacle no. 1

The introduction of lantern projections into academic lecture halls was not at all a matter of course even though the lantern had been born in the laboratory. The lantern bore a blemish from the very start as Christiaan Huygens, a mathematician interested in optics and watchmaking, and who constructed the apparatus for his father in 1659, apparently did not want his reputation as a scientist tarnished by being associated with an instrument of entertainment. An instrument which, as Deirdre Feeney mentions, demanded long hours of the lens maker's craft. In a letter to his brother he declared that he was '[...] ashamed that people will know that [such trifles] came from me'.[5]

In his widely read writings on the magic lantern the Jesuit Athanasius Kircher associated it with demonic and fantastic apparitions, and the lantern's dissemination by itinerant scholars such as the Dane Thomas Rasmussen Walgenste(i)n added to the instrument's general negative reputation as a 'lantern of horror'.[6] Instead of using it as a teaching tool as Johannes Zahn had proposed in 1685–1686, it fell into the hands of itinerant showmen who in the following century earned their living by showing crude self-painted pictures of foreign countries or the Seven Wonders of the World.[7] The same goes for the commercial exploitation of phantasmagoria spectacles by 'physicists' such as Paul de Philipstal and Étienne-Gaspard Robert aka Robertson.[8] From *c.* 1800 onward, toy lanterns made of tin with poor lighting systems and mediocre lenses produced by manufacturers such as Johann Wolfgang Rose in Nuremberg turned the device into an entertainment for children.[9]

After two centuries, during which the projection lantern mainly served popular entertainment, it slowly gained a new reputation.[10] By the mid-nineteenth century science-themed shows for large audiences were being organized by, among others, Henri Robin in Paris and

5 Quoted in Feeney, Deidre, 'The Magic Lantern as a Creative Tool for Understanding the Materiality and Mathematics of Image-Making', in *The Magic Lantern at Work. Witnessing, Persuading, and Connecting*, ed. by Martyn Jolly and Elisa deCourcy (New York, London: Routledge, 2020), pp. 16–31 (pp. 25–26). On the construction of the 'first' Dutch telescope in 1608 and the role of lens makers such as Evert Harmansz and Johan van der Wyck, cf. van den Berg, Rob, 'Het geheim van de Delftse brillenmaker', *NRC*, 16–17 May 2015, quoting historian Huib Zuidervaart from Huygens Instituut (KNAW) in The Hague.

6 Walgensten showed terrifying images which were believed to have caused the sudden death of the Danish king in 1670.

7 For more on the lantern's history, cf. chapter IV of Ruchatz, Jens, *Licht und Wahrheit. Eine Mediumgeschichte der fotografischen Projektion* (München: Wilhelm Fink, 2003). See also Vermeir, Koen, 'The Magic of the Lantern (1660–1700): On Analogical Demonstration and the Visualization of the Invisible', *The British Journal for the History of Science*, 38.2 (2005), pp. 127–59.

8 See, for example, the report by David Brewster on the effect of Robertson's show: Brewster, David, *Letters on Natural Magic, addressed to Sir Walter Scott, Bart. By Sir David Brewster, K. H.* (London: John Murray, 1832), pp. 80–82.

9 On the Rose workshop, see Scholze, Bernd, 'The Beginning of the Modern Toy Industry in Germany: Peter Friedrich Catel and the First Toy Lanterns', *The Magic Lantern*, 1 (December 2014), pp. 1–10 (pp. 1, 8–10). David Brewster in his letters to Walter Scott confirms this and gives as main reason the badly executed images whose flaws appear clearly when enlarged on the screen. Cf. Brewster, *Letters on Natural Magic*, pp. 78, 85.

10 Attested educational use of the lantern occurred at the court of the Freemason Louis XVI where the dauphin was taught with the aid of slides, cf., e.g., Mannoni, Laurent, *Le Grand art de la lumière et de l'ombre. Archéologie du cinéma* (Paris: Nathan, 1994), pp. 86–88; as well as in the pedagogical practice of Madame de Genlis around 1780, cf. Ruchatz, *Licht und Wahrheit*, pp. 155–57.

Brussels[11] and at respectable venues such as the Royal Polytechnic Institution in London with its famous large-size slides.[12] Despite their educational aspirations, however, these were commercial ventures that first and foremost sought to entertain.

Later on educational lectures again often served commercial interests, albeit driven by a pedagogical vocation. Between the 1880s and the 1920s a wave of travelling performers (often presenting medical topics) travelled North America, Australia and New Zealand, Southeast Asia, South Africa, and the United Kingdom.[13] For Belgium this occurred between 1890 and 1930.[14] They toured with slides and a narrative, stayed only for one or two lectures, employed an agent to rent the biggest premises[15] and advertised their visits in advance in the local newspaper like circus people. This contributed to a discrediting of the medium in the eyes of the academic elite,[16] who regarded such performers as little better than hucksters. Even so serious travelling lecturers needed self-promotion to assure their income, an activity which, as Joe Kember noticed in respect of the lecturing physician Anna Mary Longshore Potts, '[...] created frictions with the dignity and formality customarily accorded to the latter [as medical professional]'.[17] Besides, the sheer number of touring lecturers[18] could discredit the medium, as Elias Barker, an experienced lanternist and distributor of slides from Salisbury, complained in 1898, 'Third and fourth-rate exhibitions [...] have no doubt more than anything else brought the lantern into disrepute.'[19] His hope was that the lantern would be used for the dissemination of science and 'become a distinctive feature in the educational routine of the future'.[20]

This quick look back at the period between 1659 and 1900 points up the optical lantern's association with (itinerant) showmanship, (children's) entertainment and, the problematic popularization of knowledge for the masses. The resulting reputation was a

11 On Robin cf. Vanhoutte, Kurt, and Nele Wynants, 'On the Passage of a Man of the Theatre Through a Rather Brief Moment in Time: Henri Robin performing Astronomy in Nineteenth Century Paris', *Early Popular Visual Culture*, 15.2 (2017), pp. 152–74.

12 Cf. Brooker, Jeremy, *The Temple of Minerva: Magic and the Magic Lantern at the Royal Polytechnic Institution, London 1837–1901* (London: The Magic Lantern Society, 2013). Cf. also Ruchatz, *Licht und Wahrheit*, p. 164.

13 Cf. Kember, Joe, 'Anna Mary Longshore Potts and the Anglophone Circuit for Lantern Lecturing in the Late Nineteenth Century', in *The Magic Lantern at Work. Witnessing, Persuading, and Connecting*, ed. by Martyn Jolly and Elisa deCourcy (New York, London: Routledge, 2020), pp. 138–56 (pp. 138–39).

14 Cf. Buelens-Terryn, Margo, Iason Jongepier, and Ilja Van Damme, 'Shine a Light: Catholic Media Use, Transformations in the Public Sphere, and the Voice of the Urban Masses (Antwerp and Brussels, *c.* 1880 — *c.* 1920)', in *Faith in a Beam of Light*, ed. by Sabine Lenk and Natalija Majsova (Turnhout: Brepols Publishers, 2022).

15 According to Angela G. Ray, after 1850 in the US a 'popular lecturing system' began with 'the "creation of public lecturer" as an occupational category'. Ray, quoted in Kember, Joe, 'The Lecture-Broker: The Role of Impresarios and Agencies in the Global Anglophone Circuit for Lantern Lecturing, 1850–1920', *Early Popular Visual Culture*, 17.3 (2019), pp. 279–303 (p. 282).

16 For an illustrative example of 'infotainment' for the masses, cf. Kember, 'Anna Mary Longshore Potts', pp. 148–49.

17 Kember, 'Anna Mary Longshore Potts', p. 142.

18 Kember, 'The Lecture-Broker', p. 284, speaks of 'thousands of engagements across the United States' for the 1890s.

19 Barker, E[lias], 'The Lantern as an Educator', in *The Magic Lantern Journal and Photographic Enlarger, Almanac & Annual, 1898–9*, ed. by J. Hay Taylor (London: Magic Lantern Journal Company Ltd, [1898]), pp. 92–94 (p. 92).

20 Barker, 'The Lantern as an Educator', p. 93.

considerable obstacle to overcome if the lantern was to be deemed worthy to enter the hallowed halls of academe.

The Lantern's Power to Enchant and Educate as Pretence, or Obstacles no. 2 and no. 3

As Jens Ruchatz observed, since its inception the projection lantern was positioned between the two poles of magic and illusion on the one hand and its potential for science and education on the other, '[...] the sober description of the Laterna magica as an instrument of teaching and enlargement does not ignore the power of the projected image to fascinate'.[21]

The fascination produced by the picture on a screen in a dark environment on any kind of audience could work against the lantern's use as a teaching tool, as magic and sobriety are hardly compatible (obstacle no. 2). Even scientific representations such as an immensely enlarged drop of water, to show what is invisible to the naked eye, often appeared as a spectacle of curiosity.[22]

The same goes for the exoticizing representations of geographical, botanical, or zoological phenomena that were presented in venues ranging from fairgrounds to public lecture halls. Mid-nineteenth century showmen presented (often self-made) painted slides and dissolving views. Similarly from the 1820s onwards the catalogue of the British manufacturer of scientific instruments Carpenter & Westley[23] offered depictions of animals in vivid colours that appeared fantastic rather than drawn from nature (Fig. 1.1). Mass audiences were no doubt delighted and fascinated by them (until public zoos made it possible to see what the animals really looked like) and the accompanying talk may indeed have provided correct information. The zoologist would, however, have felt such illustrations to be nothing short of offensive (obstacle no. 3) and incompatible with the maxim widely applied to teaching children at the time that says, 'He who sees correctly, reasons well.'[24]

The problem was one of accuracy, not the fact that these were hand-drawn images. In the 1920s, for example, biologists used photographic images alongside drawings showing an idealized version of the plant.[25] (Fig. 1. 2) Photography provided the opportunity to create a new repertory offering the audience a different kind of magic: the illusion of being 'true to the original'.

---

21 Ruchatz, *Licht und Wahrheit*, p. 124.

22 Micro-cinematography, too, still fascinated the audience, e.g., THE UNSEEN WORLD (Charles Urban Trading Co., UK, 1903) was praised as 'The Micro-Bioscope Series is undeniable interesting and of real educational value, and should draw crowds wherever exhibited.' <https://www.scienceandmediamuseum.org.uk/what-was-on/when-camera-beats-eye-f-percy-smith-archive#&gid = 1&pid = 3> [accessed 25 May 2021]. Cf. also Gaycken, Oliver, '"The Swarming of Life": Moving Images, Education, and Views through the Microscope', *Science in Context*, 24.3 (2011), pp. 360–81.

23 Carpenter & Westley started mass producing slides, but the process was not yet entirely mechanized. Cf. Roberts, Philip, 'Philip Carpenter and the Convergence of Science and Entertainment in the Early-nineteenth Century Instrument Trade', *Sound and Vision* (spring 2017), no pag.

24 'Qui voit juste raisonne bien.' This maxim is quoted in the book by the French zoologist and minister of education, Paul Bert, *Premières notions de zoologie. Lectures à l'usage des élèves des établissements d'enseignement secondaire, des écoles normales primaires et des écoles primaires supérieures, 4th ed.* (Paris: G. Masson, 1885), p. vii.

25 Cf. the slide collection at the *Université libre de Bruxelles*. Cf. also Daston and Galison, *Objectivity*, pp. 161–72.

Fig. 1.1. 'Hippopotamus', possibly Carpenter & Westley, *c.* 1850. Robert Vrielynck Collection, courtesy of MuHKA.

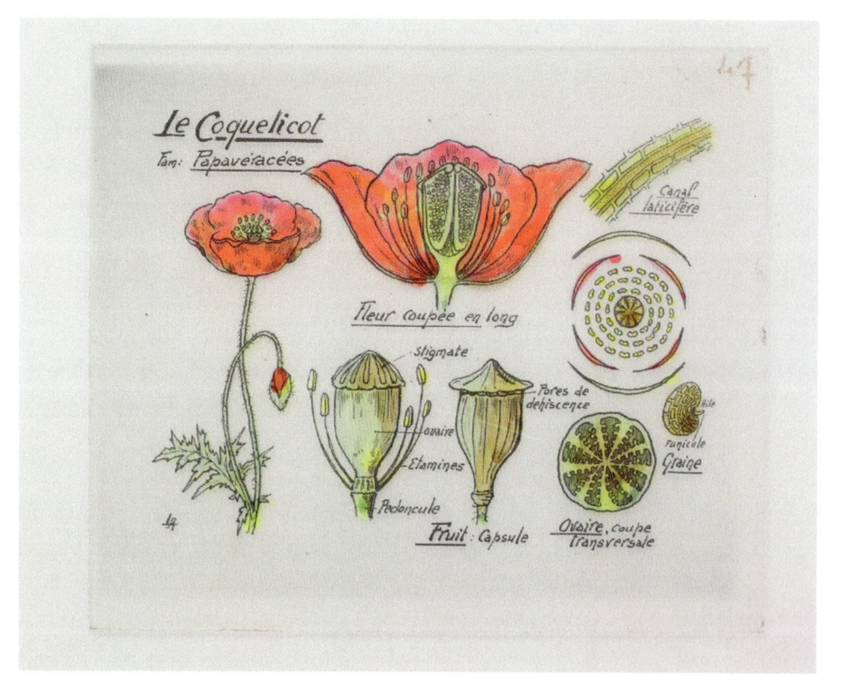

Fig. 1.2. 'Le Coquelicot', Léon Roup, *c.* 1925–1926. Fonds Léon Roup, courtesy of Archives, Patrimoine et Réserve précieuse, Université libre de Bruxelles.

As long as manufacturers such as Carpenter & Westley considered their product as 'a desirable consumer object' and, as Philip Roberts remarks, associated the lantern with education solely for marketing reasons and as a means of making it 'acceptable' to their 'middle class' clients,[26] the fascinating aspect of the screen image would take precedence over its potential for science and education.

26 Roberts, 'Philip Carpenter and the Convergence of Science'. The catalogue for 1850 contained series entitled 'Natural History, & c.' and 'Astronomical Diagrams'.

The Academic Habitus, or Obstacle no. 4

One of the problems in Western European countries such as Germany, Austria, Switzerland as well as the United Kingdom[27] lay in the attitude of the potential users: a professional snobbishness — in German *Standesdünkel* — that had to be overcome before a serious discussion of the possibilities offered by the lantern could take place. The prefix 'Standes-'[Standesdünkel] refers to a closed social group, defined by common social functions and externally perceptible distinguishing characteristics. [...] An example of the intellectual demarcation of a professional group [... is] the seclusion of the higher civil service [...].' According to Christoph Pazdzior, to keep a group exclusive requires 'processes of closure, segregation and selection' ('ständischen Schließungs-, Segregations- und Ausleseprozessen', p. 226) and a compliant behaviour as well as a 'deliberate demarcation towards the lower levels of society' ('gewollte Abgrenzung nach unten', p. 304).[28] This is an adequate characterization of the general mentality of the academic world around 1900.

The self-image and *habitus* of many teachers at institutions with long-standing traditions and the value attributed to their hierarchical position often outweighed pedagogical insights that were critical of traditional teaching methods and favoured forms of visual instruction.

> Overall, habitus is described as 'a stable system of interiorized rules of action that not only serve to enable adjustment to practical requirements but also self-interpretation and the interpretation of social relations'. [Rehbein] Such patterns of action acquired through learning and education are essential for the communication with one's social peers, i.e. representatives of the same habitus, when choosing and practicing a profession.[29]

One good example of the hierarchical difference in *habitus* is the so-called German 'Kinodebatte', which first started in the 1900s. Whereas numerous teachers familiar with the sciopticon and active in primary education (*Volksschule*) saw the potential of film as

27 The exclusive Royal Society (which admitted new members only by election) was for instance much less interested in photography than the BAAS which was open to 'generalists and specialists', according to Tucker, Jennifer, 'Magical Attractions: Lantern Slide Lectures at British Association for the Advancement of Science Annual Meetings, ca 1850–1920', in *The Magic Lantern at Work. Witnessing, Persuading, and Connecting*, ed. by Martyn Jolly and Elisa deCourcy (New York, London: Routledge, 2020), pp. 67–87 (p. 68).

28 'Das Präfix "Standes" verweist auf eine geschlossene soziale Schicht, definiert durch gemeinsame gesellschaftliche Funktionen und nach außen wahrnehmbare Distinktionsmerkmale. [...] Ein Beispiel für die dünkelhafte Abgrenzung einer Berufsgruppe [... ist] die Abgeschlossenheit der höheren Beamtenschaft [...].' (p. 23–24). Pazdzior, Christoph, *Understatement oder Standesdünkel? Hanseatisches Selbstverständnis und Kaufmannsbildung im 19. Jahrhundert. Ein Beitrag zur berufspädagogischen Regionalgeschichte* (doctoral thesis, University of Hamburg, 2016; online version at http://opus.ub.hsu-hh.de/volltexte/2017/3151/pdf/Dissertation_Pazdzior.pdf [accessed 8 March 2021]), pp. 23–24, 226, 304. Cf. also Pierre Bourdieu's theory of habitus in his *Outline of a Theory of Practice* (Cambridge: Cambridge University Press, 1977).

29 'Der Habitus wird insgesamt als "stabiles System verinnerlichter Handlungsregeln, die nicht nur der Anpassung an die Arbeitsanforderungen, sondern auch der Selbstinterpretation und der Deutung gesellschaftlicher Verhältnisse dienen" [Rehbein] beschrieben. Für Ergreifen und Ausübung eines Berufs sind diese in Erziehung und (Aus-)Bildung erworbenen Handlungsmuster, die die Kommunikation mit sozial Gleichgestellten, damit Vertretern des gleichen Habitus, erleichtern, unerlässlich.' Pazdzior, *Understatement oder Standesdünkel*, p. 51, with a quotation by Rehbein, Boike, *Die Soziologie Pierre Bourdieus* (Konstanz: UVK, 2006), p. 128.

a didactic tool and favoured *Anschauungsunterricht* both and inside and outside schools,[30] secondary school teachers (*Gymnasium, Lyceum*) insisted on front-of-class teaching, knowledge from books and 'pedantische Zucht und äußeren Drill' (strict school discipline and external drill), as the reformer Konrad Lange critically remarked.[31] After World War I, *Volksschulen* and *Realschulen* were the first to introduce film into the classroom. By virtue of their 'Herkunft und Berufsfunktion' (origin and professional position)[32] teachers in the *Volksschulen* were close to the working class and saw teaching as a vocation with the aim of giving children a chance to improve their position in society.[33] The social status of *Lyceum* teachers in Germany was much higher: since 1810 they had been expected to pass university examinations (*Staatsexamen*), which made them part of the 'aristocracy of the educated' (*Bildungsaristokratie*).[34] Their status was close, but not equal, to university professors who for a long time stood 'in unerreichbarer Höhe über der Welt' (at an unattainable height above the world).[35]

Switzerland, too, had a strictly hierarchical school system.[36] Institutions such as the Art Academy in Lausanne or secondary schools such as the Lausanne 'Gymnasium' were open to new methods whereas the introduction of the sciopticon met with some resistance in the university. Although more research is required, we see indications that this was due above all to the academic hierarchy of professorial ranks. Aloys de Molin, a young teacher of archaeology and Latin, who had studied with Jakob Burckhardt and Herman Grimm,[37] used slides when he lectured on the history of Roman Art in 1886 at the Academy. When he taught the same subject as 'Privat-Docent' in the Department of Art and Archaeology (ex-Art Academy) in the winter of 1890–1891 he did so without

---

30 Cf. Kessler, Frank, and Sabine Lenk, 'The Kinoreformbewegung in Germany: Creating an Infrastructure for Pedagogical Screenings', in *The Institutionalization of Educational Cinema*, ed. by Marina Dahlquist and Joel Frykholm (Bloomington: Indiana University Press, 2019), pp. 36–54; cf. also Ruchatz, *Licht und Wahrheit*, pp. 400–01.

31 Lange, Konrad, 'Das Wesen der künstlerischen Erziehung', in *Kunsterziehung. Ergebnisse und Anregungen des Kunsterziehungstages in Dresden am 18. und 19. September 1901*, 2nd edn. (Leipzig: R. Voigtländers Verlag, 1902), pp. 27–38 (p. 34).

32 Wilkending, Gisela, 'Volksbildung und Pädagogik "vom Kinde aus"' (1980), quoted in Schmerling, Alice, *Kind, Kino und Kinderliteratur. Eine Untersuchung zum Medienumbruch in der Kinderkultur der Kaiserzeit und der Weimarer Republik* (unpublished doctoral thesis, University of Cologne, 2007), p. 32.

33 On the self-image of *Volksschullehrer*, cf. the talk by teacher Pretzel (Berlin) in [Anon.], *Kunsterziehung. Ergebnisse und Anregungen des Kunsterziehungstages in Dresden am 18. und 19. September 1901*, 2nd edn (Leipzig: R. Voigtländers Verlag, 1902), pp. 22–24. On the role of *Volksschullehrer* and their professional association in the acceptance of living pictures as a didactic tool, cf. Schmerling, *Kind, Kino und Kinderliteratur*, p. 61.

34 Cf. Siegert, Paul Ferdinand, *Bürgerliches Selbstverständnis, Kinoreform und früher Schulfilm. Eine kulturwissenschaftliche Analyse* (unpublished doctoral thesis, Leuphana University Lüneburg, 1995), pp. 39–41, 101.

35 Lichtwark, Alfred, 'Der Deutsche der Zukunft', *Kunsterziehung. Ergebnisse und Anregungen des Kunsterziehungstages in Dresden am 18. und 19. September 1901*, 2nd edn (Leipzig: R. Voigtländers Verlag, 1902), pp. 39–57 (p. 44). Lichtwark describes the hierarchical organization of German society as a 'caste system', p. 50.

36 Cf., e.g., the list of the different types of schools in Lausanne, their various orientations and the diplomas to be obtained in the *Programme des Cours de l'Université de Lausanne. Semestre d'hiver 1905–1906* (Lausanne: Ch. Pache, 1905, and other editions).

37 Burckhardt still worked with photographic paper prints while Herman Grimm used projection. Cf. Schlick, Wilhelm, 'Herman Grimm (1828–1901) Epigone und Vorläufer', in *Aspekte der Romantik. Zur Verleihung des 'Brüder Grimm-Preises' der Philipps-Universität Marburg im Dezember 1999*, ed. by Jutta Osinski and Felix Saure (Kassel: Brüder-Grimm-Gesellschaft, 2001), pp. 73–93.

projection, although he continued to teach successfully with the apparatus at *Gymnase* and the museum he directed.[38]

At the Lausanne Arts Faculty (*Faculté des Lettres*), the first to lecture with the aid of the lantern seems to have been Ernest Chatelanat in the winter term of 1905–1906. Formerly teaching Latin before he was promoted to 'Privat-Docent' in the same Department, Chatelanat had also been the first to take the students on an archaeological excursion in 1897–1898.[39] Although De Molin had become *professeur extraordinaire* in Art and Archaeology by 1906, according to the course catalogue it was only in the winter of 1916–1917 that Albert Naef, also a *professeur extraordinaire*, taught two classes with the aid of projection.[40] Other departments adopted the lantern earlier: in 1904, Henri Meylan-Faure, professor of Greek and Antiquity, showed his own slides of a trip to the Aegean.

With respect to the Art and Archaeology department of Lausanne University, our preliminary conclusion is that, although lecturers received some money from the Faculty for the purchase of slides and the existence of a photographic laboratory for photography classes,[41] only a few lecturers used the lantern as teaching tool prior to 1916. Despite the fact that slide manufacturers had built a large catalogue of slide series on art historical and archaeological topics since the 1860s, and even though De Molin had a slide collection at his disposal[42] and used it outside the university,[43] there seems to have been little motivation to use the lantern for academic teaching. Further studies are required if our assumption that this is due to academic *habitus* is to be disproved or confirmed. In some respects, there may have been little difference between scientist Huygens being ashamed of his association with the lantern in the 1660s and the resistance of certain academics to the medium two centuries later.[44]

Abbé Moigno — A 'Change Agent' for Popular Visual Instruction, Not a Role Model for Academic Teaching

According to several studies of media history, the French former Jesuit and mathematician Abbé François-Napoléon-Marie Moigno (1804–1884) played a significant role in establishing

38 On Aloys de Molin, cf. Blancardi, Nathalie, 'Archives de verre. La Photothèque d'art et d'archéologie de la Faculté des Lettres de Lausanne', *Monument vaudois*, 6 (2015), pp. 45–55 (pp. 45–46).

39 On the art department and Chatelanat, cf. Blancardi, 'Archives de verre', p. 47. Information on the lectures comes from a list established by Blancardi based on the course catalogs of the Art History Department since 1890.

40 As his colleague, Naef organized an excursion in summer 1917. We still have to examine whether this combination is accidental or indicates an openness to other teaching methods.

41 Photography classes started in 1895. A photographer from Lausanne had offered his services in 1894, however a competitor (possibly one with an academic diploma) was appointed Privat-Docent for this position. Cf. Dossier Début Fonds, PV_CU_1880–1910 Chatelanat and Molin, pp. 57–58, 62–64.

42 Cf. Blancardi, 'Archives de verre', p. 49.

43 Cf. *Journal de Genève* 5 March 1907, p. 4; 8 March 1903, p. 4; 13 July 1907, p. 4. His talk at the *Société de Géographie* was highly praised as were the slides taken by his brother-in-law, J.-J. Mercier, on their joint trip along the Nile.

44 Preliminary research in the Netherlands indicates that the Polytechnical School in Delft and the Veterinary School in Utrecht adopted the lantern as a teaching tool earlier than the universities. Cf. *Verslag van den Staat der Hooge, Middelbare en Lagere Scholen in het Koninkrijk der Nederlanden* for the years 1896–1901 <https://www.dbnl.org/titels/tijdschriften/tijdschrift.php?id=_ver042vers01>.

the lantern as an efficient instrument of knowledge transmission. Laurent Mannoni considers him the importer of British projection techniques to France and stresses his influence on numerous French manufacturers.[45] According to Jacques Perriault, Moigno was, along with Jacques Dubosq and François Soleil, one of the founders of a 'pédagogie des projections lumineuses' (pedagogy of light projection) around 1839.[46] Jens Ruchatz identifies him as one of the 'change agents' who in the mid-nineteenth century paved the way for the educational projection of photographic slides.[47] We will show that this is the case only to a certain extent. As far as popular education in and around Paris was concerned, Moigno was indeed an early popularizer and an efficient organizer of illustrated lectures. He had, however, no influence on academic teaching at all.

Moigno was a fervent advocate of the educational value of the lantern, stating that it had allowed him to capture the attention of an audience of 'plus de deux milles [*sic*] personnes, de toutes les classes de la société' (more than two thousand people of all social classes) when lecturing at the *Salle de Bal Mérot* in Saint-Denis just outside Paris.[48] In his handbook *L'Art des projections*, Moigno detailed his ideal programme for an instructional evening as follows:

1. Opening — a musical masterpiece to initiate the audience into the world of melody and harmony.
2. Overview of scientific innovations with models, slide projections, and explanations.
3. Illustrated science demonstration (*c.* one hour).
4. Interlude (fifteen minutes) with songs, recitals from literary works or music.
5. Illustrated lecture on a historical or geographical subject.
6. Optical entertainments (chromatrope, fantascope, eidotrope, etc.)
7. Audience leaves to the accompaniment of patriotic music and the songs of various peoples.[49]

Moigno thus aimed at presenting a balanced combination of listening and watching, intellectual concentration and relaxation, technology (projector) and human performance (lecturing, physical demonstrations, music), and of education and entertainment. The total duration of such a programme would probably have been about two hours. The *Matinées scientifiques* that he had planned to organize in his *Salle du Progrès* for specific target groups, including school children, might have been shorter.[50]

Although Moigno advocated illustrated science teaching with the aid of experiments and projections[51] and was aware of how to use didactic instruments to the learner's best advantage, he does not appear to have studied pedagogical theory. He makes no mention of

45 Cf. Mannoni, *Le Grand art*, p. 249. Although Moigno's earliest attempt in 1852 failed, he succeeded in 1864. For more on Moigno' activities, cf. Mannoni, *Le Grand art*, pp. 253–57; Mannoni, Laurent, 'The Magic Lantern Makers of France', *Optical Magic Lantern Journal*, 5.2 (August 1987), pp. 3–7 (p. 5).

46 Perriault, Jacques, *Mémoires de l'ombre et du son. Une archéologie de l'audiovisuel* (Paris: Flammarion, 1981), p. 94.

47 Cf. chapter V.3.1. 'Change Agents' in Ruchatz, *Licht und Wahrheit*, pp. 209–25.

48 Moigno, François-Napoléon-Marie, *Enseignement de tous. L'Art et la pratique des projections. Les sciences, les industries, les arts enseignés et illustrés par quatre mille cinq cents photographies sur verre. Catalogue des tableaux et appareils* (Paris: Bureau du Journal Cosmos-Les-Mondes, Billon-Daguerre, [1882]), p. iii.

49 Cf. Moigno, *L'Art des projections*, pp. ix–x.

50 Cf. Moigno, *L'Art des projections*, p. xii.

51 Cf. numerous examples in *L'Art des projections*.

any (classic) texts on visual education although he could have encountered such when he had lived in Switzerland and travelled to England, Germany, Belgium, and the Netherlands.[52] One reason might be that Moigno was mainly concerned with adult education (which at the time meant from the age of thirteen and above) and theories about 'object lessons' were primarily addressed to the teachers of children. Also '[...] Moigno's mind was of a very practical cast and he was not immersed in the consideration of theories to the neglect of what is more useful'.[53] He apparently preferred to prepare boxed sets containing fifty to a hundred slides with an accompanying lantern reading which could then be rented from the Paris headquarters of his organization,[54] to writing a treatise on education.

Although he became a corresponding member of the international British Association for the Advancement of Science (BAAS) in 1855 and translated or edited the writings of numerous scientists, he seems to have conducted his practical educational work in relative isolation. Given the tensions between the Catholic Church and secular political forces in France, Moigno did not seek contact with the *Ligue de l'Enseignement*, which had been founded in 1866 by Jean Macé.[55] The *Ligue* was seen by the Catholic Church as a competitor in the struggle for the minds and the souls of young adolescents. Even more so, as the *Ligue* had a strong influence on the policies of the French Ministry of Education.[56]

As a mathematician of high repute and supporter of BAAS's mission to popularize the results of scientific research,[57] Moigno must have felt at home in the ranks of the organization's membership of academic researchers and specialized amateurs. Reading about BAAS's annual meetings, where lectures with projection were relatively common in the second half of the nineteenth century,[58] may have served to confirm his ideas. Thanks to his connections and his editorial work for the popular science journals *Cosmos* and *Les Mondes*, Moigno was well-aware of international scientific developments, but he was a savant, not an academic. His aversion to positivism,[59] his violent attacks on freethinkers, his diatribe on the 'Splendours of Faith' attacking Enlightenment conceptions of science[60] as well as his radical Catholicism in an age of secularization led to him making many enemies.[61]

---

52 Cf. [Anon.], 'L'abbé Moigno', *Nature* (24 July 1884), p. 291.

53 [Anon.], 'L'abbé Moigno', p. 292.

54 Cf. Moigno, *L'Art des projections*, pp. xi–xii.

55 Macé was a Freemason and as such was undoubtedly considered an 'enemy' by Moigno. Cf. Deshogues, Yannick, 'Jean Macé, un Franc-maçon' <http://yannickdeshogues.free.fr/pdf/7MACON.pdf> [accessed 9 November 2020].

56 On this point, cf. Quillien, Anne, 'Les plaques photographiques du Musée pédagogique. Constitution et diffusion d'un fonds pour l'enseignement', in *La plaque photographique. Un outil pour la fabrication et la diffusion des savoirs (XIXe — XXe siècle)*, ed. by Denise Borlée and Hervé Doucet (Strasbourg: Presses universitaires de Strasbourg, 2019), pp. 40–54. Unfortunately, Quillien does not say when the *Ligue*'s first illustrated lecture took place.

57 Cf. Tucker, 'Magical Attractions', p. 68.

58 Cf. Tucker, 'Magical Attractions', pp. 70–73.

59 The anonymous author of the obituary in *Nature* also mentioned Moigno strictly rejecting the positivist approach of a BAAS fellow (p. 291).

60 Cf. Moigno, François-Napoléon-Marie, *Les Splendeurs de la Foi. Accord parfait de la révélation et de la science, de la foi et de la raison, par M. l'abbé Moigno, Chanoine de Saint-Denis, Fondateur — Directeur du Journal ΚΟΣΜΟΣ — LES MONDES*, 5 vols (Paris: Blériot Frères, 1879), I: *La foi* (1879).

61 Cf. Mannoni, *Le Grand art*, p. 254; cf. also lemma 'Moigno' in *Encyclopedia of the Magic Lantern*, ed. by David Robinson, Stephen Herbert, and Richard Crangle (London: Magic Lantern Society, 2001), pp. 196–97.

Moigno saw himself first and foremost as a '*prêtre de Jésus-Christ*' (priest of Christ).[62] Faith and science were to him but one, 'Science, like faith, can only reach a young soul when it passes through a docile ear, *fides ex auditu* [...].'[63] If he really was a model of teaching as Perriault argues, it was a model primarily adopted by Catholic educators and teachers. The publishing house *Maison de la Bonne Presse* was to follow his lead later on by establishing a projection service.[64] A fervent popularizer of knowledge, Moigno wanted to reach a broad audience that encompassed all ages, social backgrounds and levels of education but was not interested in teaching the academic elite.

It is his 1882 slide catalogue, however, that reveals Moigno as a visionary. Although he saw himself firmly within the tradition of the Magic Lantern,[65] his catalogue, as Mannoni notes, may be regarded as a model for subsequent commercial slide manufacturers. It covered astronomy, physics, mechanics, obstetrics, anatomy, pathology, geology, chemistry, zoology, botany as well as photomicrography, not to forget art and architecture, literature, history, and of course religion. All of these incidentally were fields in which slides were used in academic teaching.

Moigno had the pictures for the slides taken by the photographer Armand Billon.[66] Billon explicitly stated that 'as certain publishers did not want to grant us the right to reproduce their illustrations [...] we were obliged to make a large number of drawings'.[67] Photographic reproductions of drawings or diagrams from books were already being used for popularizing science in the early 1880s and by the 1890s the practice was adopted by academics as well.[68]

The Optical Lantern and Photography, or 'Fidelity Truly Astonishing'

According to Jens Ruchatz, 'Photography played an important part in rendering projection a rational medium'.[69] It became technically feasible in the mid-nineteenth century and contributed to overcoming the problem of inaccurate representation. Neither Daguerreotypes

62 Moigno, *Enseignement de tous*, p. vi.

63 'La science, comme la foi, ne peuvent pénétrer dans une jeune âme que par une oreille docile, *fides ex auditu* [...].' 'M. L'Abbé Moigno à l'éditeur' in *La Clef de la science ou Les phénomènes de tous les jours expliqués par le Dr E. C. Brewer, membre de l'Université de Cambridge, du Collège des Précepteurs de Londres, etc. auteur de plusieurs ouvrages littéraires, historiques, scientifiques, mathématiques, etc.*, 3rd edn, rev. by Abbé Moigno (Paris: Vve Jules Renouard, 1858), pp. vii–ix (p. ix).

64 Cf. on this point Mannoni, 'The Magic Lantern Makers', p. 7.

65 Moigno, *Enseignement de tous*, p. v: 'Tel que je l'ai créé, l'enseignement par les projections n'est qu'une exhibition à la lanterne magique.' (As I created it, teaching by means of projection is simply a magic lantern performance.)

66 The catalogue does not distinguish between slides that were photographically produced and those that were reproduced by photographic means.

67 'Quelques éditeurs n'ayant pas voulu nous accorder le droit de reproduction de leurs illustrations [...] nous avons été obligés de faire faire un grand nombre de dessins [...].' Billon in Moigno, *Enseignement de tous*, p. 9.

68 Cf. the slide collections of professors at the university's *Cité scientifique*, preserved at the *Université libre de Bruxelles*.

69 Ruchatz, Jens, 'The Magic Lantern in Connection with Photography: Rationalisation and Technology', in *Visual Delight: Essays on the Popular and Projected Image in the 19th Century*, ed. by Simon Popple and Vanessa Toulmin (Trowbridge: Flicks Books, 2000), pp. 38–49 (p. 41). Another important aspect, which we cannot discuss in this chapter, was the development of lighting systems that made it possible to show slides in large

(in use *c.* 1840–1865) nor Ambrotypes (1855–1895) were practicable for projection: the former were opaque and the latter were collodion negatives that needed to be viewed against a black background. According to most histories of photography the wet collodion process was predominant from 1851 into the 1890s but required a skilled photographer to handle the plates. From 1871 on the dry plate slowly started to displace the earlier processes.[70] The brothers William and Frederick Langenheim first projected photographs in 1849 on the basis of a process developed by Abel Niépce de Saint-Victor two years earlier, patenting their own process under the name 'hyalotype'.[71] Projected photographs thus offered an unparalleled richness in detail as the Langenheims made clear in 1851:

> [...] to throw the old style of magic lantern slides into the shade, and supersede them at once, on account of the greater accuracy of the smallest details which are drawn and fixed on glass from nature [...] with a fidelity truly astonishing [...] omitting all defects and incorrectness in the drawing, which can never be avoided in painting a picture on the small scale required for the old slides.[72]

Like the cinematographic images that appeared less than half a century after the Langenheims' demonstration, projected photographs were seen as a way of creating new possibilities for the exploration of nature, including microscopy, which now could be made visible in all its details. The idea that photography was faithful to nature was echoed in many treatises by scientists during the second half of the nineteenth century.[73] It referred to 'reproduced patterns of texture', but 'not to colour or depth', as Scott Curtis emphasizes,[74] for the Lumière autochrome plate only became commercially available after the turn of the century.[75]

The Optical Lantern and Photography, or Commercial and Self-Made Slides

The emergence of photography for scientific purposes set off another international discussion: which of the available processes was the most practical? This was of interest to those who practiced photography and concerned their preferences, which in turn was of key importance to the progressively developing photographic industry, which was to turn the photographic slide into a mass-market product in the 1880s.

---

auditoriums. A newspaper article announcing the inauguration of Moigno's *Salle du Progrès*, for instance, specifies that the slides were to be projected by electric or oxyhydrogen light (*La Petite Presse*, 17 October 1872).

70 Cf. Kennel, Sarah, Diane Waggoner, and Alice Carver-Kubik, *In the Darkroom. An illustrated Guide to Photographic Processes before the Digital Age* (London: Thames & Hudson, 2010).

71 Cf. Ruchatz, *Licht und Wahrheit*, pp. 73–75.

72 Langenheim brochure quoted in Ruchatz, *Licht und Wahrheit*, p. 75.

73 On the complex and ambivalent role of photography as a faithful recording of the real in scientific discourse, cf. Daston and Galison, *Objectivity*, pp. 125–38.

74 Curtis, Scott, *The Shape of Spectatorship. Art, Science, and Early Cinema in Germany* (New York: Columbia University Press, 2015), p. 179.

75 Lumière autochromes could be projected, but the process was not well-suited to teaching, because it was expensive compared to black and white glass positives and because its materiality (thousands of coloured non-transparent grains) meant that the slides absorbed too much light.

Paper manufacturer and amateur astronomer Warren de la Rue,[76] member of BAAS and who lectured with the lantern on astronomical photography, stated in 1872: 'It has been proposed [...] to use the Daguerréotype instead of the collodion process. The former, however, is so little practiced and, moreover, so much more troublesome that it does not seem to be advisable to adopt it [...].'[77] De la Rue also referred to attempts to work with dry plates in astronomy, which signals a gradual shift towards this process, although the speaker himself declared that he believed 'that the wet collodion is preferable'.[78]

The quality of the photographs depended on the equipment and materials used, as both emulsions and lenses could cause distortion[79] as could the glass plate.[80] Those wishing to make slides themselves had to pay attention to numerous details until the launch of the Kodak camera in 1888 under the famous slogan, 'You Press the Button, We Do the Rest'. Capturing images on a photographic plate required time, skill, patience, and money with no guarantees regarding the results. Professional photographers could be employed to take pictures, but many scientists documenting their own discoveries in the field or in the laboratory elected to acquire the necessary knowhow. Courses in photography had been available since the 1860s.[81] By the end of the nineteenth century, when slide makers had built up catalogues based on standard curricula, it was possible to use commercially produced slides for academic lectures for introductory and general courses. By contrast when Warren de la Rue presented the photographs he had taken of the solar eclipse in Spain, he had to turn to professionals to have them made into slides.[82]

In this context an important role was played by the numerous photographic societies and their publications, which served as forums for the discussion of technical developments and useful practices, including how to make photographic slides. Companies selling equipment and accessories advertised their products in specialized journals and handbooks. Optical lanterns were also part of this equipment, because photographers used projection to present their work to other members and the general public.[83] In the last quarter of the nineteenth century information on photographic techniques, cameras and projection equipment, light sources, etc. as well as information about where to obtain

---

76 'Astro-photography' started with De la Rue, cf. Coppens, Jan, Laurent Roosens, and Karel van Deuren, '... *door de enkele werking van het licht...*' (n.p. [Brussels]: Gemeentekrediet, 1989), pp. 254–55.

77 Address by Warren de la Rue in *Report of the Forty-Second Meeting of the British Association for the Advancement of Science held at Brighton in August 1872* (London: John Murray, 1873), 'Notices and Abstracts', section 'Mathematics and Physics', p. 4, footnote <https://www.biodiversitylibrary.org/item/94434#page/519/mode/1up> [accessed 8 March 2021].

78 De la Rue, 'Address', p. 3.

79 'The distortion of a photographic image, if such exist, may be either extrinsic or intrinsic — that is, either optical or mechanical. The instrumental apparatus for producing the image may produce optical irregularities before it reaches the sensitive plate: or an image optically correct may by irregular contraction of the sensitive film in the process of drying, and other incidents of the process, present a faulty delineation on the plate.' De la Rue, 'Address', p. 4.

80 Cf. Coppens, Roosens, and van Deuren, '... *door de enkele werking van het licht...*', p. 268.

81 Cf. Coppens, Roosens, and van Deuren, '... *door de enkele werking van het licht...*', p. 275.

82 Cf. Tucker, pp. 70–71.

83 Cf. Ruchatz, *Licht und Wahrheit*, pp. 272–90.

such equipment was easily available. Those interested could build their own collections and show their own slides.[84]

Membership of these societies was not restricted to professional photographers, many amateurs, including scientists, swelled their ranks. One example is the Utrecht professor of Physical Chemistry, Ernst Cohen, who as an adolescent discovered photography and became a member of a photographic society. During his studies he worked as a laboratory assistant for the Amsterdam professor of physics Johannes Diderik van der Waals. In this capacity and thanks to his knowhow he took photographs of liquid jets. In 1889 Cohen published a short article in the *Revue Scientifique*, in which he gave details of his procedure. Later on he was to use these photographs in an illustrated lecture on the scientific applications of photography.[85] Cohen thus had acquired his photographic skills as an amateur, applied them to scientific research and shared his knowledge in the pages of a scientific publication. Other scientists may have followed a similar trajectory, leading them to use the optical lantern for science communication and perhaps also for teaching.

Like many other universities the *Université de Lausanne* and the *Université libre de Bruxelles* established their own photographic labs around 1895 to produce graphic materials for their teaching staff. (Fig. 1.3) Another option was to turn to the professional photographic firms to reproduce graphic material from books, postcards, photographs, or glass negatives. A third option was to have a photograph taken of an object, event, phenomenon, etc. and have it transferred to a glass slide. Yet another possibility was to draw or paint with special ink or paint on glass plates pre-processed for that purpose.[86]

The Optical Lantern and the Academics: Astronomy, Photomicrography, Art History

The question remains of how academics adopted the optical lantern as an instrument for research and lecturing and how knowledge of the possibilities offered by photography for research and teaching circulated. We will now look at three academic fields that were 'early adopters' of lantern projection. Our hypothesis is that learning from examples and the formal and informal communication networks between scholars and scientists were crucial to this process.

Astronomy

As we have seen, the BAAS offered a forum for encounters between savants and academics. Membership of a learned society and even more so a leading position in one was a mark of

84 This included collections for teaching and lecturing. Cf., for example, the numerous articles on collections in *La plaque photographique. Un outil pour la fabrication et la diffusion des savoirs (XIXe — XXe siècle)*, ed. by Denise Borlée and Hervé Doucet (Strasbourg: Presses universitaires de Strasbourg, 2019).

85 Cf. Cohen, Ernst, *Na driekwart eeuw. Levensherinneringen* (Utrecht: Matrijs, 2013), p. 40. See also Cohen, Ernst, 'La photographie des jets de liquides', *Revue Scientifique*, 7 (1889), pp. 252–53; and 'Natuurkundige voordrachten. Prof. dr. Ernst Cohen: De fotografie en haar toepassingen in de wetenschap', *De Nieuwe Courant*, 602 (1902).

86 About eight hundred items of this kind can be found in the ULB biology collection.

Fig. 1.3. Scan of a positive for a lecture in physiology, probably by Paul Héger, *c.* 1895–1899. Massaert Collection, courtesy of Archives, Patrimoine et Réserve précieuse, Université libre de Bruxelles.

distinction and often listed in publications. Warren de la Rue, for instance, was introduced in the BAAS report of 1872 as F. R. S. (Fellow of the Royal Society), Vice President of the Chemical Society (V. P. C. S.), Vice President of the Royal Astronomical Society (V. P. R. A. S.), and President of the Section Mathematics and Physics at the BAAS. The regular use of slide projections at BAAS meetings and the academic standing of its members could thus indeed help overcome the obstacle of the lantern's problematic reputation. Astronomy is a case in point.

The aforementioned address to the Mathematics and Physics section by Warren de la Rue at the 1872 BAAS meeting indicates that astronomers were among the first to adopt photography as a tool for their research. In contrast to the use of slides in popular science performances such as those by Henri Robin or the long tradition of rackwork slides illustrating the revolution of the planets — maintained over a period of more than 150 years from Adam Walker's Eidouranion in 1770 well into the twentieth century — photographs such as those taken by De la Rue had an epistemic rather than a demonstrative function.[87] We follow Lorraine Daston's definition of an 'epistemic image' as an image,

> [...] made with the intent not only of depicting the object of scientific inquiry, but also of replacing it. A successful epistemic image becomes a working object of science, a stand-in for the too plentiful and too various objects of nature, and one that can be shared by a dispersed community of naturalists.[88]

In his address, Warren de la Rue referred to photographs as epistemic images that became objects of analysis and which made it possible to confirm hypotheses. De la Rue had obtained a collodion print of the moon in 1851 using a reflecting telescope, and later used photography to document the 1860 solar eclipse.[89] He was well aware of similar initiatives elsewhere:

> It will be recollected that in 1860, for the first time, the solar origin of the prominences was placed beyond doubt solely by photography, which preserved a faithful record of the moon's motion in relation to these protuberances. The photographs of Tennant at Guntour, and of Vogel at Aden, in 1868, and also those of the American astronomers at Burlington and Ottumwa, Iowa, in 1869, under Professors Morton and Mayer, have fully confirmed those results.[90]

Those attending this section of the BAAS meeting thus learned about various options for the use of photography in their scientific studies and also about the results of such photography-based studies at the international level. De la Rue's lecture did not explicitly mention the possibility of projecting photographic glass slides, although he had done so himself at, for instance, the 1861 BAAS meeting with his photographs of the 1860 eclipse. Moigno's catalogue, published a decade later, listed about forty slides representing solar

87 On popular astronomy presentations, cf. Bush, Martin, 'The astronomical lantern slide set and the Eidouranion in Australia', *Early Popular Visual Culture*, 17.1 (2019), pp. 9–33.

88 Daston, Lorraine, 'Epistemic Images', in *Vision and Its Instruments. Art, Science, and Technology in Early Modern Europe*, ed. by Alina Payne (University Park: The Pennsylvania State University Press, 2015), pp. 13–35 (p. 17).

89 Cf. Rothermel, Holly, 'Images of the Sun, Warren de la Rue, George Biddell Airy and Celestial Photography', *The British Journal for the History of Science*, 26.2 (1993), pp. 137–69 (pp. 143, 154–58).

90 De la Rue, 'Address', p. 6. Moigno knew the article, because he published a French translation as 'Discours d'ouverture par le président, M. Warren de La Rue, D. C. L., R. R. S.', in the section 'Association britannique pour l'avancement des sciences. Réunion de Brighton', *Les Mondes. Revue hebdomadaire des sciences et leurs applications aux arts et à l'industrie par Abbé Moigno*, 29.1 (Paris: Bureaux des Mondes, September-December 1872), pp. 12–34 (p. 17, footnote) <https://babel.hathitrust.org/cgi/pt?id = hvd.hn4k8g&view = 1up&seq = 40>.

phenomena, although from the catalogue titles it is difficult to say whether these were actual photographs or photographic reproductions of drawings based on photographs or observation. They were nonetheless quite distinct from the rackwork slides of popular astronomy presentations and might be seen as signalling a shift to epistemic images in the teaching of astronomy. As Holly Rothermel argues, photography established itself as an 'integral part of astronomical work' in the last two decades of the nineteenth century.[91] By then, it had become easy to reproduce photographs on glass slides and use them for academic teaching as well, as is demonstrated for instance by the surviving lantern slide sets preserved in the former Sonnenborgh observatory at Utrecht University.[92] Photographs of astronomical phenomena thus became a standard form of research in the field of astronomy and the same goes for the projection of slides for the international presentation of the results of this research, as well as for teaching.

Photomicrography[93]

As early as 1866, Albert Montessier, a member of the Faculty of Medicine at Montpellier, published a book on photography applied to micrographic studies.[94] The Berlin chemist Paul Jeserich stated in his 1888 book on photomicrography that the parallel progress of photography and microscopy had led to the introduction of this new technology in a number of branches of science. When it became possible to reach the same photographic quality with the dry plate as with the wet collodion process and British, French, and German scientists published the results of their photomicrographic research, 'it was obvious that microphotography [*sic!*] was *destined* to serve medicine and the sciences as an *important* aid'.[95] The rapid technical developments along with the numerous manuals published by scientists in various countries boosted the adoption of this new tool. In the 1880s the 'drop of stagnant water magnified on the screen' was no longer a spectacular curiosity but an educational image to be shown in class.[96]

Photomicrographic slides were already available in the early 1870s. In 1872 Jules Girard published a book on photomicrography in Moigno's book series, which listed a

---

91 Rothermel, 'Images of the Sun', p. 138.

92 Twenty-five of these sets can be consulted at 'Lucerna — The Magic Lantern Web Resource' <http://lucerna.exeter.ac.uk/index.php> [accessed 8 March 2021]. The slides range from images of astronomical phenomena to portraits of famous astronomers and photographs of expeditions by Utrecht astronomers.

93 The term 'photomicrography' refers to microscopic phenomena reproduced by means of photography, which then could be projected, not to the projection of specimen by means of an optical lantern or a projection microscope. The word 'microphotography' generally denotes the process of reducing the size of pictures and documents to a microscopic scale, and to then project them in order to read them, as practiced during the Franco-Prussian War 1870/71.

94 Montessier, Albert, *La Photographie appliquée aux recherches micrographiques* (Paris: J.-B. Baillière et Fils, 1866).

95 '[…] wurde es auf's Deutlichste ersichtlich, dass die Mikrophotographie *berufen* sei, der Medizin und den Naturwissenschaften als *wichtiges* Hilfsmittel zu dienen […].' Jeserich, Paul, *Mikrophotographie auf Bromsilbergelatine bei natürlichem und künstlichem Lichte unter ganz besonderer Berücksichtigung des Kalklichts* (Berlin: Verlag von Julius Springer, 1888), p. 6. (Emphasis in the original text).

96 Thomson, John, 'The Magic Lantern', in *Science for All*, II, ed. by Robert Brown (London, Paris, New York: Cassell, Petter, Galpin & Co., 1883), pp. 208–14 (p. 214).

hundred slides that could be acquired.[97] A year later, Gustave Le Bon, who had lectured at the *Salle du Progrès* in November and December 1872 respectively on physiology and the 'microscopic' world, wrote a brochure for the same series, which also listed, among other things, photomicrographic slides that were available for purchase.[98] By 1880 Alfred Molteni's catalogue listed over four hundred and fifty photomicrographs, including some from Le Bon's collection, covering a variety of scientific domains such as medicine, zoology, entomology, botany, geology, etc.[99]

Interestingly, in the case of photomicrography, Moigno's *Salle du Progrès* and his book series seem to have acted as an interface between knowledge popularization and academia in France. Both Girard and Le Bon's books addressed first and foremost academic circles and encouraged them to use slides for their teaching. Le Bon explicitly addressed the problem facing histology lecturers, limited by the fact that a tissue sample viewed with a microscope could only be seen by one person at a time, whereas Le Bon had been able to lecture three hundred people at once. According to Le Bon each and every one of them could see all the details of the photomicrographic slides that he projected during his lecture.[100] Given the wide-spread adoption of photomicrographic slides, the promotion of this method of teaching appears to have been successful, arguably because on the one hand photomicrography became an important part of research in several fields and because projection allowed the efficient organization of the teaching material of the subject concerned.

Art History

Art history was among the disciplines to proceed to the wholesale adoption of the projected image relatively early on. Heinrich Dilly remarked on the interesting role it played even in 1975.[101] The improvement in photographic quality made it possible to obtain reproductions of paintings in black and white of such quality that they were judged acceptable, even by the artists themselves.[102] In a parallel development that first started in the 1840s, increasing numbers of photographers like Adolphe Braun in Dornach and the brothers Giuseppe, Leopoldo, and Romualdo Alinari in Florence had been reaching agreements with museums on the photographic reproduction of works of art and were often granted exclusive rights.

---

97 Cf. Girard, Jules, *Photomicrographie en cent tableaux pour projection* (Paris: Au bureau des *Mondes*, Gauthier-Villars, 1872).

98 Le Bon, Gustave, *L'Anatomie et l'histologie enseignées par les projections lumineuses* (Paris: Au bureau des *Mondes*, Gauthier-Villars, 1873).

99 Cf. section 'Photomicrographies', in *Vues sur Verre pour Projection* (Paris: A. Molteni, 1880), pp. 50–54 <http://archive.org/details/MolteniCatalogue32VuesSurVerre/page/n. 49/mode/2up?view = theater> [accessed 26.2.2021]).

100 Cf. Le Bon, Gustave, *L'Anatomie et l'histologie enseignées par les projections lumineuses*, pp. 8–9.

101 Cf. Dilly, Heinrich, 'Lichtbildprojektion — Prothese der Kunstbetrachtung', in *Kunstwissenschaft und Kunstvermittlung*, ed. by Irene Below (Giessen: Anabas, 1975), pp. 153–72.

102 'Von Jahr zu Jahr hat die Zahl der Künstler, welche ihre Erlaubniß zum Copiren gegeben, zugenommen [...].' ('Every year the number of artists, who give permission for their work to be copied, increases [...].') [Anon.], 'Die photographische Gesellschaft in Berlin', *Die Gartenlaube*, 2 (1878), p. 40 <https://de.wikisource.org/wiki/Die_photographische_Gesellschaft_in_Berlin> [accessed 8 March 2021]).

Nonetheless the process of building slide collections for academic institutions was slow and often hampered by a lack of finance. One early, albeit rare example was the German art historian Bruno Meyer (1840–1917) who used the lantern at the *Polytechnische Hochschule* in Karlsruhe for a decade between 1874–1884 and published on his experiences.[103] Art history collections in universities were created somewhat later, for instance in 1890 for Bordeaux and Naples.[104] The first illustrated lectures at Lausanne for example were given in 1887 and its first budget for the acquisition of slides was granted in 1903.[105] Strasbourg (then: *Kaiser-Wilhelms-Universität*) followed suit only during the winter term of 1902–1903.[106] The aforementioned Willem Vogelsang, who worked in the Netherlands, was one of the proponents of the use of the lantern in the lecture hall.

An invaluable contemporary source on lantern practice comes from the reflections of the German art historian Herman Grimm, published originally as a series of articles entitled, 'The transformation of university lectures on modern art history through the use of the sciopticon' of 1892 and subsequently included in Grimm's *Beiträge zur Deutschen Culturgeschichte*.[107] Grimm paid particular attention to the didactic changes that the use of the lantern occasioned.

First he emphasized the fact that the projected image could be seen from everywhere in the auditorium, 'die zu nahe gelegenen beiden ersten Sitzreihen vielleicht ausgenommen' (except perhaps for the too closely positioned first two rows of seats).[108] Furthermore, whereas photographs, etchings, and drawings generally reduce the size of a painting (without this giving rise to pause), lantern projection enlarges the image. Second Grimm described the possibilities offered by the sciopticon for a comparative approach, by showing two paintings representing the same subject side by side. (Fig. 1.4) He declared that 'As the images appear simultaneously, comparative examination takes place immediately.'[109] He particularly stressed how much he could now show to a large auditorium in only a brief period of time, which made the optical lantern a most efficient teaching aid.[110]

103 Cf. Männing, Maria, 'Bruno Meyer and the Invention of Art Historical Slide Projection', in *Photo-Objects: On the Materiality of Photographs and Photo Archives*, ed. by Julia Bärninghausen and others (online version at <http://mprl-series.mpg.de/studies/12/>), pp. 275–90 (p. 277). Cf. also Meyer, Bruno, 'Die Photographie im Dienste der Kunstwissenschaft und des Kunstunterrichts', *Westermanns illustrierte deutsche Monatshefte*, 47 (1879), pp. 196–209.

104 Cf. Miane, Florent, 'L'enseignement de l'histoire de l'art à Bordeaux', in *La plaque photographique*, ed. by Denise Borlée and Hervé Doucet (Strasbourg: Presses universitaires de Strasbourg, 2019), pp. 73–91 (p. 73); on Napoli Frederico II in 1890, cf. Rossella Monaco, 'Les plaques photographiques des archives Giovanni Previtali de l'université de Naples Frédéric II', in *La plaque photographique*, ed. by Denise Borlée and Hervé Doucet (Strasbourg: Presses universitaires de Strasbourg, 2019), pp. 93–105 (p. 95).

105 Cf. Blancardi, 'Archives de verre', p. 46.

106 Cf. Borlée, Denise, and Hervé Doucett, 'Les plaques de projection de l'Institut d'Histoire de l'Art de l'Université de Strasbourg. Un objet de recherches', in *La plaque photographique*, ed. by Denise Borlée and Hervé Doucet (Strasbourg: Presses universitaires de Strasbourg, 2019), pp. 108–22 (p. 113).

107 Grimm, Herman, 'Die Umgestaltung der Universitätsvorlesungen über Neuere Kunstgeschichte durch die Anwendung des Skioptikons', in *Beiträge zur Deutschen Culturgeschichte*, ed. by Herman Grimm (Berlin: Verlag von Wilhelm Hertz, 1897), pp. 276–395.

108 Grimm, 'Die Umgestaltung der Universitätsvorlesungen', p. 281.

109 'Indem die Bilder zu gleicher Zeit sichtbar gemacht werden, tritt die vergleichende Betrachtung sofort in Wirksamkeit.' Grimm, 'Die Umgestaltung der Universitätsvorlesungen', p. 282. Here Grimm may be referring to slides on which two paintings were reproduced side by side.

110 Grimm, 'Die Umgestaltung der Universitätsvorlesungen', pp. 283–85.

Fig. 1.4. Three busts, self-made slides in the collection of Alfred A. Schmidt, professor of art history in Fribourg, possibly 1960s. Courtesy of Diathèque, Section d'histoire de l'art, Université de Lausanne.

Of greatest interest perhaps is Grimm's stated view that before the introduction of the projected image, the lecturer's words were authoritative but as a result of the encounter with the art works themselves [*sic!*] students could get to know the masters by looking at them. Their own judgments now became part of the lecture, and they could check for themselves, whether the lecturer's observations were correct.[111] According to Grimm, the focus on the works, which the lantern had made possible, relegated the biographical dimension of art history instruction to the background and fostered the contemplation and examination of the individual works.[112] The optical lantern thus made a work-centred approach to art history possible.

Although commercial distributors offered a wide choice of art history slides, lecturers would still not always find the requisite image and would have a slide made from a reproduction in a book or another source by the university's photographic service. This applies to about 90 per cent of the slides preserved at the University of Bordeaux.[113] 65 per cent of the items in the database of the *Diathèque* at the *Université de Lausanne* (UNIL) lack a manufacturer's label, which indicates a large proportion of self-made slides.[114]

The diverse provenance of slides is reflected in particular in the collections held in countries such as Belgium and Switzerland, which lacked large producers or distributors of slides, in contrast to Britain, France, and Germany. Illustrative is the study made by the UNIL *Diathèque* of 10,000 glass positives acquired between *c.* 1900–1940 by different institutes within the university. Apart from the 65 per cent of unlabelled slides from the section on art history, archaeology, and architecture, about a quarter came from the German firm of Dr Franz Stoedtner (23.2 per cent), the others from E. A. Seemann, Germany (3.7 per cent), Fratelli Alinari, Italy (2.1 per cent),[115] Projections Molteni, France (1.9 per cent), and Lichtbeelden Instituut Amsterdam, the Netherlands (1.2 per cent). UNIL's lecturers thus ordered slides mainly from neighbouring countries. The presence of Swiss firms such as Fred. Boissonnas & Cie in Geneva and the Zurich-based company of J. Ganz & Co. is negligible. Furthermore the art historians at UNIL apparently preferred the larger firms such as Stoedtner, Seemann, and Molteni, while for example the smaller company of Ad. Braun et Cie,[116] which in 1920 offered only 543 items, is all but absent in the UNIL collection.

A study of existing slide collections can thus further our understanding of the material infrastructures that facilitated the adoption of the optical lantern. The UNIL case shows that commercial suppliers of slides did play an important role but that the ability to have images reproduced locally broadened the range of illustrations available for use by lecturers.

---

111 Cf. Grimm, 'Die Umgestaltung der Universitätsvorlesungen', pp. 307–08.

112 Cf. Grimm, 'Die Umgestaltung der Universitätsvorlesungen', p. 318.

113 Cf. Miane, 'L'enseignement de l'histoire de l'art à Bordeaux', p. 79.

114 It is possible that in a certain number of cases the label was covered or removed.

115 After being taken over in 1920, the company's name was changed to Fratelli Alinari I. D. E. A. (Istituto di Edizione Artistiche), slides from a later period were labelled 'Istituto Micrografico Italiano'. Cf. also Paoli, Silvia, 'Fratelli Alinari', in *Encyclopedia of Nineteenth-Century Photography*, ed. by John Hannavy, 2 vols (New York: Routledge, 2008), pp. 23–27.

116 Cf. Kempf, Christian, 'Les vues de projection et la maison Braun de Dornach', in *La plaque photographique*, ed. by Denise Borlée and Hervé Doucet (Strasbourg: Presses universitaires de Strasbourg, 2019), pp. 177–85 (p. 181).

Conclusion

For the optical lantern to be adopted as an academic teaching tool, an adequate technical infrastructure (including electricity as a safe light source for the projection lantern), practical knowhow and the availability of visual material for projection were undoubtedly a necessary but not in all cases a sufficient condition. Furthermore the discipline itself, supported by its institution status and its financial resources, had to take an active part in the implementation of the medium. Scott Curtis's observations concerning the introduction of moving pictures as a didactic tool in Germany after 1900 are no doubt equally applicable to the optical lantern, for which it took

> [...] the correspondence and mutual accommodation between the logic of a discipline — its problem-solving pattern, its investigatory methods, its ideological assumption — and [the medium's] characteristics [as well as the] match — between the formal features of the representational technology and the investigatory presumptions [...] because it provides the researcher, community, or discipline with the reassuring sense that the tool will fit the task to which it was assigned.[117]

To put it otherwise: the introduction of the optical lantern as a didactic tool was rather improbable when in a given discipline traditional teaching methods were not questioned and photography did not seem to contribute anything to resolving the problems on which the discipline was working. In 1897 *Photographische Mitteilungen* referred to the rejection of the projected image by 'ein berühmter Professor für Maschinenlehre' (a famous professor of practical mechanics) as a curiosity. The professor had stated that 'solche hingezauberten Bilder könnten das vor dem Zuhörer gezeichnete Bild des Lehrenden nie ersetzen' (pictures that are conjured up in this way could never replace a drawing made by the lecturer before the eyes of the audience). To follow this process was considered more effective for the student 'als [...] das plötzliche Erscheinen des fertigen Bildes auf dem Schirm' (than the sudden appearance of the completed picture on the screen).[118]

During the second half of the nineteenth century the projection lantern succeeded in overcoming its more or less exclusive association with travelling showmen and phantasmagoria performances. It gradually entered academic teaching after having been adopted for popular education by secular organizations and a group of enthusiasts within the Catholic Church. As we have shown, apart from considering the infrastructural and technological premises that made these developments possible, it is important to delve deeper into the specific conditions in which this process took place in a variety of disciplines to understand why in 1900 the archaeologist Holwerda and the art historian Vogelsang felt obliged to apologize for the absence of projected images in their lectures.

117 Curtis, *The Shape of Spectatorship*, p. 23.

118 Quoted in Vogel, Hermann Wilhelm, 'Über die Bildlaterne', *Photographischen Mitteilungen*, 1 (April 1897), pp. 1–3 (p. 2).

MARGO BUELENS-TERRYN

Taking the University to the People. The Role of Lantern Lectures in Extramural Adult Education in Early Twentieth-Century Brussels and Antwerp

Introduction

Public lectures have been one of the most dominant popular educational practices from the second half of the nineteenth century onwards.[1] Rooted in the Anglo-Saxon world, the culture of the popular lecture became a widespread phenomenon which also proved an important tool for non-formal adult education outside the walls of universities, hence the term 'extramural'.[2] Based on the examples of English university extensions (or 'community engagement') and those of French popular universities, similar initiatives were developed in both the Dutch- and French-speaking parts of Belgium (Flanders and Wallonia respectively) at the turn of the century. Although differences can be noted between university extensions and popular universities, their similarities are in fact more interesting and, in this chapter, the use of both terms side-by-side will be justified. Both types organized activities that

* This chapter was written as part of the research project 'B-magic. The Magic Lantern and Its Cultural Impact as Visual Mass Medium in Belgium (1830–1940)'. 'B-magic' is an Excellence of Science project (EOS-contract 30802346, 2018–2023), supported by the Research Foundation Flanders (FWO) and the Fonds de la Recherche Scientifique (FNRS). I would like to thank Kaat Wils and Nelleke Teughels, my supervisors Ilja Van Damme, Kurt Vanhoutte and Iason Jongepier, as well as Kristof Loockx, for providing such useful feedback and suggestions on earlier versions of this article. A special note of gratitude goes to Wout Vande Sompele, for his help with the very brief introduction to the economic history of (bread) prices.

1 van Damme, Dirk 'The University Extension movement (1892–1914) in Ghent, Belgium in comparative perspective', in *Adult Education between Cultures. Encounters and Identities in European Adult Education since 1890*, ed. by Barry J. Hake and Stuart Marriott (Leeds: University of Leeds, 1992), p. 13.

2 On the rise and development of popular lectures in the Anglo-Saxon world, see, for instance: Cunningham, Peter, Suzan Oosthuizen and Richard Taylors (eds), *Beyond the Lecture Hall: Universities and Community Engagement from the Middle Ages to the Present Day* (Cambridge: University of Cambridge, 2009); Eastman, Carolyn, 'Oratory and Platform Culture in Britain and North America, 1740–1900', *Oxford Handbooks Online*, (2016), <https://doi.org/10.1093/oxfordhb/9780199935338.013.33>; Wright, Tom F., 'The Lyceum Movement', in *The Cosmopolitan Lyceum: Lecture Culture and the Globe in Nineteenth-Century America*, ed. by Tom F. Wright (Amherst and Boston: University of Massachusetts Press, 2013).

Margo Buelens-Terryn • University of Antwerp, margo.buelens-terryn@uantwerpen.be

Learning with Light and Shadows: Educational Lantern and Film Projection, 1860-1990, ed. by Nelleke Teughels and Kaat Wils, TECHNE-MPH, 8 (Turnhout, 2022), pp. 51-75
© BREPOLS PUBLISHERS 10.1484/M.TECHNE-MPH-EB.5.131494

brought higher education to the people and in which lantern lectures played a central role.[3] This chapter aims to examine the use of the projection lantern in Belgian extramural education as an instrument to disseminate (scientific) knowledge and accommodate an emerging ideal of visual higher education for people from all walks of life. As a medium of mass communication, the lantern was believed to reach even those who were either illiterate or barely literate.[4] Moreover, the combination of word and image used in lantern lectures was seen as an important tool in 'the battle for attention' which reflected broader trends in the highly mediatized Belgian society.[5]

Despite the recent 'visual turn' or 'pictorial turn' in education and pedagogical historiography, the precise role of the projection device in extramural adult education remains understudied.[6] The literature about university extensions and popular universities focuses almost exclusively on the relationship between the emergence of extramural adult education on the one hand, and the changing societal frameworks of that time on the other. The need for new forms of (scientific) popular education fitted within broader contemporary evolutions in, among others, pedagogical reforms. The *Education Nouvelle*-movement (New Education), historically situated between the mid-nineteenth and mid-twentieth centuries, stood for a more scientific understanding of education and society, and considered this approach as the engine of emancipation and progress.[7] A new pedagogical ideology emerged at the turn of the century, supported by a significant part of the Belgian bourgeoisie and intellectual elite. They linked the perceived intellectual inferiority of the workers to the 'social question', which connected social issues with intensified industrialization.[8] As a result, new forms of popular education were created within the new industrial, proletarian urban centres in the hope of uplifting the masses, while maintaining the existing social order.[9] Another stimulus for popular education came with the introduction of plural male suffrage (1893) and single male suffrage (1919) in Belgium, a time when the 'ordinary' people entered the political arena as active players and needed to be convinced through word and images.[10] In a society that was under increasing pressure from tensions between secular

3 van Damme, Dirk, *Universiteit en volksontwikkeling: het hooger onderwijs voor het volk aan de Gentse universiteit (1892–1914)*, (Gent: Archief RUG, 1983), p. 37.

4 Lenk, Sabine, and Nelleke Teughels, 'Spreken met licht. Magische projectieplaatjes', *Koorts*, 1 (2020), pp. 4–9 (pp. 5–6).

5 Kember, Joe, 'The "Battle for Attention" in British Lantern Shows, 1880–1920', in *A Million Pictures: Magic Lantern Slides in the History of Learning*, ed. Sarah Dellmann and Frank Kessler (New Barnet: John Libbey Publishing Ltd, 2020), pp. 52, 58–59; Jonckheere, Evelien, *Aandacht! Aandacht! Aandacht en verstrooiing in het Gentse Grand Théâtre, Café-concert en Variététheater, 1880–1914* (Leuven: Universitaire Pers Leuven, 2017), pp. 9, 91.

6 Catteeuw, Karl, 'Als de muren konden spreken … Schoolwandplaten en de geschiedenis van het Belgisch lager onderwijs' (unpublished doctoral thesis, Katholieke Universiteit Leuven, 2005), pp. 11–26.

7 Grootaers, Dominique, 'Belgische schoolhervormingen in het licht van de "éducation Nouvelle" (1870–1970)', *Jaarboek voor de geschiedenis van opvoeding en onderwijs* (2001), pp. 9–33 (pp. 9–10).

8 van Damme, *Universiteit en volksontwikkeling*, pp. 7–11, 40; Elwitt, Sanford, 'Education and the Social Questions: The Universités Populaires in Late Nineteenth-Century France', *History of Education Quarterly*, 22 (1982), pp. 55–72 (p. 55).

9 van Damme, *The University Extension movement*, p. 12.

10 De Vroede, Maurice, 'Hogeschooluitbreidingen en Volksuniversiteiten', *BTNG-RBHC*, 10 (1979), pp. 225–78 (pp. 258–59, 267–68).

and religious groups, and with increasing class conflict, both the transfer of knowledge and class integration were seen as essential weapons.[11]

The relationship between the 'social question' and popular universities in France has been analysed by historian Sanford Elwitt.[12] For Belgium it is almost exclusively from the perspective of (the contribution to) popular education that university extensions and popular universities have been studied. Pedagogue Maurice De Vroede, active until the late 1980s, explored the emergence of both forms of extramural education and their success in terms of reaching the lower classes of society in Flanders and Wallonia respectively, while his students elaborated studies on specific popular universities in Brussels.[13] Pedagogue Dirk van Damme mainly discussed the university of Ghent's extension programme in its international context and mentioned lantern slides in the changing didactic framework of the university extensions to help workers overcome difficulties in attending lectures and courses. Still, the manner in which the lantern was incorporated is not discussed.[14] If mentioned in the literature, the projection lantern is mainly associated with lectures on geography, ethnography, and travel.[15] Furthermore, Van Damme focused on the university extension from the point of view of '*cultuurflamingantisme*' ('cultural flamingantism'), as Greet Nuyts did for the Antwerp *Katholieke Vlaamsche Hogeschooluitbreiding* (Catholic Flemish University Extension).[16] The Flemish movement strived for a 'Dutchification' of education and administration in function of the emancipation of Flanders which was regarded by the Flemish-minded as being a part of the country that was oppressed by the French-speaking people. 'Cultural flamingantism' reformulated these objectives in a cultural-political way.[17]

This chapter will demonstrate where, why, and in what specific way the projection lantern was used for the communication of all types of knowledge. Based on case studies of *hogeschooluitbreidingen* (university extensions) and *universités populaires* (popular universities) in two of the most important cities in Belgium, Brussels and Antwerp, I aim to highlight the added value brought by employing visual media to reach out to the masses. Moreover, I will demonstrate how this medium contributed to the popularization of scientific knowledge which was an important goal within both forms of extramural adult education. In what follows, I will first describe the sources and methodology I

---

11 van Damme, *Universiteit en volksontwikkeling*, pp. 7–11, 40; de Vries, Boudien, 'De waarde van kennis bij arbeiders en de kleine burgerij in de tweede helft van de negentiende eeuw', *De Negentiende Eeuw*, 33 (2009), pp. 53–70 (p. 55).

12 Elwitt, 'Education and the Social Questions', pp. 55–72.

13 De Vroede, 'Hogeschooluitbreidingen en Volksuniversiteiten', pp. 225–78; Bossaerts, Beatrijs, 'Drie Volksuniversiteiten uit het Brusselse, begin Twintigste Eeuw' (unpublished master thesis, Katholieke Universiteit Leuven, 1979); Van der Wee, Annemie, 'De Volksuniversiteit te Sint-Gillis, 1901–1914' (unpublished master thesis, Katholieke Universiteit Leuven, 1978); Couvreur, Luc, 'De Volksuniversiteit te Schaerbeek' (unpublished master thesis, Katholieke Universiteit Leuven, 1977).

14 van Damme, *Universiteit en volksontwikkeling*, p. 42; van Damme, *The University Extension movement*.

15 Couvreur, 'De Volksuniversiteit te Schaerbeek', p. 36; Bossaerts, 'Drie Volksuniversiteiten uit het Brusselse, begin Twintigste Eeuw', pp. 34, 47, 57; Van der Wee, 'De Volksuniversiteit te Sint-Gillis, 1901–1914', pp. 33–41.

16 Nuyts, Greet, 'De Katholieke Vlaamse Hogeschooluitbreiding, Antwerpen 1898–1914' (unpublished master thesis, Katholieke Universiteit Leuven, 1974).

17 Flemish is a derivative of Dutch that is spoken in Flanders. van Damme, *The University Extension movement*, pp. 19, 37; Witte, Els, and Jan Craeybeckx, *Politieke geschiedenis van België sinds 1830. Spanningen in een burgerlijke democratie*, (Antwerpen: Standaard Wetenschappelijke Uitgeverij, 1985), p. 149.

used, and provide context. Subsequently the public aspect of the lectures is addressed by looking at attendance numbers and entrance fees, and at the geographical spread of lantern lectures within Antwerp and Brussels, with particular attention given to the socio-economic characteristics and population density of the municipalities in relation to the intended audience. In the final sections, I will examine how the lantern was used as a tool in the battle for attention and consider what knowledge was transferred by means of the projection lantern.

Sources, Methodology and Context

The role of the projection lantern in extramural adult education will be examined using a corpus of sources that is twofold. The first part is composed of publications and archival material of the university extensions and popular universities.[18] These include annual reports, publications on their wider operation, and material relating to specific topics such as how to select suitable subjects for lectures. These sources have already been explored by researchers in the past, primarily for insights into the operation and purpose of the organizations. The sources have not yet been studied from the point of view of the use and role of the lantern. In this chapter, the main focus will be on the Antwerp Catholic Flemish University Extension, (hereafter CFUE), in comparison to the broader university extension- and popular university-movement in Brussels and Antwerp. Both cities are approached broadly as there is far too little historical reflection in the literature on how the (lantern) lecture circuit transcended the boundaries of the major (capital) cities. However, this tells us much about the ubiquity of extramural adult education whose founders wanted to be present everywhere in order to get their message across. The surrounding suburban municipalities of Brussels and Antwerp are therefore included.[19]

The second part of the sources consists of an extensive database of Belgian announcements and reviews of lantern lectures in newspapers for the sample periods 1902–1904 and 1922–1924. In this way two interesting time frames are included: around 1900, Belgium was still largely a reflection of nineteenth-century civil society. The turn of the century was marked by the clash between established powers and emerging emancipatory movements along with a gradual opening up to 'the masses'. Despite the *belle époque*'s belief in progress, the spectre (or fear) of 'the masses' loomed. In the 1920s, Belgium had

18 Since the focus in this chapter is on the Catholic Flemish University Extension, the collections of KADOC (Documentation and Research Centre on Religion, Culture and Society) in Leuven proved to be especially useful.

19 The surrounding communes of Brussels have been included: Anderlecht, Auderghem, Berchem-Sainte-Agathe, Brussels, Etterbeek, Evere, Forest, Ganshoren, Ixelles, Jette, Koekelberg, Molenbeek-Saint-Jean, Saint-Gilles, Saint-Josse-ten-Node, Schaerbeek, Uccle, Watermael-Boitsfort, Woluwe-Saint-Lambert and Woluwe-Saint-Pierre. (Leblicq, Yvon, and Machteld De Metsenaere, 'De Groei' in *Brussel. Groei van een hoofdstad*, ed. by Jean Stengers (Antwerpen: Mercatorfonds, 1979), pp. 167–71.) For Antwerp, the suburban municipalities Berchem, Borgerhout, Deurne, Hoboken, Merksem and Mortsel are included. (May, Laura, 'Suburban place-making. Political economic coalitions and 'place distinctiveness' (Antwerp, *c.* 1860-*c.* 1940)' (unpublised doctoral thesis, Universiteit Antwerpen, 2020), pp. 18, 45, 63; Vrints, Antoon, *Bezette Stad. Vlaams-nationalistisch collaboratie in Antwerpen tijdens de Eerste Wereldoorlog*, (Brussel: Algemeen Rijksarchief, 2002), p. 19.)

emerged as a mass society, with an intensified Flemish emancipation and an increasingly expanded welfare state.[20]

Since Belgian society was divided ideologically into three main groups, that is, socialist, catholic and liberal (a phenomenon also known as 'pillarization'), I have collected announcements and reviews of lantern lectures in newspapers from the three main pillars for both Brussels and Antwerp and, if available, in both Dutch- and French-speaking newspapers. For Brussels, I have also analysed one neutral and one communist newspaper (both in French).[21] As such, these methodological considerations provide a representative sample of newspapers to enable the investigation of general trends in lantern practices within extramural education in both Belgian cities. Although the literature indicates that lantern lectures were extensively reported, these advertisements and reviews have rarely, if ever, been researched to uncover spatial or other substantive patterns in the lantern lecture circuit.

Information about the date, hours, speakers, topics, locations and (reactions of) the (intended) audience can all be extracted from the newspaper announcements (see Fig. 2.1). This allows for instance to analyse the themes discussed and the geographical spread of lantern lectures. Even so, the collected references inevitably constitute an underrepresentation of the total number of lantern lectures given; lantern lectures that were announced outside the mainstream daily press are not included. Moreover, lectures were also advertised through word-of-mouth, street posters, or via other advertising channels. The daily press, however, had a broader geographical reach and thus provides a good picture of how a broad, interested readership was informed about public lectures through general media channels. The picture portrayed here therefore does not seek to recreate a historical reality in absolute numbers but should be understood instead as an indication of the familiarity of the projection device among the urban masses around 1900–1920. When we look at the overall picture of public lantern lectures, it immediately becomes clear how widespread the practice of illustrated lectures organized by extramural adult education was. Of the 5774 unique lantern lectures announced in the Brussels and Antwerp newspapers for both sample periods together, almost a quarter were organized by university extensions and popular universities. Other organizers included cultural and scientific associations, alumni unions, and other societies.

---

20 Deneckere, Gita, *1900: België op het breukvlak van twee eeuwen* (Tielt: Uitgeverij Lannoo, 2006), pp. 7–16; Strikwerda, Carl, *A House Divided. Catholics, Socialists and Nationalists in Nineteenth-Century Belgium*, (Lanham, Boulder, New York e.a.: Rowman and Littlefield Publishers, 1997), p. 280; Van Ginderachter, Maarten, *The Everyday Nationalism of Workers: A Social History of Modern Belgium*, (Stanford: Stanford University Press, 2019), pp. 167–69.

21 The newspapers included for Antwerp are: *De Nieuwe Gazet* (Liberal, Dutch), *Le Matin, Le Précurseur and Le Nouveau Précurseur* (Liberal, French), *Het Handelsblad* and *Gazet van Antwerpen* (Catholic, Dutch), *La Métropole* (Catholic, French), *De Werker* and *De Volksgazet* (Socialist, Dutch). For Brussels: *Het Laatste Nieuws* (Liberal, Dutch), *L'Indepéndance Belge* (Liberal, French), *Het Nieuws van den Dag* and *De Standaard* (Catholic, Dutch), *Journal de Bruxelles* (Catholic, French), *Le Peuple* (Socialist, French), *Le Drapeau Rouge* (Communist, French) and *Le Soir* (Neutral, French). For more information about the exact selection and processing of these newspaper(s) (articles) see: Buelens-Terryn, Margo, Iason Jongepier and Ilja Van Damme, 'Lichtbeelden voor de massa. Toe-eigening en gebruik van de magische lantaarn in Antwerp en Brussel (*c.* 1860-*c.* 1920)', *Stadsgeschiedenis* 14 (2019), pp. 122–36 (pp. 130–32).

Universités populaires

— Université Populaire Nord-Est. (Ecole n° 9, rue des Eburons, 50.)

Samedi 19, à 8 h. 1/4, cours de l'extension de l'Université libre. M. Demoor : Bases scientifiques de l'éducation physique. (4e leçon.)

— Le Foyer Intellectuel. (Local : Ecole rue du Fort, 80, Saint-Gilles.)

Samedi 19, à 8 h. 1/2, M. G. Hallut, conférence avec exécution et audition musicales : Jean Sébastien Bach.

Dimanche 20, visite des antiquités égyptiennes au musée du Cinquantenaire. Départ du local à 9 h. très précises. Réunion à 10 h. au Musée.

Lundi 21, à 8 h. 1/2, M. Charles Buls, ancien bourgmestre de Bruxelles, conférence avec projections lumineuses : Le forum romain.

A 8 h., section de chimie expérimentale par MM. Ruelle et Vidal.

Fig. 2.1. Example of an announcement of lantern lectures by Brussels popular universities. Le Peuple, 19 November 1904, p. 3.

	Brussels	Antwerp
1902-1904	634	10
1922-1924	254	66

Table 2.1. The number of unique lantern lectures in extramural education mentioned in the newspapers that took place in Brussels and Antwerp (1902–1904 and 1922–1924). Author's database.

For both sample periods, Table 2.1 shows the number of unique lantern lectures of extramural education announced in newspapers that took place in Brussels and Antwerp. At first glance, it is already clear that Brussels was making earlier and more intensive use of the projection lantern in the context of university extensions and popular universities.

French-speaking extramural initiatives were mainly active in Brussels and were inspired by the popular universities in France.[22] As the bilingual capital of the Kingdom of Belgium,

22 van Damme, *The University Extension Movement*, pp. 11–14.

Brussels had strong international ties and was the beating heart of Belgium's largely French-speaking financial circles as well as the hometown of the monarch, nobility, and the liberal *haute bourgeoisie*.[23] In the historiography the popular university movement is mostly described as being less influential than the university extension movement. However, based on the analysis of lectures advertised in newspapers, this does not apply to Brussels, where socialists and social-liberals strongly supported adult education.[24] The pioneering role in the use of the educational lantern by the Brussels socialists is related to their specific characteristics. Carl Strikwerda describes this socialism as 'virtually working-class liberalism', as the Brussels' workers 'possessed the highest wages and literacy, had the longest tradition of organization and had the greatest access to resources such as printing presses, meeting halls, and middle-class assistance.'[25]

Shortly after the first popular universities were established in Brussels and its municipalities, they experienced a rapid growth both in terms of activity and membership. Their original aim (the intellectual, moral, artistic, and social uplifting of the working class) was strongly supported by the Belgian Workers Party, that is, pursuing the emancipation of the ordinary people more widely. Announcing a lantern lecture at a popular university in the socialist newspaper *Le Peuple*, an unknown editor added: 'We can't tell our friends enough how much regular attendance at the popular university is necessary for their intellectual emancipation.'[26]

By contrast, it were the university extensions that were the most popular in Flanders, where the Flemish-speaking variants looked at Britain as a model.[27] Before the First World War the Antwerp socialist movement lagged behind other major Belgian cities, which partly explains the large difference in the numbers of lantern lectures that took place between Brussels and Antwerp. Only after the First World War did the more Flemish-oriented Antwerp socialists take over the leading role from their Ghent colleagues, even though this did not immediately translate into a use of the lantern that was as enthusiastic as their Brussels counterparts.[28] It was the Catholic Flemish University Extension (CFUE, °1898) that dominated Antwerp extramural education during the period under study. Only in the 1920s, liberal and socialist initiatives were developed that are worth mentioning. The founders of CFUE deliberately chose a 'Catholic' and 'Flemish' approach. They wanted to demonstrate that religion and science were not opposed, but could be considered as two sides of the same coin. Unlike most popular universities, lectures were consistently held in Flemish: 'Because by using the Flemish language we wanted to contribute to refuting

23 Wagenaar, Michiel, *Stedenbouw en burgerlijke vrijheid. De contrasterende carrières van zes Europese hoofdsteden*, (Amsterdam: Uitgeverij Thoth, 2001), pp. 54–70; Kenny, Nicolas, *The Feel of the City. Experiences of Urban Transformation* (Toronto, Buffalo and Londen: University of Toronto Press, 2014), p. 32; Strikwerda, *A House Divided*, pp. 36–39, 57.

24 Elwitt, 'Education and the Social Questions', p. 57.

25 Strikwerda, *A House Divided*, pp. 53, 86.

26 Translated from the French: « Nous ne pouvons assez redire à nos amis combien est nécessaire à leur émancipation intellectuelle la fréquentation régulière des universités populaires. » *Le Peuple*, 3 April 1902, p. 2.

27 van Damme, *The University Extension Movement*, pp. 11–14.

28 De Munck, Bert, and Maarten Van Ginderachter, 'Over drinkebroers en vechtersbazen. Sociale organisaties, arbeidersverhoudingen en collectieve actie', in *Antwerpen. Biografie van een stad*, ed. by Inge Bertels, Bert De Munck and Herman van Goethem (Antwerpen: De Bezige Bij, 2010), pp. 245–73 (pp. 256–57, 261).

the false assertion that our language is unsuitable for science'.[29] The Flemish university extensions were thus employed both for raising Flemish awareness and for intellectually and culturally developing Flanders, an enterprise which gained momentum in the 1920s as illustrated by Table 2.1.

It is no coincidence that the CFUE came into being in Antwerp in an era of strong Catholic self-awareness in the Scheldt city. Antwerp represented the old counter-reformatory stronghold of the past and was known as the beating heart of the Flemish movement, in which Catholicism played an important role. As the largest Belgian city without its own university, the Scheldt city with its harbour and the particular dynamics that existed between its port and trade, demanded a well-developed professional education.[30] The lantern activities of the CFUE increased over time both in the centre of Antwerp and the surrounding municipalities. Various (local) sub-sections were established during the first two decades of the twentieth century, focusing on specific contemporary socio-political issues which were dealt with using the projection lantern. These sub-sections tended to meet at fixed times and locations throughout the week, excluding weekends.

Who Comes at What Price: Attendance Numbers and Entrance Fees

Both university extensions and popular universities tried to spread knowledge and higher culture to every layer of society. The founders of the CFUE in Antwerp saw a need for this in their own time, as described in 1907 on the occasion of their upcoming tenth anniversary:

> The twentieth century in particular will be the century of one's own work and one's own education; and to be able to fight the ever fiercer 'struggle for life' one needs weapons, for which one has to look particularly to science and scientific knowledge. The *University Extensions* are therefore called upon to be spread everywhere and to be followed, they will be like armories for the workers of the twentieth century.[31]

The popular universities, for their part, defined their overarching goal as being to lead the workers on the path of emancipation through intellectual and moral education.[32] Whether

29 Translated from the Dutch: '*omdat door het gebruik der Vlaamsche taal wij wilden bijdragen de valsche beweering te weerleggen, dat onze taal ongeschikt is voor de Wetenschap.*' (*Hooger Onderwijs voor't Volk. Hoogeschooluitbreiding*, Katholieke Vlaamsche Hoogeschooluitbreiding: Antwerpen, 1901–1902, pp. 16–17); De Vroede, 'Hogeschooluitbreidingen en Volksuniversiteiten', p. 259.

30 de Smaele, Henk, *Rechts Vlaanderen. Religie en stemgedrag in negentiende-eeuws België* (Leuven: Universitaire Pers Leuven, 2009), p. 361; Beyen, Marnix, Luc Duerloo and Herman Van Goethem, 'Het calimerocomplex van de stad. Een politieke cultuur van klagen en vernieuwen', in *Antwerpen. Biografie van een stad*, ed. by Inge Bertels, Bert De Munck and Herman van Goethem (Antwerpen: De Bezige Bij, 2010), pp. 67–108 (pp. 84–86); Vrints, *Bezette Stad*, pp. 19–23; Lenders, Piet, '150 jaar hoger handelsonderwijs in Antwerpen' in *Antwerpen en de Jezuïeten, 1562–2002*, ed. by Herman Van Goethem (Antwerpen: UFSIA, 2002), pp. 163–84 (p. 165).

31 Translated from the Dutch: '*De XXe zal bijzonder de eeuw van eigen werk en eigen vorming zijn; en om den steeds hevigeren "struggle for life" te kunnen strijden heeft men wapenen noodig, die men grootendeels in bijzondere wetenschap en hoogere kennis zal moeten zoeken. De Hoogeschooluitbreidingen, geroepen om overal verspreid en gevolgd te worden, zullen als wapenhuizen zijn voor de zwoegers der XXe eeuw.*' *Gedenkboek van de Katholieke Vlaamsche Hoogeschooluitbreiding van Antwerpen* (Katholieke Vlaamsche Hogeschooluitbreiding: Antwerpen, 1907), p. 5.

32 Bossaerts, 'Drie Volksuniversiteiten uit het Brusselse', p. 27.

or not they actually reached their intended audience, and thus their goals, is often difficult to determine in the field of lantern research. The sources give only indirect information about the audience and we therefore have to rely on indications, such as entrance fees, periodic reviews about attendance and reactions of the audience, as well as the relation between the intended audience as described in publications and the locations of lectures.

The rare reviews published in newspapers sometimes emphasize that 'the room was packed' or that there was 'a very large audience' present.[33] On occasion, it is declared in organizations' reports that the size of this audience fluctuated from a couple of dozen to as many as three hundred people at their peak, and consisted of both women and men as the former were regularly encouraged to accompany their husbands to lantern lectures.[34] It was also sometimes stated in newspaper announcements that the organizers expected a large turnout. This prediction may or may not have been prompted by the success of a previous lecture by the same speaker. In the context of lantern lectures for religious and socio-political education in the United Kingdom, Karen Eifler states that 'the many statements about enthusiastic audience reactions testify to the power that social organizations attributed to their lantern performances'.[35] The same critical reflection must be made when analysing newspaper reviews and articles. Testimonies about audience turnout must be read first and foremost from the perspective of the value attached to these lantern lectures by the organizers; a discrepancy undoubtedly existed between expected and actual figures. After all, they did not always achieve the desired results. A disappointed voice in the Catholic newspaper *Het Handelsblad* stated:

> It is regrettable that so few people are interested in the remarkable lectures given by the CFUE. These very instructive lectures should surely be attended by all those who, by virtue of their position in society, are called upon to promote the cultural development of our people. Please, leading Catholics of Berchem, shake off your indifference and come to the CFUE that are held every 14 days in the *St-Stanislasgesticht* and support through your presence the very laudable efforts of the organizers.[36]

Attendance numbers are not particularly meaningful until we know who exactly had (financial) access to the lectures. Entrance fees can provide more insight into this. From the beginning of the university extension movement in Belgium, the issue of entrance fees was discussed. Some people followed the English model, advocating the introduction of attendance as a sign of mutual respect between speaker and listener. Moreover, it would motivate students, it was felt. The Ghent socialists also supported this point of view in

33 Translated from Dutch to English: '*de bovenzaal was proppensvol*' and '*een zeer talrijk publiek*'. *Het Handelsblad*, 11 March 1922, p. 3 and *Het Handelsblad*, 31 October 1924, p. 3.

34 Bossaerts, 'Drie Volksuniversiteiten uit het Brusselse', p. 34; Van der Wee, 'De Volksuniversiteit te Sint-Gillis, 1901–1914', pp. 30, 35; KVHU, *Hooger Onderwijs*, Antwerpen, 1901–1902, p. 18.

35 Eifler, Karen, 'Sensation — Intimacy — Interaction: Lantern Performances in Religious and Socio-Political Education', *Early Popular Visual Culture*, 17 (2019), 45–70 (p. 54).

36 Translated from the Dutch: '*Spijtig is het te moeten bestatigen dat zoo weinig belangstellenden de merkwaardige voordrachten der K. V. H. U. volgen. Deze zeer leerrijke spreekbeurten zouden toch zeker moeten bijgewoond worden door al de personen die, door hunne positie ib de samenleving, geroepen zijn om de kultureele ontwikkeling van ons volk te bevorderen. Toe, vooraanstaande katholieken van Berchem, schudt de onverschilligheid van u af, en komt in't vervolg naar de voordrachten der K. V. H. U. welke om de 14 dagen in het St-Stanislasgesticht gehouden worden en steun door uwe aanwezigheid het zeer loffelijk pogen der inrichters. Het Handelsblad*, 10 December 1922, p. 4.

relation to the Ghent extension: it would humiliate the working class if popular education had to be offered as a form of charity. On the other side was a group led by Professor Paul Frédericq of the Ghent department of Dutch philology and literature, a dominant figure in the central committee of the Ghent university extension, who strived for free admission to the lectures.[37]

Based on newspaper advertisements, the organizers of extramural adult education in Brussels and Antwerp seem to have sided with the latter group and made their lectures 'free and without charge'.[38] In contrast, for lectures given within a course often an entrance fee was asked. For example, in February 1904 the CFUE established a lecture series titled 'On Congo and Colonial Sciences', in response to 'the systematic campaign of the English press against Congo Free State, the purpose of which is only too apparent.'[39] In doing so, they countered the exploding criticism, following the report of the British consul in Boma, of Belgian King Leopold II's colonial project.[40] This course included five lectures, for which a total of 1 franc was charged.[41] To give an idea of the value of this amount: in 1896, an adult male in manufacturing earned an average of 3.28 Belgian francs a day (expressed in the franc of 1914).[42] In 1904, the price of wheat stood at 0.174 francs per kilo, and one paid 0.26 francs for a loaf of white bread in Ghent.[43] A single lecture within this course therefore cost just under one loaf of bread, but the attendee did have to pay for the five lectures in one go (1 franc). Overall, this relatively low entrance fee leads me to conclude that lantern lectures were affordable and could therefore reach more people through the communication of science and general higher knowledge. This is confirmed by the existing literature which has demonstrated that on a national level expenditure by Belgian households on 'shows and entertainment' increased significantly in the first decades of the twentieth century (from approximately thirty-five million francs in 1900 to fifty million in 1910). Moreover, the law of compulsory Sunday rest (approved on 26 July 1905) also allowed an increasing amount of time to be spent on leisure activities.[44]

The ability to afford potential entrance fees is clearly no hard evidence that the people actually showed up at these lantern lectures. Moreover, statements about audience turnouts almost never make clear what the (socio-economic) background of the attendees were.

---

37 van Damme, *The University Extension Movement*, p. 20.

38 *Katholieke Vlaamsche Hoogeschooluitbreiding te Antwerpen, 15de jaar*, Katholieke Vlaamsche Hoogeschooluitbreiding: Antwerpen, 1912–1913, p. 17. See also announcements such as *Het Handelsblad*, 30 January 1904, p. 1 (Translated from the Dutch: *'vrij en kosteloos'*).

39 Translated from the Dutch: *'De stelselmatige veldtocht der Engelsche pers tegen Belgisch Congo, waarvan het doel maar al te duidelijk wordt.' Het Handelsblad*, 7 February 1904, p. 2. The newspaper article speaks of 'Belgian Congo', although at this time the colony was still the personal property of Leopold II and thus called Congo Free State. It was not until 1908 that the colony was transferred to the Belgian state. Stanard, Matthew G., *Selling the Congo: A History of European Pro-Empire Propaganda and the Making of Belgian Imperialism*, (Lincoln: University of Nebraska Press, 2011), p. 44.)

40 Stanard, *Selling the Congo*, pp. 27, 30–31; Van Reybrouck, David, *Congo. Een geschiedenis*, (Amsterdam: De Bezige Bij, 2010), pp. 110–11.

41 *Het Handelsblad*, 7 February 1904, p. 2.

42 Scholliers, Peter, 'Honderd jaar koopkracht in België (1914–2014)', *Brood & Rozen*, 20 (2015), 5–21 (p. 7).

43 Avondts, Gerda, and Peter Scholliers, *Gentse prijzen, huishuren en budgetonderzoeken in de 19^{e} en 20^{e} eeuw* (Brussel: Centrum voor Hedendaagse Sociale geschiedenis, 1981), p. 50.

44 Segers, Yves, *Economische groei en levensstandaard. De ontwikkeling van de particuliere consumptie en het voedselverbruik in België, 1800–1913*, (Leuven: Universitaire Pers, 2003), pp. 205–06.

The exact location (neighbourhood, district) in which the lantern lectures were organized may reflect the type of the audience that extramural adult education sought to reach and the likely success in doing so. The reasons for the establishment of lantern lectures in certain places can, in other words, be linked to the intended audience and the character of the neighbourhood, municipalities, or suburbs in question. It is thus worthwhile considering here the geographical spread of lectures in Brussels and Antwerp in relation to the intended audience.

The Right People at the Right Place: Location and Intended Audience

Extramural adult educational activities in the capital were distributed spatially and linked to the characteristics of the different municipalities. As the popular universities especially were spread throughout Brussels, it is worth noting the differences between the communes (Table 2.2). It was not until December 1900 that the first Belgian popular university was launched in Schaerbeek. During the first sampling period (1902–1904) several followed in its wake. However, some came into being only after 1904 and disappeared again before the second sample period (1922–1924) such as *Oeuvre Nouvelle* (the popular university of Brussels, 1907–1919) and the popular university of Watermael-Boitsfort (1905–1911).[45] It was mainly the popular universities in the socio-economically wealthier communes who had already opened their doors in 1902–1904, like Ixelles, Etterbeek, Saint-Josse-ten-Node, and Schaerbeek. It is probably no coincidence that they were at the forefront of setting up this form of extramural adult education and lantern lectures. After all, the popular universities were usually financially dependent on subsidies from, among others, their municipality, which was also able to provide rooms such as school buildings where the lectures could be held. Richer municipalities probably had the means to do this more quickly, which could explain the more rapid establishment of popular universities in these areas. At the same time the intended audience were often workers, so it was not by chance that the choice for the venue in Schaerbeek was 'Brasserie Chevalier', a café in a densely populated and working-class suburb.[46] In the industrial communes such as Anderlecht and Molenbeek-Saint-Jean, popular universities with a great deal of lantern activity also emerged at the beginning of the twentieth century.[47]

Yet the record holder of lantern lectures announcements in the daily press in this first sample was *Le Foyer Intellectuel*, the popular university of Saint-Gilles (founded in 1901). This commune had a double profile from a socio-economic point of view. Around the middle of the nineteenth century, the poorer blue-collar workers had been pushed out of the city centre into areas such as Molenbeek-Saint-Jean, Anderlecht, and certain parts of

---

45 Couvreur, 'De Volksuniversiteit te Schaerbeek', p. 1; Bossaerts, 'Drie Volksuniversiteiten uit het Brusselse', pp. 13, 29–30, 60.

46 Couvreur, 'De Volksuniversiteit te Schaerbeek', pp. 2, 51–52; Bossaerts, 'Drie Volksuniversiteiten uit het Brusselse', p. 31; Van der Wee, 'De Volksuniversiteit te Sint-Gillis, 1901–1914', p. 24; Leblicq and De Metsenaere, 'De Groei', p. 174.

47 Leblicq and De Metsenaere, 'De Groei', p. 174; Dupont, Wannes, 'Free-floating evils. A Genealogy of homosexuality in Belgium' (unpublished doctoral thesis, Universiteit Antwerpen, 2015), p. 377–78.

	Population 1900	**Total unique announcements of extramural education lectures (1902-1904) (n=634)**	*Population 1920*	**Total unique announcements of extramural education lectures (1922-1924) (n=254)**
Anderlecht	47,929	41	67,038	0
Auderghem	4685	0	9108	0
Berchem-Sainte-Agathe	1845	0	3851	0
Brussels	183,686	72	154,801	53
Etterbeek	20,838	50	39,813	9
Evere	3892	0	7,192	0
Forest	*9509*	*0*	*31,152*	*16*
Ganshoren	2872	0	4451	0
Ixelles	*48,615*	*26*	*81,245*	*42*
Jette	*10,053*	*0*	*16,109*	*31*
Koekelberg	*10,650*	*0*	*12,502*	*19*
Molenbeek-Saint-Jean	58,445	23	71,225	19
Saint-Gilles	51,763	255	64,814	21
Saint-Josse-ten-Node	32,140	70	31,843	0
Schaerbeek	63,508	74	101,526	7
Uccle	*18,034*	*24*	*32,056*	*37*
Watermael-Boitsfort	6520	0	10,096	0
Woluwe-Saint-Lambert	3468	0	11,300	0
Woluwe-Saint-Pierre	2686	0	8072	0

Table 2.2. The number of unique extramural education lantern lectures mentioned in the newspapers that took place in Brussels and its communes (1902–1904 and 1922–1924), and the official population numbers (according to general censuses, 1900 and 1920). The municipalities in which lantern activity grew between the first and the second sample are shown in italics. Author's database; Vrielinck, Sven, *De territoriale indeling van België (1795–1963). Bestuursgeografisch en statistisch repertorium van de gemeenten en de supracommunale eenheden (administratief en gerechtelijk). Met de officiële uitslagen van de volkstellingen*, 3 vols (Leuven: Universitaire Pers Leuven, 2000), III, pp. 1667–1777.

Saint-Gilles. At the turn of the century, these working-class neighbourhoods became the fourth largest industrial and working-class communes in the Brussels region. At the same time, an artistic concentration arose. In contrast, wealthy citizens moved to other, more residential parts of Saint-Gilles, after which it would eventually develop into a bourgeois suburb of Brussels.[48] In short, it can be said that both the (financial) means, as well as the intended audience, were present.

48 Leblicq and De Metsenaere, 'De Groei', p. 174; Debroux, Tatiana, 'Des artistes en ville: géographie rétrospective des plasticiens à Bruxelles, 1830–2008' (unpublished doctoral thesis, Université libre de Bruxelles, 2012), pp. 141, 188; Dupont, p. 377.

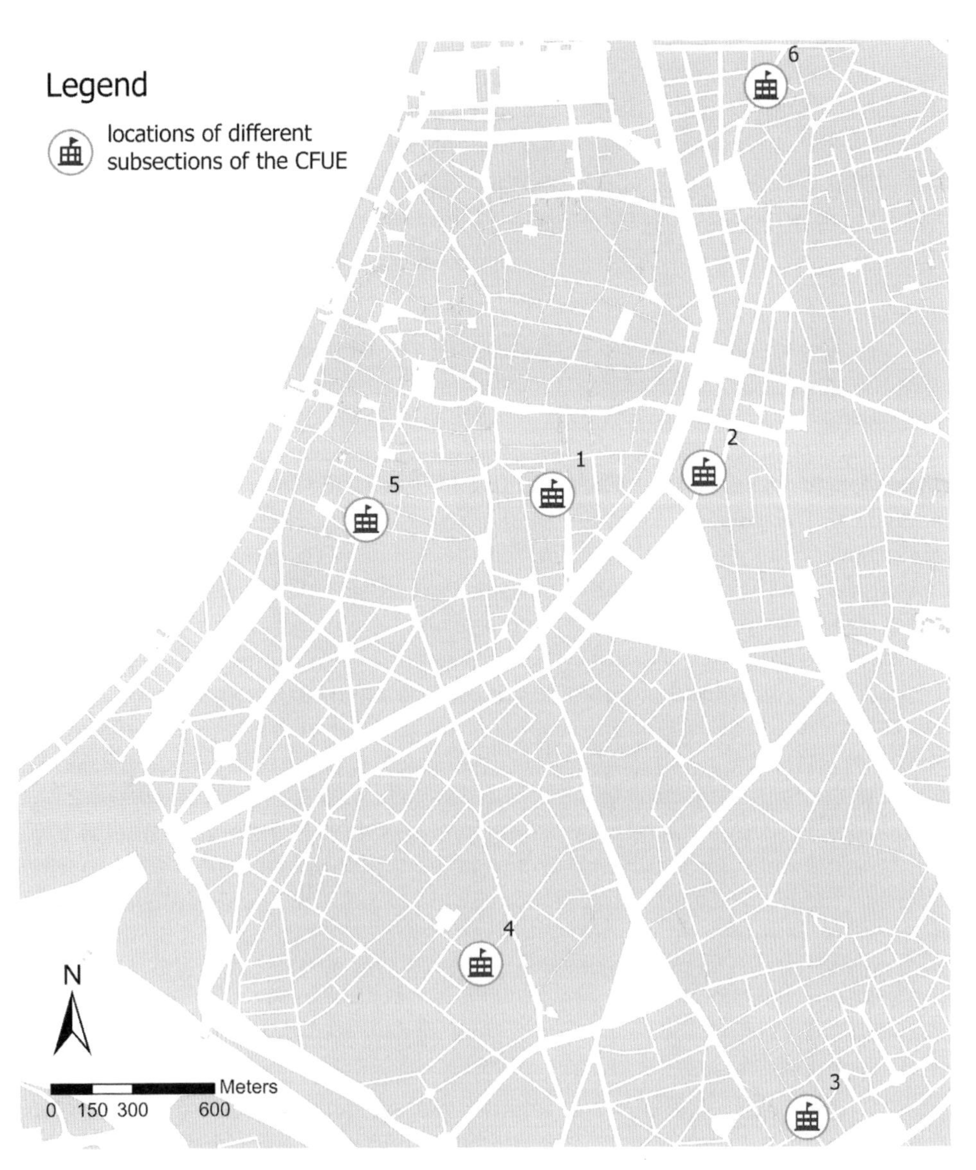

Fig. 2.2. Locations of the subsections of the Catholic Flemish University Extension held in Antwerp (1902-1904 and 1922–1924). Author's database.

The abovementioned sub-sections of the Catholic Flemish University Extension were each aimed at a specific audience, something which manifested itself in Antwerp spatially with a greater spread of, and differentiation between, the lecture halls used for illustrated lectures. In 1902–1904, the CFUE in Antwerp only reached the press with lantern lectures in *Zaal Anthonis* (Room Anthonis, number 1 on Fig. 2.2). In the second sample period, by contrast, the lectures in the centre of Antwerp took place in the rooms of the '*Stedelijke Normaal- en Oefenschool voor jongens*' (Urban Ordinary and Training School for Boys, number 2). In addition to the centrally located sites, several satellites emerged that were

	Population 1900	**Total unique announcements of extramural education lectures (1902-1904) (n=10)**	*Population* 1920	**Total unique announcements of extramural education lectures (1922-1924) (n=66)**
Antwerp	272,831	9	302,058	51
Berchem	*19,962*	*1*	*32,115*	*6*
Borgerhout	37,693	0	52,860	1
Deurne	8517	0	15,337	0
Hoboken	10,202	0	21,006	0
Merksem	*11,648*	*0*	*20,173*	*8*
Mortsel	3463	0	7866	0

Table 2.3. The number of unique extramural education lantern lectures mentioned in the newspapers that took place in Antwerp and its suburban municipalities (1902–1904 and 1922–1924), and the official population numbers (according to general censuses, 1900 and 1920). The municipalities in which lantern activity grew between the first and the second sample are shown in italics. Author's database; Vrielinck, *De territoriale indeling van België (1795–1963)*, pp. 1669–1735).

more scattered throughout the city and increasingly nearer the outskirts of the city (see Fig. 2.2). The lecture series and classes of the '*Katholieke Vlaamsche Volkshogeschool'* (Catholic Flemish University College), for example, were given in the *Sint-Norbertusgesticht* (number 3). Other sub-sections of the CFUE were an '*Afdeeling voor Dames-Onderricht* (Department for Ladies' Education), a '*School voor Onderricht en Verdediging van den Godsdienst'* (School for Teaching and Defending Religion, number 4), an '*opleidingschool voor Sociale Werkers'* (Social School, number 5) and *Vak- en Volksonderwijs* (Professional and Popular Education in *Vrede Sint-Amands*, number 6).[49] Unlike the other halls, this last one, so it was explicitly stated, was equipped with a Dussaud projector.[50] It is not clear whether or not the other rooms had fixed projectors or not, or whether the Dussaud projector was so exceptional that it was worth mentioning. Apart from these regular venues, the CFUE also held lectures 'on request and where it proves effective' at other organizations in Antwerp or elsewhere.[51]

The CFUE's work was not limited to Antwerp's city centre. Indeed, the surrounding municipalities became part of their field of action (see Table 2.3). Due to the construction of the *Brialmont* bulwark after 1859, two of the earliest suburban municipalities of Antwerp, Borgerhout and Berchem, both municipalities with a strong population and close to the Antwerp city centre, came to lie *intra muros*. Deurne, Merksem (north-east of Antwerp), Hoboken, and Mortsel (south of Antwerp) were the early growing suburban municipalities

49 *Katholieke Vlaamsche Hoogeschooluitbreiding te Antwerpen, 24ste jaar*, Katholieke Vlaamsche Hoogeschooluitbreiding: Antwerpen, 1921–1922, pp. 4–25.

50 *Katholieke Vlaamsche Hoogeschooluitbreiding te Antwerpen, 25ste jaar*, Katholieke Vlaamsche Hoogeschooluitbreiding: Antwerpen, 1922–1923, p. 21.

51 Translated from the Dutch: '*op aanvraag en waar het doelmatig blijkt'* (*Hooger Onderwijs*, p. 39.)

extra muros. Merksem, one of the municipalities with a strong CFUE lantern activity in the 1920s, was known for its industrial developments. The workers related to this economic activity could have been an important factor in establishing and maintaining a focal point of CFUE's activity.[52]

Although the literature situates the peak of extramural adult education before the First World War, my results show that the frequency of lantern lectures organized by university extensions and popular universities experienced both downward and upward trends in activity.[53] The CFUE in Antwerp gained momentum in the interwar period as a result of the growing Flemish movement and several popular universities in Brussels also managed to stand their ground and continued, or even expanded, delivering their lectures using lanterns. The popular universities of Forest, Ixelles, Uccle, Koekelberg, and Jette gained power in the 1920s, compared with the first sample period. The latter two communes retained their strong labour class profile during this period. Generally speaking, no clear patterns are discernible in Brussels: newspapers indicate that extramural education in both higher and labour class communes experienced decreases as well as increases, so these evolutions are probably due to local factors such as financial support or the agency of the different boards and their adaptability to the real audience, which did not always overlap with the expected audience.

During the starting period at least, the popular universities explicitly focused on the working class. After a few years, several popular universities observed that this public was not in fact being reached, probably because, in the early twentieth century, working hours were not yet regulated, resulting in long working days and limiting the free time (and presumably also the energy) for workers to attend additional evening activities, let alone weekday activities. Although the nineteenth century brought many French-speaking Walloon immigrants to Brussels, the (almost exclusive) use of French still excluded a large part of the working class. Despite the French language spread in the communes of Brussels in the *fin-de-siècle*, at the beginning of the twentieth century the majority of the industrial population in Molenbeek-Saint-Jean, Anderlecht, Koekelberg, Jette, downtown Brussels, and the wider Brussels agglomeration were monolingually Flemish. Nevertheless, Francophone leaders often refused to transcend this language barrier, missing the opportunity to reach two-thirds of the working masses that spoke Flemish (numbers from the 1880s). Many artisanal workers, in contrast, spoke French or were bilingual.[54]

While boards of some popular universities continued to try and reach the originally intended audiences, others argued that their activities were not exclusively intended for workers but for people from all walks of life. Hence, servants and members of the middle and upper classes increasingly made up most of the audience.[55] '*Extension Universitaire de Bruxelles*' and '*Extension de l'Université libre de Bruxelles*' were related to the Brussels

52 May, 'Suburban place-making', pp. 18, 45, 63.

53 De Vroede, 'Hogeschooluitbreidingen en Volksuniversiteiten', p. 271; Couvreur, 'De Volksuniversiteit te Schaerbeek', p. 63.

54 Couvreur, 'De Volksuniversiteit te Schaerbeek', pp. 3–4, 59–69; Bossaerts, 'Drie Volksuniversiteiten uit het Brusselse', pp. 12, 22–29, 46, 52, 59–69; Van der Wee, 'De Volksuniversiteit te Sint-Gillis, 1901–1914', pp. 8–9; Strikwerda, *A House Divided*, pp. 56–57; Van Ginderachter, *The Everyday Nationalism of Workers*, p. 146.

55 Couvreur, 'De Volksuniversiteit te Schaerbeek', pp. 3–4, 59–69; Bossaerts, 'Drie Volksuniversiteiten uit het Brusselse', pp. 12, 22–29, 46, 52, 59–69; Van der Wee, 'De Volksuniversiteit te Sint-Gillis, 1901–1914', pp. 8–9.

University and, not surprisingly, attracted mostly teachers, the middle classes, and members of the bourgeoisie.[56] In addition, the university extension movement turned out to be largely middle-class in orientation. The CFUE, for example, aimed at a more or less educated audience with their regular lectures and courses, while they looked for the 'less educated man himself in their own midst' by holding their guest lectures in, for example, workmen's circles.[57]

As a result both the popular universities and the university extension programmes attributed an intermediary role to certain middle-class groups, such as teachers, who were believed to have a 'multiplier effect'. When these groups and the upper layer of the working classes were (re)integrated into the dominant civic culture, they could act as intermediaries towards (the rest of) the working class. In late nineteenth- and early twentieth-century Belgium, the uncertain political climate thus made the middle classes a decisive factor in the possible outcome.[58] In other words, by investing in the popular university movement, the Brussels socialists especially tried to break down the gap between 'high culture' and the lower classes.[59]

The shifts over time, already visible in Table 2.1, are now explained in more detail by looking at the socio-economic characteristics of the different suburban municipalities and the (shift in the) intended audience. The population density of these suburban municipalities (see also Table 2.2 and Table 2.3), helps explain why, on the one hand, the differences in terms of numbers of lantern lectures announced in the press are so significant between the two cities and, on the other hand, why the choice was made to establish extramural education in certain municipalities and not in others. As part of the objective of spreading knowledge and science among the people, it was important that as many people as possible were reached. In municipalities with a higher population density the support for this must have been greater. In the list outlining the ten most important urban centres' populations in Belgium in 1900, the city of Antwerp was undoubtedly in first place. Brussels ranked second, but several of the capital's suburban municipalities were also included: Schaerbeek (fifth place), Ixelles (sixth place), Molenbeek-Saint-Jean (seventh place), and Sint-Gillis (ninth place).[60] Half of the top 10 was therefore represented by Brussels and its suburbs while population numbers in Antwerp's suburban municipalities remained significantly lower. Nevertheless, there are clear trends in both cities: as a rule, extramural education was established in the municipalities with the highest populations. The municipalities in which the number of lantern lecture announcements increased from the first to the second sample period were also generally those that experienced (strong) population growth.

---

56 Bossaerts, 'Drie Volksuniversiteiten uit het Brusselse', p. 11.

57 Translated from Dutch to English: '*den minderen man (…) in zijn midden*' (*Gedenkboek van de KVHU*, p. 10.)

58 van Damme, *The University Extension Movement*, pp. 33–34, 38; van Damme, *Universiteit en volksontwikkeling*, pp. 11, 42.

59 Strikwerda, *A House Divided*, p. 145.

60 Greefs, Hilde, Bruno Blondé and Peter Clark, 'The Growth of Urban Industrial Regions: Belgian Developments in Comparative Perspective, 1750–1850', in *Towns, Regions and Industries. Urban and Industrial Change in the Midlands, c. 1700–1840*, ed. by John Stobart and Neil Raven (Manchester and New York: Manchester University Press, 2005), pp. 223–23.

Something for the Eye: Visual Methods

Wherever it settled, the Catholic Flemish University Extension described itself as 'an essential school whose doors are wide open to anyone who wants to broaden his horizons'.[61] The organization therefore set high standards for its lantern lecture speakers. While at many university extensions abroad, only academics took to the podium for these lectures, this was not the case at the CFUE. Although they certainly welcomed these sorts of speakers, in practice they preferred people from 'real life', 'because they are more in touch with the people, and therefore know them better, and know better what type of knowledge they lack most.'[62] In the Antwerp context of port and trade, this fitted in seamlessly with the aforementioned increasing need for professional (higher) education. Nevertheless, with a few exceptions, lecturers were recruited among people with a university degree or a comparable level of education.[63] In Brussels, on the other hand, initially mainly students or professors from the liberal University of Brussels gave these lectures, but this soon evolved into a situation where guest speakers belonged to a more general intellectual elite.[64] Academics took part in both forms of extramural adult education, but in addition people from the medical professions, as well as engineers, lawyers, and even religious people, (former) military personnel and politicians, all took the floor. All had in common that they were seen as specialists in the subject they were talking about, which in several cases is illustrated in recent publications by the speakers on the subject.

Speakers taking part in extramural adult education were at the service of their listeners 'to give further information and explanations and to show the way to further study'.[65] Although ideally the content of lectures was adapted to the intellectual level of the audience, the audience often turned out to be insufficiently prepared.[66] After all, primary education only became compulsory in Belgium on the eve of the First World War in 1914.[67] According to the CFUE itself, many people simply came by to spend an evening outside even though their lectures were not intended as a place where a lovely time could be spent indoors during cold winter nights.[68] Even those with a genuine interest in line with the CFUE objectives often lacked the necessary levels of concentration, as the lectures usually took place in the evening (often starting at around 8 p. m. or 8.30 p. m.),

61 Translated from the Dutch: *'eene wezenlijke school waarvan de deuren wagewijd openstaan voor iedereen, die zijn gezichtsveld wil verruimen.'* Lefevre, G., *De Taak der Kath. Vlaamsche Hoogeschooluitbreiding te midden de andere Katholieke Vlaamsche Inrichtingen* (Katholieke Vlaamsche Hoogeschooluitbreiding: Antwerpen, 1912), p. 5.

62 Translated from the Dutch: *'personen uit het werkelijk leven', 'omdat zij meer met het volk in betrekking komen, en dus beter kennen en beter weten aan welke kennis het volk het meest gebrek lijdt.' Hooger Onderwijs*, p. 21.

63 Lefevre, *De Taak der Kath. Vlaamsche Hoogeschooluitbreiding*, p. 6.

64 Bossaerts, 'Drie Volksuniversiteiten uit het Brusselse', p. 48; Couvreur, 'De Volksuniversiteit te Schaerbeek', pp. 33–36; Van der Wee, 'De Volksuniversiteit te Sint-Gillis, 1901–1914', pp. 44–45.

65 Translated from the Dutch: *'om verdere inlichtingen of verderen uitleg te verstrekken en den weg tot verdere studie aan te duiden.'* (*KVHU te Antwerpen, 15de jaar*, p. 4.)

66 Leo Van Puyvelde and Aug. Van Roey, *Aard en Inrichting. Welke onderwerpen worden best behandeld*, Katholieke Vlaamsche Hoogeschooluitbreiding: Antwerpen, 1910, p. 8–11.

67 Van Ginderachter, *The Everyday Nationalism of Workers*, p. 93.

68 Van Puyvelde and Van Roey, *Aard en Inrichting*, p. 8; Lefevre, *De Taak der Kath. Vlaamsche Hoogeschooluitbreiding*, p. 5.

after a long working day.[69] In the battle for attention, the projection lantern was thus seen as an important weapon. How to use 'the lantern with which you can lead your listeners through the dark unknown' was discussed at length within extramural adult education.[70] Light was brought into the darkness by, amongst other things, the method used, the way the speaker's thoughts were ordered, the language used, and the subject matter dealt with.

An important aspect in keeping the audience's attention focused on the lantern lecture was clearly the implementation of 'visual methods'. Certainly, in the interwar period, this should be seen in terms of broader pedagogical reforms and debates about new methods of teaching and learning which assigned a central role to visual instruction, new media technologies, sensory aids, and visual devices. Known as the 'New Educational Movement', 'Reform Pedagogy' or 'progressive education', this new movement emphasized the child itself and the importance of an 'intuitive' way of acquiring knowledge, shifting the focus from text to (sensory) experiences with concrete objects.[71] It was, however, not always possible to present what was being talked about in a real, concrete form. The best alternative was to let the audience see the subject indirectly. The projection lantern was an ideal technique with which to make clear at a single glance to all the spectators what the speaker was talking about. In this way, the subject became more accessible.

University extensions and popular universities already used this medium from their foundation on. By the 1910s it was observed by a member of the CFUE that more and more use was made of the lantern, something which was strongly encouraged. It was emphasized that these images should not be shown quickly at the end of the lecture but should be projected simultaneously with accompanying words. Sufficient time should be taken to allow them to sink in. The use of the lantern, in this way, was both useful and pleasant as 'it gives rest to weary ears and some diversion to eager eyes'.[72] The popular universities tried to combine the pleasant with the useful, which sometimes led to the former obtaining the upper hand. The CFUE faced the same dangers, stressing in multiple publications that the way in which knowledge was transferred should be attractive. Still, at the same time, voices within the CFUE continued to emphasize that intellectual development remained the fundamental objective. By starting with less serious and more attractive yet still scientific material, they tried to awaken the desire for learning among the lower strata of society.[73]

For these reasons projection lanterns, scientific experiments, demonstrations, and even music were often incorporated into lectures.[74] Exact figures on the ratio of lectures

---

69 Couvreur, 'De Volksuniversiteit te Schaerbeek', p. 52; Bossaerts, 'Drie Volksuniversiteiten uit het Brusselse', pp. 31–36; Van der Wee, 'De Volksuniversiteit te Sint-Gillis, 1901–1914', p. 15.

70 Translated from the Dutch: *'de lantaarn waarmee ge uw toehoorders in het donkere onbekende kunt voortleiden.'* Joos, Amaat, *Wetenschappelijke voordrachten voor het volk*, Katholieke Vlaamsche Hoogeschooluitbreiding: Gent, 1910, pp. 3–15.

71 Reese, William J., 'The Origins of Progressive Education', *History of Education Quarterly*, 41 (2001), vi, 1–24 (pp. v, 2, 16); Depaepe, Marc, and Angelo Van Gorp, 'The Canonization of Ovide Decroly as a 'Saint' of the New Education', *History of Education Quarterly*, 43 (2003), 224–49 (p. 224.)

72 Translated from the Dutch: *'omdat ze de vermoeide ooren laat rusten en wat bezigheid geeft aan de begeerige oogen.'* (Joos, *Wetenschappelijke voordrachten voor het volk*, pp. 3–15.)

73 Van Puyvelde and Van Roey, *Aard en Inrichting*, pp. 3–6; Bossaerts, 'Drie Volksuniversiteiten uit het Brusselse', p. 33; Van der Wee, 'De Volksuniversiteit te Sint-Gillis, 1901–1914', pp. 30, 48; Couvreur, 'De Volksuniversiteit te Schaerbeek', p. 60.

74 van Damme, *The University Extension Movement*, p. 30.

with and without slides are difficult to obtain. Not all lantern lectures were announced in newspapers. Not all other advertising practices have stood the test of time. In addition, the same lectures can be found in different sources some of which were accompanied with slides, some without. A 1903 *Report of the Commissioner of Education* on American university extensions consistently made the comparison between the domestic and foreign extension movements. Belgium was not discussed, but it was reported that in France 'as a rule the lectures are illustrated by magic lantern views, of which the government furnished 29,000 collections for the use of lecturers. The *Ligue de l'Enseignement*, a private society engaged in the work, distributed 44,986 views.'[75] Since the Belgian popular universities kept their eyes firmly fixed on France, this suggests that this was also the case in Brussels and the French-speaking part of Belgium.[76] Literature on the establishment and activities of some popular universities in Brussels sporadically lists the lectures given, both with or without slides, within certain years of operation, based on the few surviving periodicals of these organizations. Although these lists are not exhaustive or sequential, making an evolution in time difficult to observe, the data from the popular university in Saint-Gillis from 1902 and 1903 seem to confirm the widespread use of the projection lantern in their lectures.[77]

At the peak of the university extension- and popular university-movement, averages of one to five lectures per week were reached.[78] Because of this frequent use at their peak, we can assume that the individual popular universities and university extensions probably considered it worthwhile to invest in one or more lantern projectors themselves. The CFUE annual report of 1912–1913, for example, mentions the purchase of a new device with which 'the image of photographs and objects could be projected directly onto the screen, without the use of diapositives' to 'clarify the lectures by means of beautiful images'.[79] In the 1920s, however, many extramural education organizations appeared to be in financial trouble. It was therefore announced in 1922 that organizations for popular development, including organized lantern lectures, amongst others, could apply for state grants. The motivation for this was expressed as follows: 'Such organizations deserve our warm affection and the fullest support because, especially through after-school studies, one can complete one's development and education and become responsible citizens.'[80] To receive this grant, the organizations had to:

> Exist for at least one year, be directed by a committee of at least five members, be completely not-for-profit and therefore ask no (or very small) entrance fees, not carry

75 *Report of the Commissioner of Education for the Year 1902*, Washington Government Printing Office, 1903, pp. 679–80.

76 van Damme, *The University Extension Movement*, p. 37; De Vroede, 'Hogeschooluitbreidingen en Volksuniversiteiten', p. 265.

77 See for example: Van der Wee, 'De Volksuniversiteit te Sint-Gillis, 1901–1914', pp. 32–35.

78 Couvreur, 'De Volksuniversiteit te Schaerbeek', pp. 18, 60; Van der Wee, 'De Volksuniversiteit te Sint-Gillis, 1901–1914', p. 31; *Hooger Onderwijs*, p. 14.

79 Translated from the Dutch: *'het beeld van foto's en voorwerpen rechtstreeks op het doek werpen zonder behulp van diapositieven'*; *'door prachtige lichtbeelden te verduidelijken' KVHU te Antwerpen, 15de jaar*, p. 6.

80 Translated from the Dutch: *'Zulke kringen verdienen onze warme genegenheid en de volledigste ondersteuning, daar men vooral door de naschoolse werken de opvoeding en het onderwijs kan volledigen en verstandige burgers vormen.' Het Laatste Nieuws*, 15 May 1922, p. 1.

out any political, religious and anti-religious propaganda, be accessible to all, and ultimately accept the supervision of the state.[81]

Since both the university extensions and the popular universities declared themselves as politically neutral, at first glance, both seem to have been able to meet all the conditions. The newspaper article announcing the new state support explicitly mentioned that this money could be used for the rental of projection equipment. In addition, 'it will finally be possible in Belgium, as in other countries, to compensate lecturers, who are currently sometimes difficult to find and who are sometimes discouraged.'[82] The financial difficulties which extramural education was clearly experiencing at the time may help to explain the downward trend (in Brussels at least) of lantern lectures in Table 2.1.

The sources generally remain silent as to where the slides came from. Speakers who frequently gave lectures beyond this kind of educational institutions would probably have been able to provide their own slides. However, not all speakers could do this, and it seems unlikely that the university extensions and popular universities all had their own slides for so many different topics (see Table 2.4 and Table 2.5). Another option could have been to turn to the city's various slide-lending services, as was done by primary and secondary schools. Ongoing research by Wouter Egelmeers shows that the *Werk der Lichtbeelden* (Catholic Lending Service for Slides, °1908) in Antwerp, but also the Brussels City Depot (°1906) and the slide lending service of the liberal *Ligue de l'Enseignement* (Educational League) in Brussels, could have offered a solution.[83] Only through these scraps of information can we catch a glimpse of the omnipresence of the lantern medium in extramural adult education and the kinds of technical materials used.

The battle for attention also had to be fought by carefully considering the construction of argumentation and the broader knowledge required. The audience therefore needed to be given the opportunity, either during the lecture or afterwards, to ask questions about aspects they did not understand so that they could be clarified. Using the inductive method to teach scientific subjects was also recommended. Amaat Joos, administrator of the Episcopal Normal School in St Nicholas and member of the Royal Flemish Academy, tried to pre-empt the criticism that this method was also used for primary school children by stating that, in contrast to the deductive method, the inductive way was the only way to reach the diverse audience to whom the doors of the lecture hall were open. The CFUE

81 Translated from the Dutch: *'ten minste één jaar bestaan, bestuurd zijn van een komiteit van ten minste 5 leden, alle winstbejag vermijden en van hun toehoorders geene of heel kleine ingangsrechten eischen, geen politieke, godsdienstige of anti-godsdienstige propaganda voeren, en toegankelijk zijn voor iedereen en uiteindelijk het toezicht van den Staat aanvaarden.' Het Laatste Nieuws*, 15 May 1922, p. 1; De Vroede, 'Hogeschooluitbreidingen en Volksuniversiteiten', pp. 258, 265–68, 274–75.

82 Translated from the Dutch: *'zal men eindelijk in België, net als in andere landen, de voordrachtgevers kunnen vergoeden, die nu soms moeilijk te vinden zijn en die wel eens ontmoedigd worden.' Het Laatste Nieuws*, 15 May 1922, p. 1.

83 This relates to unpublished ongoing research by Wouter Egelmeers presented on 23 May 2021 at the digital conference 'Sound and Vision. Exploring the role of audio and visual technologies in the history of education'. More information on this conference can be found at: https://www.uantwerpen.be/en/projects/b-magic/events/events-archive/sound-and-vision-ex/. For more information about these lending services, see: G.-Michel Coissac, 'Les projections en Belgique', *Le Fascinateur. Organe des récréations instructives de la Bonne Presse* 90 (June 1910), 147–51 and E. Guyot, *Bulletin communal. Première partie. Compte rendu des séances.* (1909), II, pp. 720–21.

also suggested giving the audience something to hold onto by handing them a short table of contents beforehand to make it easier to follow the lecture and to be able to go over the material later at home. This could be distributed in the CFUE's magazine, on a board in the lecture hall, or on paper, which could then be handed out at the entrance.[84]

In theory, the popular universities preferred a more democratic method of *éducation mutuelle* (mutual education), where knowledge was not just passed from person A to person B, but a mutual exchange of knowledge and experience was pursued. In reality, however, most lecturers spoke *ex-cathedra*. Shortly after its establishment and during the first great success of the popular universities, there would have been too many people present to allow a real discussion to take place. In other words, there was little, or no, room left for bidirectional learning which led to a situation probably very similar to that of the university extensions where the transfer of knowledge in itself was central.[85]

Furthermore, the manner and language used were also essential for communicating the message properly. Projected images in combination with the proficiency with which the lecturer spoke, helped retain the attention of the audience. Importance was attached to alternating strength and tone of voice. Dry reasoning had to be alternated with questions and exclamations. Another recommendation read: 'When the occasion arises, introduce a joke or a surprising twist so that people can laugh heartily and recharge themselves. The small loss of time that might result is more than compensated for by the renewed attention.'[86] To achieve the above-mentioned objectives, the CFUE deliberately chose 'the language of the people that everyone in Flanders understands', namely Flemish.[87] On the other hand, the popular universities placed strong emphasis on sound knowledge and use of French to further promote the spread of this language as part of the 'Frenchification' process.[88]

The Spotlight on Society: Themes Discussed

Now that it is clear what audience the university extensions and popular universities were trying to reach by using visual methods, it is time to take a closer look at the knowledge they wanted to disseminate. After all, the organizers of lantern lectures were mindful of the fields of interest of the audience and the usefulness of certain themes for them. In its publications, the Catholic Flemish University Extension stated that it deliberately chose subjects that were related to current events: 'after all, issues of current significance arouse the interest of the public and a University Extension must not fail to utilize this interest to disseminate new

84 Joos, *Wetenschappelijke voordrachten voor het volk*, pp. 3–15.

85 Couvreur, 'De Volksuniversiteit te Schaerbeek', pp. 9, 66, 70; Van der Wee, 'De Volksuniversiteit te Sint-Gillis, 1901–1914', pp. 5, 90.

86 Translated from Dutch to English: *'Haalt bij passende gelegenheid ook een klucht of een fijnen zet uit, zoo kunnen de menschen eens hartelijk lachen en zich verkwikken. Het klein tijdverlies dat er zou kunnen uit voortspruiten, wordt ruim vergoed door vernieuwde aandacht.'* Joos, *Wetenschappelijke voordrachten voor het volk*, p. 15.

87 Translated from the Dutch: *'de taal van het volk die iedereen in Vlaanderen begrijpt'* Van Puyvelde and Van Roey, *Aard en Inrichting*, p. 6); *Gedenkboek van de KVHU*, p. 6.

88 Strikwerda, *A House Divided*, p. 37; Van der Wee, 'De Volksuniversiteit te Sint-Gillis, 1901–1914', p. 23; Couvreur, 'De Volksuniversiteit te Schaerbeek', p. 5.

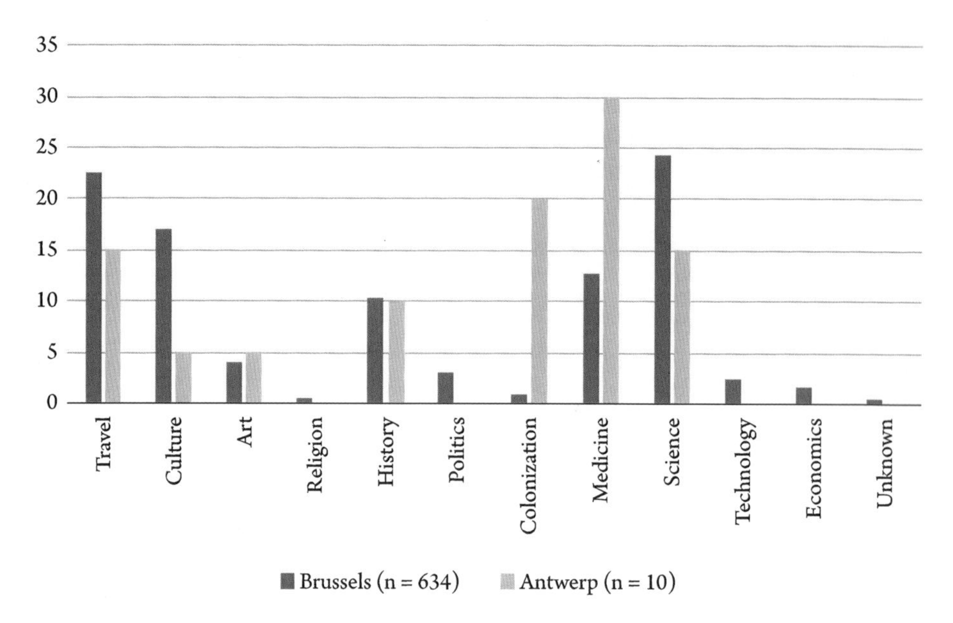

Table 2.4. Themes discussed during extramural adult education lantern lectures held in Brussels and Antwerp (1902–1904). Author's database.

knowledge and science'.[89] In the first decades of the twentieth century the CFUE tried to teach young people, workers and the bourgeoisie about religion through lantern lectures, so that the audience was 'able to respond to the innumerable errors that are spread everywhere and incessantly in pamphlets and newspaper articles'.[90] Through lantern lectures the CFUE also very consciously responded to broader contemporary domestic socio-political situations, for example, in the 1910s, by addressing the 'women's question' and 'women's development'.[91]

More broadly, my analysis demonstrates that the lantern was used by university extensions and popular universities to disseminate all sorts of knowledge as vividly and practically as possible. Although there are some differences in emphasis between Brussels and Antwerp, both are largely characterized by the same trends (see Table 2.4 and Table 2.5). As the literature has already suggested, travel lectures were indeed ideal as they could be supported by slides, although cultural and historical lectures also made extensive use of this visual aid. In contrast to the Ghent university extension, which was mainly established by members of the university's department of Dutch philology and literature, literature and language did not make up the bulk of lantern lectures in either Brussels

89 Translated from Dutch to English: *'Punten van actueel belang maken immers de belangstelling van het publiek gaande en eene inrichting voor Hoogeschooluitbreiding mag niet nalaten die belangstelling ten nutte te maken, om weeral wat kennis en wetenschap te verspreiden.' Katholieke Vlaamsche Hoogeschooluitbreiding van Antwerpen*, Katholieke Vlaamsche Hoogeschooluitbreiding: Antwerpen, 1903–1904, p. 11.

90 Translated from the Dutch: *'in staat gesteld te antwoorden op de ontelbare dwalingen die in vlugschriften en dagbladartikels overal en onophoudend verspreid worden.'* Lefevre, *De Taak der Kath. Vlaamsche Hoogeschooluitbreiding*, pp. 6–13.

91 Lefevre, *De Taak der Kath. Vlaamsche Hoogeschooluitbreiding*, pp. 6–13.

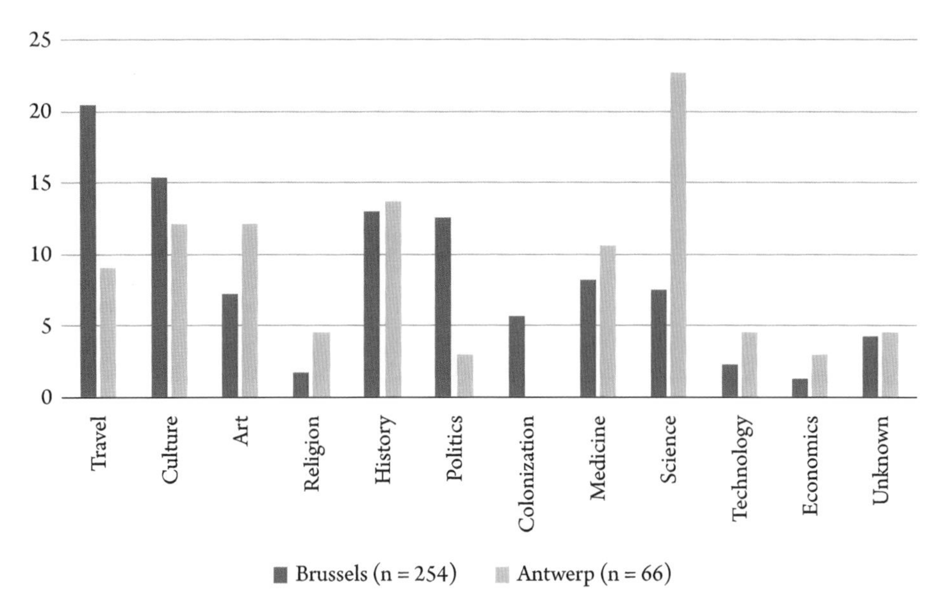

Table 2.5. Themes discussed during extramural adult education lantern lectures held in Brussels and Antwerp (1922–1924). Author's database.

or Antwerp.[92] This connection with what is today known as the Arts Department was less strong in the Brussels extensions and Antwerp did not even have its own university. Moreover, the university extension and popular universities in Brussels were not driven by 'cultural flamingantism' (with a strong focus on Dutch). In addition, lectures on literature required less visual support. Instead, cultural topics such as architecture and (city or rural) landscapes provided more suitable subjects for slides, as did historical subjects such as art history, ancient history, or (older or recent) episodes of Belgian history.

As van Damme has demonstrated in his research, the extension movement was characterized by the admiration for 'the triumphs of the age in scientific discovery', which translated into a considerable part of the lectures being devoted to scientific and medical subjects. It was precisely for these topics that slides could provide the necessary clarification for the less well-educated audience. After all, the abstract thinking often associated with scientific lectures was seen as too demanding. Visual education was a way to respond to this: 'observation, which happens almost without effort, instantly supplies an image which otherwise would be painfully birthed by imagination.'[93] This explains why many of the lantern lectures were about science and medicine, whether or not combined with experiments and demonstrations. Raising awareness about the expansion of natural sciences was supposed to result in an understanding of the benefits of modern industrial society and the acceptance of the social system. Furthermore, the projection lantern was

92 van Damme, *The University Extension Movement*, pp. 20, 24.

93 Translated from the Dutch: *'door de aanschouwing, vermist de waarneming, die haast zonder moeite gebeurt, oogenblikkelijk het beeld geeft, dat anders pijnlijk door de verbeelding gebaard wordt.'* (Joos, *Wetenschappelijke voordrachten voor het volk*, pp. 13–15.)

also used to popularize modern medical findings and sciences through the promotion of hygienic habits among the (working) people and improving health standards.[94] In this way, lectures responded to the greater societal concern about degeneration, the causes of which were sought in the triangular relationship between alcohol, syphilis, and tuberculosis. Furthermore, these lectures also had a strong moralizing and civilizing component: after all, prostitution was considered the main source of contagion of venereal diseases, but other forms of 'sinful' behaviour, such as masturbation and tobacco use, were also associated with these diseases.[95]

Besides scientific and medical lectures on various topics, including anthropology, the care of the mentally ill, and bacteriology and its applications, which were all considered useful, politically oriented lectures also began to increase. This did not include political propaganda as such, rather themes such as the Great War or (international) contemporary issues such as, for instance, the contemporary situations in Ireland and Hungary. Several children's trains had been running from Hungary to Belgium and the Netherlands since the early 1920s and children were placed within Catholic families in order to recover from the hardships of the First World War and the subsequent revolutions in their homeland.[96] More broadly this also characterized the 'internationalist turn' of Catholicism and the Catholic university extensions whereby changes in Belgium were linked to the wider world.[97]

Conclusion

The projection lantern turned out to be the ideal medium through which to spread all sorts of knowledge among ordinary people in an accessible way. The medium was, therefore, deployed *en masse* at public lectures that were part of extramural adult education in Brussels and Antwerp during the timeframe studied (1900s–1920s). Based on archival material and mainly the Catholic Flemish University Extension (CFUE) publications, on the one hand, and an extensive research into Belgian daily newspapers on the other, I have highlighted four important aspects of lantern lectures in the context of university extensions and popular universities in this chapter: entrance fees charged and attendance numbers, the geographical spread of the lectures in relation to the intended audience, the way in which the projection lantern was deployed and finally, the themes discussed when using slides.

In this way, I was able to demonstrate the added value of the projection lantern in the battle for the attention and the souls of ordinary people. Since the intended audience might be misinformed about the nature of the lecture, tired after a long working day, or not

94 van Damme, *The University Extension Movement*, pp. 28–30.

95 Nys, Liesbet, 'De ruiters van de Apocalyps: 'alcoholisme, tuberculose, syfilis' en degeneratie in medische kringen, 1870–1940', in *Degeneratie in België 1860–1940: een geschiedenis van ideeën en praktijken*, ed. by Johan Tollebeek, Geert Vanpaemel and Kaat Wils (Leuven: Universitaire Pers Leuven, 2003), pp. 11–12, 20–22.

96 Aalders, Maarten J., Gábor Pusztai and Orsolya Réthelyi, 'Honderd jaar geleden. Internationaal licht op de kindertreinen als verbinding tussen Hongarije, Nederland en België', in *De Hongaarse kindertreinen: een levende brug tussen Hongarije, Nederland en België na de Eerste Wereldoorlog*, ed. by Maarten J. Aalders, Gábor Pusztai and Orsolya Réthelyi (Hilversum: Uitgeverij Verloren), p. 7.

97 Laqua, Daniel, *The age of internationalism and Belgium, 1880–1930: Peace, progress and prestige* (Manchester and New York: Manchester University Press, 2013), pp. 85–93.

adequately prepared due to their level of education, great value was placed on the images that supported and reinforced the speaker's words. The value of slides is also confirmed by the government grants given to organizations that strived for popular development through, among other things, lantern lectures from the 1920s onwards. By using visual methods, in line with pedagogical trends in the 1920s, abstract knowledge could be conveyed in a more accessible way. 'Training with the eye' was a central idea in extramural education as the world was brought to students rather than vice versa. No distinctions were made and, in addition to scientific themes, topics were also covered that were intended to shape the audience morally, culturally, and socially. The themes were therefore often chosen with current events in mind. The audience was approached in specific urban spaces and their surrounding suburban municipalities. Typically, the more affluent, but also densely populated municipalities with some concentration of workers, became the home of frequent extramural adult education lantern lectures.

The literature, however, only sporadically mentions the projection lantern, as the scientific approach and methods of the new forms of extramural education were seen as the means by which class reconciliation and solving the social problems caused by intensified industrialization could be achieved. The role, and the added value, of the lantern for the functioning and the realization of the objectives of the university extensions and popular universities has, until now, remained understudied. Nonetheless, the goals of extramural education in relation to the society in which they operated cannot be considered independently from the use of the projection lantern. Regardless of its success, the projection lantern proved to be an essential tool simply at the level of actively trying to reach less educated people and to convey scientific knowledge.

Neglecting the lantern projection in the context of popular development and popularization of science pursued by the university extensions and popular universities inevitably leads to a misunderstanding and/or underestimation of the impact of these forms of extramural education. This is confirmed by the fact that the literature generally argues that extramural education lost momentum in the interwar period. However, my research shows that although the number of announcements for lantern lectures in Brussels seem to have been declining, they nonetheless remained significant. In fact, the importance of lectures within extramural adult education context actually increased in Antwerp, a response to the fact that Belgium was now a mass society, one in which the Flemish movement was increasingly important and in which the welfare state was further expanded, amongst other things. More than ever the transmission of knowledge was important in terms of reaching and convincing the (urban) masses about various issues. Although the power of the image may not have been sufficient to reach the lowest strata of society, it did reach the middle classes. As a politically important group during a time defined by social changes, they were assigned an intermediary role and might as well have been able to take advantage of the opportunities resulting from the transmission of knowledge enabled through the lantern. By lighting up the room with the lantern it became possible to bring the university, as *pars pro toto* for scientific knowledge and higher culture, to the people, many of whom did not (yet) have the opportunity to enjoy further (higher) education.

JAMILLA NOTEBAARD
NICO DE KLERK

The Photographic Turn in Visual Teaching Aids: Films and Slides for Schools in the Netherlands, 1911–1926

Introduction

According to the Dutch databases of digitized newspapers *Delpher.nl* and *Archieven.nl* reports on and advertisements for public illustrated lectures in the Netherlands began to appear in the 1870s. Initially sparse and dealing with a limited number of topics, their frequency skyrocketed as the 1890s progressed and photography-based slides became the norm. The illustrated lantern lecture had become a mass medium, not so much to entertain but rather to inform, instruct or educate. This boost coincided with a time, the last quarter of the nineteenth century, when the so-called social question demanded the concern of governments and citizens alike. The abominable working and living conditions of a fast growing proletariat, particularly in the cities, not only led to regulation and legislation, albeit slowly, but also to a host of private charity and uplifting activities, in both established and newly founded organizations (notably those inspired by the then popular Toynbee work). In order to reach their target groups collectively the illustrated lantern lecture became an important means. It remains doubtful whether these initiatives actually managed to reach the most destitute, but by the end of the 1890s the lantern lecture fulfilled a significant uplifting and educational purpose.[1] To some extent, academic outreach contributed to this social movement, too. Indeed, it is partly because of this presence in the public sphere that we know about the use of the lantern by universities. But what about schools, elementary as well as secondary?

For all we know lantern slide projections may have led a quiet, unnoticed life in school teaching, even though we suspect that they were used only sparsely. One reason

1 Not only doubt and uncertainty are characteristic of the current state of knowledge about the significance of the lantern lecture, negligence is, too. To date studies of the social question have not acknowledged the ubiquity and significance of the optical lantern's use, even after the launch of *Delpher.nl*. See for instance: Christianne Smit, *De volksverheffers: sociaal-hervormers in Nederland en de wereld 1870–1914* (Hilversum: Verloren, 2015).

Jamilla Notebaard • Utrecht University, j.notebaard@uu.nl

Nico de Klerk • Utrecht University, n.h.deklerk@uu.nl

Learning with Light and Shadows: Educational Lantern and Film Projection, 1860-1990, ed. by Nelleke Teughels and Kaat Wils, TECHNE-MPH, 8 (Turnhout, 2022), pp. 77-98

10.1484/M.TECHNE-MPH-EB.5.131495

is that their energy sources must have posed serious obstacles, not just of handling and safety—explosion hazards in gas-lit projectors. But also their very availability could cause problems; in Amsterdam, for instance, as late as 1912 not all elementary schools were connected to the electricity grid.[2] And, secondly, while schools may have reached the disadvantaged, certainly after the 1900 *Leerplichtwet* (Compulsory Education Act), their activities were by definition of course largely intramural. Insofar as there was discussion about the lantern's use in elementary and secondary teaching there seem to be no traces of it in public sources, at least in newspapers, around the turn of the twentieth century. And while an early brochure devoted to demonstrative teaching aids used a broad definition that included all the senses as well as the spoken word, it did not address photography-based media.[3] But cinema's rise to notoriety and the concern and controversy it caused would quickly change all that. And with it the lantern, too, was dragged into the spotlight.

In the following article we trace the debates on photography-based teaching aids that appeared in public discourse around 1910: their pedagogical and didactic merits, and their implementation. Clearly, reflections on film prevail, since there was much concern about its harmful effects on children. The first part of our text focuses on the rhetorical and, sometimes, anecdotal level of the debate, while the second part deals with the ways politics and science entered the fray.

I.

The value of the late nineteenth-century concept of visual education, according to media historian Jens Ruchatz, was that 'visual knowledge transmission functioned to convey a concrete, sensory impression of those, as yet unknown things for whose understanding it is a prerequisite.'[4] It is no coincidence, probably, that the earliest discussions of visual education in the Netherlands were written about the subject of geography. Geography, of course, could obviously benefit from visuals, as they brought the outside world into the classroom. But while, in the words geography teacher Henri Zondervan, modern education *is* visual education, the necessary teaching aids were in his opinion underused, and when they were used at all, it was inadequately.[5] Hence his rather general overview of various types of visual teaching aids—atlases and globes; educational walks; and pictures—and how

2 'Lichtbeelden op de Lagere Scholen te Amsterdam', *Algemeen Handelsblad*, 28 March 1912, evening edn., 1st section, pp. 1–2, <https://resolver.kb.nl/resolve?urn = ddd:010649873:mpeg21:p. 001> [accessed May 3, 2021].

3 Sikkel, J. C., *De aanschouwing in het onderwijs. Paedagogische bijdrage voor de Scholen met den Bijbel* (Amsterdam: s.n., 1902). A teacher and a reverend, Sikkel specifically meant God's word. The national Dutch (elementary and secondary) education system was officially secular, even though traditionally its curriculum was meant to educate students in both civil and Christian virtues. But during the nineteenth century there arose, besides public, government-funded schools, a variety of denominational schools with a more explicitly confessional curriculum; only in 1917 were these schools funded by government, too.

4 Ruchatz, Jens, *Licht und Wahrheit. Eine Mediumgeschichte der fotografischen Projektion* (München: Wilhelm Fink, 2003), pp. 228–29. The original German text reads: 'Der anschaulichen Wissensvermittlung kommt in diesem Kontext die Funktion zu, von noch unbekannten Dingen jenen konkret-sinnlichen Eindruck zu vermitteln, der notwendige Voraussetzung für ihr Begreifen ist.'

5 Zondervan, Henri, 'Aanschouwelijkheid bij het aardrijkskundig onderwijs', *Vragen des tijds*, 1e deel, 1896, 235. <https://resolver.kb.nl/resolve?urn = MMKB10:000736001:00007> [accessed 23 November 2021].

they could be put to optimal service. It is here that the optical lantern was first mentioned in relation to teaching, although Zondervan ranks it with wall charts and other picture materials, unconcerned about the differences between them (size, whether or not they are projected, presence or absence of colour). As well, he dismissed the costs of a lantern and the problem of darkening a room as mere temporary inconveniences.

Four years later, a more practical consideration of the projection lantern was offered by assistant teacher F. J. A. Paesi. He discussed it in the context of the 'new direction' in teaching methods, particularly those that assisted pupils in understanding the world around them in visual ways. In the example he provides the optical lantern is used to reflect on a guided tour through a rubber factory. During this 'educational walk' photographs were being made that were afterwards used to illustrate pupils' compositions; later—days or even weeks after this instructive outing—they were used as an aide-mémoire in a lantern slide presentation to refresh pupils' knowledge acquired during the event.[6] All three methods were part of the complete experience and seen as necessary for students to gain an understanding of the world: of soil, of raw materials, and how they are processed to make goods.[7] The two abovementioned texts, however, remain the only retrieved sources on the use of the optical lantern for elementary and secondary school until the 1910s in the Netherlands. How can we explain this?

Despite the lantern's wide circulation ever since the 1890s, mostly in popular science lectures or for entertainment purposes, the lack of references in newspapers suggests that schools were not yet equipped to implement this new device on a large scale. As Jennifer Eisenhauer claims in her discussion of the history of the American usage of the lantern for educational purposes in art history, a technological device acquires its cultural meaning through discursive practices: 'Embedded within the magic lantern's emergence as a scientific and educational tool is a reframing of the relationship between object and viewer from a discourse of *magic* vision to one of *scientific vision*.' Such changes, according to Eisenhauer, were part of shifting concepts of 'vision, knowledge and subjectivity'.[8] Hence she attaches more value to these discursive dimensions that caused a change in meaning than to technical improvements.

Unfortunately, however, Eisenhauer does not elaborate on the arguments that led to this new vision nor does she explicate the shifting meaning of the concepts. She accepts the shift at face value by simply observing that at the end of the nineteenth century the optical lantern was widely used as a tool for education and instruction, rather than entertainment. Despite this thin base to support her claim, her discussion of the mechanical improvements that led to the wider spread of the lantern in education at the end of the nineteenth century is very helpful with respect to the Dutch situation.[9] As Eisenhauer observed, the role played by photography in making lantern projections suitable for educational purposes not only

6 By the time of the first educational film screenings the class outing appears to have been a common phenomenon; see: 'De bioscoop als school' *Leeuwarder Courant*, 12 June 1911, 1st section, 1, <http://resolver.kb.nl/resolve?urn = ddd:010599160:mpeg21:p. 001> [accessed 3 May 2021].

7 A. F. J. Paesi, 'Naar aanleiding van een schoolwandeling', *School en Leven*, 31 May 1900, 611–13, <https://resolver.kb.nl/resolve?urn = MMKB14:001557040:00001> [accessed 3 May 2021].

8 Eisenhauer, Jennifer F., 'Next Slide Please: The Magical, Scientific and Corporate Discourses of Visual Projection Technologies', *Studies in Art Education*, 47, 3 (2006), 199–200.

9 Ibid., 198–200.

Fig. 3.1. Logo of the Lantern Slide Association on a newsletter circulated in 1911. Internationaal Instituut voor Sociale Geschiedenis. Documentatiecollectie Cultuur Nederland. Doos 12.8. Cultuur (V-Z) folder 8. Vereeniging tot het houden van voordrachten met lichtbeelden.

improved the quality of the slides, it also removed the association of imperfection due to 'human intervention' in painting or drawing images. Most relevant for the Dutch situation is her statement about the impact of the improvements of light and the availability of electricity on a wider scale.[10] For instance, it took until the 1910s before Amsterdam was connected to the electricity grid and could count on a constant and reliable energy source to teach by means of the lantern.

Another important factor was the supply of educational lantern slides, which changed significantly after 1910 with the appointment of a new director of the Dutch *Lichtbeelden-Vereeniging* (Lantern Slide Association). Founded in 1898 to propagate and enable illustrated lectures, the association created a centrally housed lending collection of slides for all kinds of organisations and associations. Its new director introduced a policy that targeted schools and universities for its activities as well. The association appointed a new board of advisors, consisting of professors, schoolteachers, and other stakeholders who were asked to advise on relevant topics, images, and literature that could be used for the production of slides and accompanying texts.[11] As well it requested Royal Approval, which it was granted in 1911. This allowed the association to set up a photo studio to produce negatives, develop photographs, and make glass slides. And by hiring three photographers it was able

10 Ibid., 200–01.

11 IISG. DCN. Box 12.8. Culture (V-Z) folder 8. *Vereeniging tot het Houden van Voordrachten met Lichtbeelden* (Association for the Delivery of Illustrated Lectures). Brochure announcing new management structure of the Lantern Slide Association; A. Dijkgraaf, 'Het Bioscoopvraagstuk', *Vragen des Tijds*, (1912), p. 116.

to create a 'photographic archive for the Netherlands', subdivided into the categories of geography and geology, art history, natural history, astronomy, medicine, agriculture, and 'stories for the young and the old'.[12] The educational orientation of the association as well as improvements in the technical set-up of the device, contributed to a series of trials to determine the feasibility of lantern projections in schools.

Initial Trial Screenings

More or less simultaneously with the first tests to establish the feasibility of the optical lantern in classrooms, a number of trial screenings were carried out with educational film projections. This circumstance allows one to see how the relative merits of each were formulated, argued—or taken for granted. The photographic turn in Dutch teaching aids took off in 1911 with trials in three municipalities: Leeuwarden, The Hague, and Amsterdam. While the first two focused exclusively on film, the one in Amsterdam was concerned with the affordances of the optical lantern. In the trials answers were sought to the question of the didactic value of the optical lantern and film.

The first initiative was undertaken by the mayor and aldermen of Leeuwarden, capital of the northern province of Friesland. On 12 and 13 June 1911, a special programme of films was screened for schoolchildren between the age of nine and twelve at the local Friso cinema theatre. The programme was put together with the help of educational experts.[13] A newspaper report the following day concluded that the production of purpose-made educational films should receive more attention, while it also commented on the possibility of showing films beforehand, in the classroom, or a teacher's explanatory lecture during the screening.[14] Three teachers, in a letter to the editor, were confident that any problem would be solved:

> There is much, very much to improve. In the first place—no matter how beautiful the films were overall—it is imperative to have other pictures, taken directly from the school's curriculum. We won't bother with other defects; they do not alter the fact that film has proven to be suitable for implementation within the school.[15]

In the same town of Leeuwarden, a year after these trial screenings, city counsellor Zandstra believed that a trial with the optical lantern in elementary schools should be conducted as soon as possible. It is interesting to read in the report of the city council meeting that

---

12 Joh. D. Hintzen, 'De Lichtbeelden-Vereeniging te Amsterdam', *De Groene Amsterdammer*, July 2, 1927, p. 4, <http://historisch.groene.nl/nummer/1927-07-02/pagina/4#4/0.18/-115.27> [accessed May 3, 2021]; 'De A. N. D. B', *Algemeen Handelsblad*, April 20, 1911, evening edn., 1st section, p. 1, <http://resolver.kb.nl/resolve?urn = ddd:010337957:mpeg21:p. 001> [accessed May 3, 2021].

13 'De bioscoop als school' (12 June 1911), p. 1.

14 'De bioscoop als school', *Leeuwarder Courant*, 13 June 1911, 1st section, p. 1, <https://resolver.kb.nl/resolve?urn = ddd:010599161:mpeg21:a0004> [accessed 3 May 3, 2021].

15 L. Bij de Leij, D. van der Schaaf, A. C. Nubé, 'De bioscoop in dienst van't onderwijs', *Leeuwarder Courant*, June 14, 1911, 2nd section, p. 7, <http://resolver.kb.nl/resolve?urn = ddd:010599162:mpeg21:p. 007> [accessed May 3, 2021]. The original Dutch text reads: 'Er is nog veel, zeer veel te verbeteren; in de eerste plaats moeten er — hoe mooi deze films meerendeels ook waren — heel andere beelden komen, rechtstreeks gegrepen uit de leerstof der school. De overige gebreken kunnen we laten rusten; ze doen niets af aan het feit, dat de Bioscoop bewezen heeft, geschikt te zijn om in de school te worden ingeburgerd.'

besides often unsubstantiated debates about these two teaching aids' relative merits, personal stakes, too, could obstruct the assessment of their value. In this case, alderman of education J. A. A. Schoondermark, the driving force behind the abovementioned film screenings, opposed this motion, citing a research report on educational film screenings that the council awaited.[16] The motion was voted down and nothing was heard of Zandstra's plan again.

In The Hague, secondly, the municipality appointed a committee, in December 1911, to investigate the significance of film for education.[17] The committee's final report, partly based on data obtained abroad, was discussed in August 1913. The general conclusion was that film should best be seen as a complement to other teaching practices. Accordingly, the optical lantern had an important role to play in preparing pupils for subsequent film screenings, since, as the committee stated, the optical lantern 'had been already used' in classrooms.[18] Again, the presence and usage of the lantern was taken for granted, its affordances or disadvantages were not part of the argument in this report. In the fall of 1913 the committee's most important conclusions were adopted: 'Attending film screenings in regular cinema theatres is not recommended'; 'It is desirable to combine the school cinema with the *Museum ten Bate van het Onderwijs* (Museum for the Benefit of Education) and other institutes that serve the general interests of education and teachers'.[19] In 1916, it established such school screenings, initially within the premises of said museum.[20] In 1918 the renamed *Gemeentelijke Schoolbioscoop* (Municipal School Cinema) relocated to a building of its own,[21] and in July 1920 the municipality took formal and financial responsibility.[22]

Notwithstanding the fact that the lantern had already been integrated in the curricula of Amsterdam schools,[23] in 1911 the municipality of Amsterdam decided to do a trial with the regular use of the optical lantern for elementary education. The central question in this trial also dealt with the matter of the space where this way of teaching should take place: within the school or in a building especially equipped for that purpose. Despite

---

16 'Rapport', *Leeuwarder Courant*, December 11, 1912, 1st section, p. 2, <https://resolver.kb.nl/resolve?urn = ddd:010679815:mpeg21:p. 002> [accessed May 3, 2021].

17 'Schoolbioscoop, *Limburger Koerier*, September 26, 1913, 1st section, p. 2, <https://resolver.kb.nl/resolve?urn = MMKB23:001174183:mpeg21:p. 00002> [accessed May 3, 2021]; David van Staveren, *De bioscoop en het onderwijs* (Leiden: A. W. Sijthoff, 1919), 16.

18 ''s-Gravenhage. De Bioscoop-Kwestie', *De Maasbode*, September 24, 1913, morning edn., p. 1, <https://resolver.kb.nl/resolve?urn = MMKB04:000185081:mpeg21:p. 001> [accessed May 3, 2021]; Van Staveren (1919), 17–18.

19 'Uit de residentie. Rapport der bioscoop-commissie', *De Nieuwe Courant*, September 23, 1913, evening edn., 2nd section, p. 9, http://resolver.kb.nl/resolve?urn = MMKB15:000785212:mpeg21:p. 00009> [accessed May 3, 2021] and similar newspaper reports over the next few days; 'De Bioscoop, II', *Het Vaderland*, October 5, 1913, morning edn. B, pp. 5–6, <http://resolver.kb.nl/resolve?urn = MMKB23:001512013:mpeg21:p. 00005> [accessed May 3, 2021]. The Museum for the Benefit of Education was founded in 1904; since 1986 it is called Museon.

20 'Museum ten Bate van het Onderwijs', *Het Vaderland*, April 19, 1916, morning edn., p. 1, <http://resolver.kb.nl/resolve?urn = MMKB23:001514134:mpeg21:p. 00001> [accessed May 3, 2021].

21 'Opening Gem. Schoolbioscoop', *De Maasbode*, August 30, 1918, evening edn., 2nd section, p. 2, <http://resolver.kb.nl/resolve?urn = MMKB04:000187986:mpeg21:p. 006> [accessed May 3, 2021] (and similar newspaper reports on this and the following days).

22 *Rapport van den Onderwijsraad* (1922), 47.

23 Schoolteacher A. Dijkgraaf bluntly stated this fact without any further validation. See: Dijkgraaf (1912), pp. 118–19.

the difficulties of actually getting this trial off the ground, a meeting was held to discuss the most pressing matters with respect to this visual teaching aid.[24] One of the aldermen of Amsterdam attending this meeting believed that the optical lantern should be set up in all schools, since he considered it an indispensable tool. However, the conditions for setting up the lantern in Amsterdam schools were problematic, because, as noted, not all of them were connected to the electricity grid. Furthermore, it proved to be extremely difficult and expensive to fully darken auditoria during the day.[25]

These initial trials focused on a single medium rather than a comparison between the two — even though there was some talk of one (usually the lantern) alongside or in the service of the other (usually film). On top of that there was little reflection on the specific affordances of each medium nor on the results of these tests. Contrary to expectation, there was no thorough analysis to understand the potential of a new teaching aid nor an evaluation of the benefits for students' memories or the knowledge they had gained. In none of these cases such reflections are presented. However, in their direct aftermath a discussion blossomed about the desirability of the combination of both media.

Projecting Conditions

The subsequent debate between supporters and opponents of both media stemmed from the broader 'cinema question' (Dutch *bioscoopvraagstuk*). As Dutch film scholar Floris Paalman has argued, in the 1910s the debate about films for educational purposes was a direct consequence of the perceived social perils of 'regular' cinema.[26] Commentaries in the months following these first trials would return throughout the following decade: Which criteria should apply to make visual teaching aids acceptable in schools? And which conditions had to be fulfilled to make them function in an educational context? We review these commentaries on the basis of the following four points.

Technical Set-Up

The example of Amsterdam had shown that much of the functioning of these media depended on technical and logistical issues and possibilities. Various positions were held on this matter. Schoolteacher Jos Veerbaldt wrote in 1912 that each school should have an optical lantern. She actually preferred the lantern over film, since it was less distressing and the slides for educational purposes were easier to make than films.[27] The same position was taken by teacher J. Kruithof, who stated moreover that the optical lantern's distinct advantage was that it could be used anytime a teacher wanted to. Despite the difficulties of

24 'Lichtbeelden op de Lagere Scholen te Amsterdam', March 28, 1912.

25 Ibid.

26 Floris Paalman, 'Far and close: the Gemeentelijke Schoolbioscoop in Rotterdam', Marina Dahlquist, Joel Frykholm (eds), *The Institutionalization of Educational Cinema. North America and Europe in the 1910s and 1920s* (Bloomington: Indiana University Press, 2019), 81.

27 Jos Veerbaldt, 'Vrouwenbeweging. Iets over bioscoopen en projectielantaarn', *Bataviaasch Nieuwsblad*, June 1, 1912, 5th section, pp. 17–18, <http://resolver.kb.nl/resolve?urn = ddd:011036017:mpeg21:p. 017> [accessed May 3, 2021]. Veerbaldt was the pseudonym of Josephine Baerveldt-Haver, a Dutch feminist who devoted herself to women's suffrage.

darkening the room, both considered the lantern an indispensable teaching aid.[28] Neither of them, however, reflected on such practical matters as the costs of darkening the room, of buying the optical lantern, and renting the slides.

Another schoolteacher, A. Dijkgraaf, argued for using both lantern and film projections in schools, preferably all schools. However, in order to fully use the potential of these media, schools needed projection equipment that could alternate between slides and film, such as the *Schulprojectionsapparat*-Porta.[29] But even though at this time projectors equipped for both media were pretty standard, all three of the abovementioned commentators failed to discuss other conditions, notably a connection to the electricity grid or having teaching staff capable of operating the lantern and/or film projector.

A final example on the reflection of film as a visual teaching aid in comparison to the optical lantern came from Dutch school inspector and later pedagogue J. H. Gunning Wzn., who was vehemently opposed to film as an educational tool, because he believed that film merely stimulated the senses. Despite possible reforms, films were continuous and therefore distressing. From a pedagogical perspective the optical lantern, because of its still images, was the right teaching aid.[30] In spite of his firm criticism, however, Gunning did not explicitly say why he preferred still over moving images. Of course, one can assume that this had much to do with the possibility of teachers' explanations of the visuals or of stimulating pupils to think for themselves. But it is striking to see how little evidence or arguments were brought into the discussion to convince their respective audiences.[31] Two years after Gunning's referenced article '*Tegen de bioscoop*' (Against cinema) he reported on the newly invented device to stop the film at any moment in mid-projection. Despite this improvement he remained unconvinced of the pedagogical value of film.[32]

Cinema Inside or Outside School?

These technical issues, in turn, led to the question whether lantern or film projections needed to take place inside school or outside, at a purpose-built, so-called school cinema. An interesting and relevant figure in this respect and especially in the further development of the debate was David van Staveren. He had been an elementary school teacher in The Hague since 1913 before he became the director of that city's Municipal School Cinema. Although he had no previous experience with or knowledge of educational cinema, he nevertheless considered it necessary to screen educational films in a school cinema: a separate establishment with professional machinery and knowledgeable staff, where films could be screened for a number of school classes at the same time.[33] However, not

---

28 J. Kruithof, 'De projectie-lantaarn in de school', *De School met den Bijbel*, November 7, 1912, pp. 301–02, <https://resolver.kb.nl/resolve?urn = MMUBVU06:001767019:00005> [accessed May 3, 2021].

29 Dijkgraaf (1912), 120–21.

30 J. H. Gunning Wz., 'Tegen de bioscoop', *Het Kind*, August 24, 1912, pp. 138–39, <https://resolver.kb.nl/resolve?urn = MMKB14:002587017:00002> [accessed May 3, 2021].

31 Paesi (1900), pp. 611–13; Ruchatz (2003), 228–29.

32 J. H. Gunning Wz., 'De bioscoop verbeterd. Een Nederlandsche uitvinding', *Het Kind*, June 13, 1914, p. 95, <https://resolver.kb.nl/resolve?urn = MMKB14:002589013:00007> [accessed May 3, 2021].

33 Bert Hogenkamp, 'Staveren, David van', *Biografisch Woordenboek van het Socialisme en de Arbeidersbeweging in Nederland (BWSA)*, <https://socialhistory.org/bwsa/biografie/staveren> [accessed May 3, 2021]; Van Staveren (1919), 7–9.

everyone agreed. In 1921, another schoolteacher, G. H. Wanink, polemicized with van Staveren about the necessity of school cinemas. Wanink argued that as the films that were shown had to be short, the trip to a separate building would be too time-consuming.[34] Moreover, unlike van Staveren, Wanink believed that the screening of films within schools was not an insurmountable problem, since teachers could be taught how to operate a projector and claimed they welcomed this medium in their classrooms.[35] No consensus was reached, but, more remarkably, nowhere in this polemic did either side substantiate why a school cinema was a good option or not.

What Did Children Get to See?

A general critique at the time was the persistently insufficient amount of proper educational films. The abovementioned schoolteacher A. Dijkgraaf argued for a central organisation entrusted with the production of educational films.[36] This future organization should take as an example the structure of the Lantern Slide Association and the support it had enlisted of a number of experts to ensure itself of high-quality slides, prints and books.[37] A board of advisors, consisting of educational specialists and teachers, among others, was indispensable. In order to emphasize the usefulness of film as a teaching aid, much attention should be directed to 'turn of the evil tide' of film and offer a clear focus on its potential as a valuable didactic tool.[38] Despite his enthusiasm for the optical lantern and its benefits for classroom teaching, Dijkgraaf appeared convinced of the added value of film, provided that it was significantly improved.[39]

And room for improvement there was, according to a number of teachers, doctors, and psychologists. Already in 1907, a committee in the German city of Hamburg had reported on a series of controlled film screenings, apparently followed by a survey, which showed that they could be painful to the children's eyes, especially of those who already had eyesight problems. Such ailments had mostly to do with the rapid succession of images as well as their flicker.[40] German psychologist Robert Gaupp argued that children should not be exposed for long periods of time to film, since this would lead to exhaustion, dizziness and dimmed eyesight.[41] For the same reason warnings were advanced about the lengthy exposure to the darkness during screenings. However, Dutch ophthalmologist D. M. Straub and an anonymous school

34 'Opbouw der maatschappij. De schoolbioscoop op nieuwe banen', *De Nieuwe Courant*, May 14, 1921, morning edn., p. 4, <https://resolver.kb.nl/resolve?urn = MMKB15:000761069:mpeg21:p. 00004> [accessed May 3, 2021].

35 G. H. Wanink, 'De Schoolbioscoop, I, *De Bode: Orgaan van den Bond van Nederlandsche Onderwijzers*, September 23, 1921, pp. 3–4, <https://resolver.kb.nl/resolve?urn = MMIISG10:000690034:00003> [accessed May 3, 2021].

36 Veerbaldt (1912); J. H. Gunning Wz. (August 24, 1912), 128; Dijkgraaf (1912), 122. See also part II below.

37 Dijkgraaf (1912), 116.

38 Ibid., 112.

39 Ibid., 120.

40 Thierry Lefebvre, 'Flimmerndes Licht. Zur Geschichte der Filmwahrnehmung im frühen Kino', transl. from the French by Sabine Lenk, *KINtop. Jahrbuch zur Erforschung des frühen Films*, 5 (Basel — Frankfurt am Main: Stroemfeld/Roter Stern, 1996), 72–73.

41 J. Kruithof, 'Bioscoop-Vertooningen', *De school met den Bijbel*, October 10, 1912, pp. 236–37, <https://resolver.kb.nl/resolve?urn = MMUBVU06:001767015:00004> [accessed May 3, 2021]; Robert Gaupp, 'Der Kinematograph vom medizinischen und psychologischen Standpunkt', *Dürerbund. Flugschrift zur Ausdruckskultur*, #100 (1912), 4; see also: Lefebvre (1996), 76.

doctor were reported as stating that as long as film screenings did not last more than five quarters of an hour no damage would be done.[42] Others were concerned about the damage film could cause to children's moral wellness. For instance, J. Kruithof, citing Gaupp and a Prussian government report, warned against the adverse stimulation of children's nerves and imagination by unseemly or hideous scenes.[43] As so often, however, substantiated arguments and research of measured dangers of film as a visual teaching aid were lacking. As a matter of fact, even as late as 1922 this was precisely the criticism of pedagogue Ph. Kohnstamm in his review of the *Staatscommissie Bioscoopgevaar* (State Committee Cinema Peril).[44]

The Role of the Teacher

As van Staveren had made obvious, he was not impressed with the technical skills of schoolteachers. S. de Jong Ezn. stated that operating the projector was deemed impossible for teachers with no experience or technical know-how. In all likelihood their comments were informed by their preference for school cinemas, in The Hague and Rotterdam, respectively.[45] But while teachers were advised to leave the projector alone, others stressed the teacher's authoritative role as the one who controls the teaching aid instead of being controlled by it.

A much-heard argument in understanding the role of the teacher had to do with the preparations that needed to be taken care of before a classroom film screening. Commenting on a trial in the Belgian town of St Gilles, Dijkgraaf stressed the importance of a lecture, illustrated or not, before taking students to see a film. In other words, a solid preparation was required to make them understand what they were going to see. Moreover, he argued in his elaborate article that it was sometimes necessary to have both media available simultaneously: in order for children to memorize certain phenomena they needed to have the possibility to look at still as well as moving images — the former to explain the specific characteristics of a waterfall, for example.[46] This had also been one of the recommendations of the Hague investigative committee's conclusion on a school cinema: 'it is desirable to frequently use an optical lantern in all schools; it is indispensable for the preparation of film-based teaching.'[47] Finally, as a sufficient amount of educational films were lacking, fears arose that the curriculum would be determined by those films that were available rather than by the teacher. As Ph. Kohnstamm stated, it cannot be that the school is in the service of the cinema.[48]

---

42 'De Bioscoop en de School', *De Preanger-bode*, November 8, 1913, evening edn., 2nd section, p. 5, <https://resolver.kb.nl/resolve?urn = MMKB08:000125191:mpeg21:p. 005> [accessed May 3, 2021].

43 Kruithof (1912), p. 236.

44 'Het bioscoopgevaar', *Zutphensche Courant*, April 8, 1921, 2nd section, p. 5, <https://resolver.kb.nl/resolve?urn = MMRAZ02:000407080:mpeg21:p. 00005> [accessed May 3, 2021].

45 S. de Jong Ezn., 'De bioscoop. IV', *De School met den Bijbel*, November 20, 1913, pp. 287–88, <https://resolver.kb.nl/resolve?urn = MMUBVU06:001768021:00008> [accessed May 3, 2021].

46 Dijkgraaf (1912), pp. 120–21; Paesi, (1900), pp. 611–61.

47 'Schoolbioscoop', *Algemeen Handelsblad*, September 24, 1913, evening edn., 2nd section, p. 7, <http://resolver.kb.nl/resolve?urn = ddd:010651316:mpeg21:p. 007> [accessed May 3, 2021]. The original Dutch text reads: 'Het is wenschelijk op alle scholen geregeld de projectielantaarn te gebruiken; bij de voorbereiding van het bioscopisch onderwijs is deze onmisbaar.'; Ph. Kohnstamm, *Bioscoop en volksontwikkeling* (Amsterdam: Nutsuitgeverij, 1922), 32.

48 Ibid., 35.

The Next Stage of the 'Cinema Question': Regulations and Pedagogical Perspectives

In 1918, the cinema question took a new and interesting turn: the government became involved. On 2 November 1918, a state committee on the peril of cinema was set up to investigate what measures should be taken to combat the peril of cinema. By that time, it was argued, film shows were not considered dangerous in themselves anymore (as being harmful to the eyes, for instance), but their content was, both morally and socially, specifically for those under the age of eighteen.[49] Moreover, in the report's conclusion its members explicitly discussed on what grounds films for this age category should be censored. Newspaper reports at the time merely commented that these films should be approved or banned on the basis of their 'appropriateness' rather than on their 'admissibility'.[50] But what that actually meant remained unclear.

Simultaneously with the felt need for censoring film for the young, film as an educational tool was being institutionalized and embraced in The Hague. Van Staveren had been appointed director of its school cinema. For him films were not problematic at all, on the contrary: 'Film fills a lacuna in education. It is all about evoking images in the children's minds that will not fade too quickly or disappear altogether.'[51] One of the members of the state committee, Andrew de Graaf, stated that films should above all be useful for 'instructive purposes'.[52] An opponent of censorship, he wrote in one of his brochures: 'Only as school cinema, at the service of education and in the hands of competent pedagogues, something good might come of it.'[53] Similar arguments had been made by van Staveren, Dijkgraaf, and others in the context of creating pedagogically valuable teaching aids.[54]

Many positions were taken, many opinions expressed, but most of them were rhetorical or anecdotal. In 1922, Dutch pedagogue Ph. Kohnstamm entered the discussion with his critical review, referenced above, of the government report on the perils of cinema. Kohnstamm was largely concerned with the lack of evidence for almost all its claims. He stated that, despite the presence of 'schoolmen' on the committee, it had lost an opportunity by not calling on pedagogues or psychologists. In his view films had no instructive value, as they

---

49 J. de R., 'Wij leven snel!', *De Bioscoop-Courant*, January 10, 1919, p. 2, <https://resolver.kb.nl/resolve?urn = MMEYE01:000815002:00004> [accessed May 3, 2021]; W. W. van der Meulen, 'Overheidstoezicht op de bioscopen', *Vragen des Tijds*, January 1, 1921, p. 436, <https://resolver.kb.nl/resolve?urn = MMKB10:000773001:00444> [accessed May 3, 2021].

50 'Staatscommissie bestrijding bioscoop-gevaar', *Algemeen Handelsblad*, October 16, 1920, morning edn., 2nd section, p. 5, <https://resolver.kb.nl/resolve?urn = ddd:010654776:mpeg21:a0129> [accessed May 3, 2021] and many similar news reports over the following two weeks.

51 Van Staveren (1919), 5–6. The original Dutch text reads: 'De bioscoop vult een leemte in 't onderwijs aan. Het gaat er toch om, bij de leerlingen beelden op te roepen die niet al te spoedig zullen vervagen of geheel verdwijnen.'

52 A. de Graaf, 'De Bioscoop en hare gevaren', *De Getuige*, April 15, 1917, pp. 3–4, <https://resolver.kb.nl/resolve?urn = MMKB13:002808032:00003> [accessed May 3, 2021]; see also: H. de Bie, 'Nog eens de bioscoop', *De School met den Bijbel*, July 4, 1918, pp. 4–5, <https://resolver.kb.nl/resolve?urn = MMUBVU06:001769001:00009> [accessed May 3, 2021]; 'Nieuwe boeken: de bioscoop', *School en Leven*, February 26, 1920, pp. 428–30, <https://resolver.kb.nl/resolve?urn = MMKB14:001608028:00006> [accessed May 3, 2021].

53 Quoted in: 'Binnenland', *Het Oosten: Weekblad gewijd aan Christelijke Philantropie*, October 22, 1919, [unpaginated], <https://resolver.kb.nl/resolve?urn = MMLIND01:001432043:00002> [accessed May 3, 2021]. The original Dutch text reads: 'Alleen als school-bioscoop in handen van bekwame paedagogen, in dienst dus van het Onderwijs, is er nog iets goeds van te verwachten.'

54 Van Staveren (1919) 7; Dijkgraaf (1912): p. 116; 'De bioscoop als school' (June 12, 1911).

invite passivity, while their tiresome flicker and glitter makes viewers forget their content.[55] Consequently, film was only functional for memory retention when students were familiar with the topic.[56] Additionally, 'precise observation' was hardly possible. In contrast, the still image was 'indispensable' for Kohnstamm, since this not only allowed closer and more sustained observation, but also enabled teachers to speak simultaneously with the visuals.[57] His other objection against film as an educational teaching aid was the fact that the school curriculum had to adjust to the available films, instead of the other way around.[58]

Kohnstamm was the only person in this choir of comments to call for a 'pedagogical psychological experiment'. Instead of just parroting other peoples' words, what was needed was systematic research into the didactical value of still and moving images, the optical lantern and film.[59] And even though his abovementioned review also contained a few unsubstantiated statements, his call was issued precisely when such research was being prepared by his colleagues of the Municipal University of Amsterdam's Pedagogical-Psychological Laboratory, which he directed at the time.[60]

II.

The debates triggered by the introduction of photography-based teaching aids, film in particular, found their provisional peak in 1919, with the brochure *De bioscoop en het onderwijs* (Cinema and education), by the director of the Hague Municipal School Cinema, David van Staveren. With his boastful language the author seemed to have taken the flight forward, an evasive move to force the matter of film in education. He quickly dismissed a few of the common objections: that it would accustom students to cinemagoing and its corrupting effects, that it infringed on teaching time, and that film's sheer visuality and pace left no time for reflection. With the last point he also skirted the issue of the relative merits of film and lantern projections. Instead, he insisted curtly and in passing that 'the living image has a much more lasting impact than the standing image'; the latter, moreover, would 'often only realize its full potential (…) alongside film', whatever that was supposed to mean — unless this was an implicit reference to one of the Hague investigative committee's recommendations that the lantern 'is indispensable for the preparation of film-based teaching'; see note 47).[61] Film, apparently, was the teaching aid everybody had been waiting for. And the school cinema was the place to see it in operation.

55 'Het bioscoopgevaar' (April 8, 1921), 5.
56 Kohnstamm (1922), 33.
57 'Het bioscoopgevaar' (April 8, 1921), 5.
58 Kohnstamm (1922), 34–35.
59 'Het bioscoopgevaar' (April 8, 1921).
60 On page 33 Kohnstamm states in a footnote: 'I am very pleased that my relevant remarks in this journal [*Volksontwikkeling*, March 1922, p. 280) have led to this investigation, the first thorough psychological research about the functioning of film that we possess.' Original text: 'Het verheugt mij zeer dat mijn desbetreffend eopmerkingen in dit tijdschrift IMaart 1921, p. 280) aanleiding hebben gegeven tot dit onderzoek, het eerste nauwkeurige psychologische onderzoek over de werking van den film dat wij bezitten.' Quoted in: Kohnstamm (1922), 33.
61 Van Staveren (1919), 6; 5. The original Dutch texts read: '…dat het levende beeld van veel blijvender invloed is dan het stilstaande.'; '…ze zal vaak pas haar volle waarde verkrijgen (…) naast de film'.

DE BIOSCOOP EN
HET ONDERWIJS
DOOR D. VAN STAVEREN.

A. W. SIJTHOFF'S UITG.-Mij. — LEIDEN.

Fig. 3.2. Titlepage of David van Staveren's brochure on cinema and education (1919). David van Staveren, *De Bioscoop en het Onderwijs* (Leiden: A. W. Sijthoff's Uitg-Mij, 1919).

Indeed, Van Staveren's self-assertive brochure not only failed to extinguish the debates, it stoked the fire under another one: would the cinema come to the school or the school to the cinema? Over the next few years controversies over photography-based teaching aids, their exhibition, and their form broadened to brochures, reports, and books, while interest shown by politics and science in the late 1910s and early 1920s lent more weight to the matter. The national government, in the shape of a mixed Catholic-Protestant coalition Cabinet, entered the debate in 1919, after calls for financial support of a future national institute to secure the manufacture and distribution of proper educational films. And in 1922, science, in the shape of the Pedagogical-Psychological Laboratory of the Municipal University of Amsterdam, examined the high-minded claims made about the 'didactic value' of the new teaching aids.

The Reluctant State

The idea of a national film institute, provisionally called State Film Archive, arose from the wide agreement, the debates notwithstanding, that an ample stock of proper educational and information films was a long way away. For slides there was no such problem. As we have seen, after the Lantern Slide Association had reoriented its policy to school education, in 1910, it steadily increased its series of lantern slides—from an already considerable catalogue that it had begun to assemble since 1898—with the creation of a photographic archive.[62] The persistent shortage of films, however, was not for lack of plans.

A few private, non-commercial attempts had been made to collect or produce films. The *Koloniaal Instituut* in Amsterdam had commissioned the making of a few dozen films in colonial Indonesia, then called the Netherlands East-Indies, in 1912–1913 and in 1917. They were explicitly meant for educational institutes at all levels in the Netherlands, besides museums, colonial exhibitions, etc. The institute also accumulated, mainly through donations, a large collection of lantern slides and photographs, from which it made more slides.[63] In 1917, moreover, it established its *Comité voor Lezingen en Leergangen* (Committee of Lectures and Courses) to train teachers of elementary and secondary schools in the use of visual teaching aids. Secondly, the *Nederlandsche Vereeniging Filmcentrale* (Dutch Association Film Centre), founded in 1917, had pledged to collect films on Dutch industry, agriculture, and other economic activities for distribution among schools, although not much was heard of it since. And in 1919 the *Vereeniging Nederlandsch Centraal Filmarchief* (Association Dutch Central Film Archive) was founded with the aim to collect films of Dutch historical and topical importance.[64] But while all three organisations focused

62 Hintzen (July 2, 1927), p. 4. At the time this article was published the Lantern Slide Association's slide collection consisted of *c.* 30,000 slides.

63 *Derde jaarverslag 1913* ([Amsterdam: Vereeniging 'Koloniaal Instituut', 1914), 15. The influential Swiss pedagogue Gottlieb Imhof considered the film and lantern slide collection of the Colonial Institute rich in content. Screened at the Schweizer Mustermesse, in Basel in 1922, a selection was provided for classroom screenings, introduced by Imhof and a representative of the Institute. Particularly the combination of still and moving images was much approved by teaching staff; see: Anita Gertiser, 'Domestizierung des bewegten Bildes. Vom dokumentarischen Film zum Lehrmedium', *Montage A/V*, 15, 1 (2006), 62–63.

64 'Ned. Centraal Filmarchief. Eerste filmkeuring: geen gelukkige keuze', *Het Vaderland*, 52, March 24, 1920, evening edn., A, p. 1, <http://resolver.kb.nl/resolve?urn = ddd:010006520:mpeg21:p. 001> [accessed May 3, 2021]. This association hoped to play a national role in the shape of a state film archive that the abovementioned discussions in 1919 hinted at.

Fig. 3.3. Royal Institute for the Tropics in Amsterdam (Mauritskade), 1936. (Royal Institute for the Tropics: <https://www.kit.nl/nl/over-ons/geschiedenis/#media-0–6017>

on specific subjects (on which only the Colonial Institute seemed to have had sizeable collections, particularly photographs and slides), more was needed to serve a school's complete curriculum.

In fact, an initiative of wider scope had been launched in February 1912 by J. A. A. Schoondermark, the alderman of education in Leeuwarden. Heartened by the abovementioned, widely praised educational film experiment of the previous summer, he invited his peers and other relevant parties to a meeting in Utrecht to discuss film as a teaching aid on a coordinated, national scale. The meeting decided to set up a committee to investigate the possibility of educational film production and estimate a budget for a partnership of participating municipalities.[65] In 1913 this initiative was embraced by the *Vereeniging van Nederlandsche Gemeenten* (VNG; Association of Dutch Municipalities). This administrative umbrella organization was willing to finance the production of educational films on condition that it was supervised by experts and that sufficient municipalities and educational institutes at every level would participate.[66] This plan came to naught due to wartime circumstances. Indeed, the amount of films the VNG had proposed to produce—ten in the first year and five during every following

65 See e.g.: 'De school en de bioscoop', *Nieuwe Rotterdamsche Courant*, 11 February 1912, morning edn., A, p. 6, <http://resolver.kb.nl/resolve?urn = ddd:010032032:mpeg21:p. 006> [accessed 3 May 2021] and similar news reports over the following days.

66 'Schoolbioscopen', *De Nieuwe Courant*, 24 July 1913, evening edn., 2nd section, p. 7, <http://resolver.kb.nl/resolve?urn = MMKB15:000785060:mpeg21:p. 00007> [accessed 3 May 2021].

year — may well have been too ambitious in an economy cut off from supplies by neighbouring warring nations.[67]

In May 1912, entrepreneur Maurits Binger launched another initiative with the foundation of two film companies: the *Maatschappij voor Wetenschappelijke Cinematographie* (Scientific Cinematography Co.) and the *Maatschappij voor Artistieke Cinematographie* (Artistic Cinematography Co.). A commercial venture, its goal was to 'unite in a regular organization the various ways of utilizing film for industry, art, education, economy, and ethics, alternated with suitable entertainment.' As well it put forward a proposal for a mobile, 1000-seat auditorium that could be erected all over the country and function as a school cinema.[68] In May 1913 the two companies were merged in the Scientific and Artistic Cinematography Co. During these two years they produced about two dozen documentary films before they became, in 1914, *Filmfabriek Hollandia* (Film Studio Hollandia). From that point onwards the company refocused on the production of feature fiction films. Its success, thanks to reduced foreign imports during the war years, lasted only until the early 1920s. The postwar removal of restrictions on the import and export of films, in March 1919, and the subsequent flood of American product sealed Hollandia's fate; in 1923 it was declared bankrupt.[69] In the meantime, its documentary production had taken a backseat while the mobile auditorium had been totally forgotten.

Despite Hollandia's failure, it became even clearer than before the war that the Dutch film market was too small to recover the costs of educational films proper. As late as 1931, historian Frances Consitt, in her investigation of film as a teaching aid in the much bigger British market, observed this problem, too: 'Producers refuse to create educational films for a non-existent market.'[70] If educational institutes wanted to employ film as a teaching aid, they had no choice but to rely on commercial production and distribution companies, whose films also must have a theatrical release — or in the terms of the mission statement quoted above, 'alternate with suitable entertainment.' This, of course, significantly compromised the ideal of an educational film.

Van Staveren's 1919 brochure clearly signalled his dependence on commercial parties: its front and back covers as well as its flyleaves carried advertising for two Dutch commercial distribution companies whose catalogues of 'educational and scientific films', of both domestic and foreign origin, served theatrical as well as nontheatrical venues. An overview of van Staveren's school cinema programs featured a considerable number of their titles (despite the municipality's investigating commission's condition at the time that it only

67 *Rapport van den Onderwijsraad* (1922), 24; E. Bonebakker, 'Utrecht', *De Kinematograaf*, 19 April 1918, p. 3588, <https://resolver.kb.nl/resolve?urn = MMEYE01:000768012:00012> [accessed 3 May 2021].

68 'Wetenschappelijke cinematographie', *Algemeen Handelsblad*, 8 May 1912, evening edn., 3rd section, p. 10, <http://resolver.kb.nl/resolve?urn = ddd:010649940:mpeg21:p. 010> [accessed 3 May 2021]. The original Dutch text reads: '…die zich ten doel stelt door een vaste organisatie te vereenigen de verschillende toepassingen der bioscoop op het gebied van industrie, kunst, onderwijs, economie en etiek, afgewisseld door gepaste ontspanning.'

69 André van den Velden, Fransje de Jong, Thunnis van Oort, 'De bewogen beginjaren van de Nederlandsche Bioscoop Bond, 1918–1925', *Tijdschrift voor Mediageschiedenis*, 16, 2 (2013), 24; Eye Filmmuseum, <http://catalogus.eyefilm.nl/ce/Corporaties/wetenschappelijke cinematographie> [accessed May 3, 2021].

70 Francis Consitt, *The Value of Films in History Teaching* (London: G. Bell and Sons, 1931), 1.

show 'purpose-made films').[71] Parenthetically, the same overview did not bear out his claim, quoted above, that lantern projections would realize their 'full potential [...] alongside cinema', as none are listed. A recent study of the Municipal School Cinema in Rotterdam, furthermore, lists a number of its programs screened in 1921 and 1922. Many of their titles, too, were commercially distributed nonfiction films, with a varying share of foreign imports.[72] No wonder that, even as late as 1924, Dutch film producer-cum-cameraman H. C. Verkruysen could write that 'the educational film is not satisfactory yet.'[73]

The abovementioned initiatives as well as the then current situation had been discussed during talks between experts and high-level government representatives, in 1919. What was clearly needed, according to the experts, was a counterbalance to the film entertainment business to make the kinds of film that would be useful for education and befit a state film archive. Moreover, given the reliance on commercial product, the government, it was argued, would lose the opportunity to assume a central role and control a field now largely left to the whims of private enterprise.[74] Eventually the government dismissed the suggestion, citing financial reasons. But in 1921 it reconsidered, possibly prodded by the recommendation to financially support associations and institutes that offer programs of instructive films or of suitable entertainment, a recommendation that was mentioned in the report by the State Committee to Combat the Perils of Cinema, published in October 1920.[75] The Elementary Education section of the Education Board, a government advisory body, was subsequently asked to appoint another commission to investigate the significance of film as a teaching aid. However, the commission's work may have been a mere political maneuver; the conclusions submitted in its report of 1922 did not lead to any measures. But the government did follow up on the 1920 report on the perils of cinema with the 1926 *Bioscoopwet* (Cinema Law) and a national censorship office in 1928.

This decision has been attributed to what was called a 'moralizing government'. A study of that title defined this term as governance based on policies of 'moral and cultural values and norms'. It claimed, moreover, that it reflected the attitude of Dutch governments from around the turn of the twentieth century until the Second World War as that of being a 'night-watchman state', concerned merely with the core tasks of foreign policy, national safety, public order, and good morals; only in case of threats to these matters would a government intervene.[76] As noted, cinema's perceived perils to public order and morals, for instance, had indeed forced it into action. This was in line with interventions made in the past, albeit mostly on a local level, to curb fairgrounds or blood sports — a type of activities that strongly suggests that moral and cultural values were heavily informed by considerations of class and taste.

---

71 *Rapport van den Onderwijsraad* (1922), 48–51; 'Uit de residentie. Rapport der bioscoop-commissie' (23 September 1913), p. 9.

72 Paalman (2019), 86–91.

73 H. C. Verkruysen, *De cultureele beteekenis van het lichtbeeld* (Amsterdam: Elsevier, 1924), 62. The original Dutch text reads: '... de onderwijsfilm deugt nog niet.'

74 Expert statement by W. H. Idzerda, privatdocent Photography at the *Technische Hoogeschool* Delft (Institute of Technology, Delft) and at that time director at the documentary department of Hollandia Film Studio; 'Appendix III', *Rapport van den Onderwijsraad* (1922), 77–78.

75 'Staatscommissie bestrijding bioscoop-gevaar' (16 October 1920).

76 J. H. J. van den Heuvel, *De moraliserende overheid: een eeuw filmbeleid* (Utrecht: Lemma, 2004), 7–8.

But there is more to be said against the notion of a night-watchman state in this context. First of all, it is not just a misnomer with regard to the period under consideration, it is obsolete, as the cited study failed to notice government legislation ever since the 1880s that addressed the excesses of the abovementioned social question by prohibiting child labour, limiting working hours, regulating housing construction or making elementary education compulsory.[77] Furthermore, the cited study's focus on censorship and exhibition, certainly a prominent issue until the late twentieth century, makes its argument self-fulfilling. It misses the view from below and the innovative ideas, particularly among a new generation of teachers around the turn of the twentieth century, that had created a new dynamism.[78] And due to its poor grasp of cinema history it misses part of the view from above, too, as Cabinets — both liberal and confessional — involved themselves from time to time in commissioning or sponsoring film *productions*, for reasons of colonial and military propaganda rather than morals or cultural values. For instance, the abovementioned films commissioned by the Colonial Institute, in 1912, had been made possible by the Ministry of Colonies, which had paid the filmmaker's salary during the year his commission lasted.[79] In 1916, the Ministry of War had commissioned a propaganda film on the Dutch army and navy, *DE LEGER- EN VLOOTFILM* (ARMY AND NAVY FILM; 1917).[80] Again the Ministry of Colonies, in 1926, commissioned a number of films about the Netherlands East-Indies.[81] Eventually, though, national safety forced the government to enter the world of media itself: in 1934, in order to counter national-socialist propaganda and serve national interest by its own information, it created a state press service. So much for the night-watchman state.

Didactic Value

In 1922 two psychologists of the then recently founded Psychological-Pedagogical Laboratory of the Municipal University Amsterdam (Gemeente Universiteit) conducted two unique experiments.[82] Unique in the Netherlands, that is. Because at the same time, and unbeknown to each other, a series of similar experiments was being conducted in the United States (in fact, its lead researcher, Frank N. Freeman, had conducted educational film experiments as early as 1918).[83] Both were concerned with the didactic value, or what the Americans called pedagogical effectiveness, of visual teaching aids. The Dutch experiments

---

77 Ruud Koole, *Twee pijlers: het wankele evenwicht in de democratische rechtsstaat* (Amsterdam: Prometheus, 2021), 184.

78 Jan Bank, Maarten van Buuren, *1900: hoogtij van de burgerlijke cultuur* (The Hague: Sdu, 2000), 229–64.

79 'Minutes of meeting board of directors Association 'Colonial Institute' (January 15, 1912)', Archive Koninklijk Instituut voor de Tropen (Royal Tropical Institute), file no. 219, quoted in: Janneke van Dijk, Jaap de Jonge, 'Johann Christian Lamster (187–1954)', van Dijk, de Jonge, Nico de Klerk (eds), *J. C. Lamster, een Vroege Filmer in Nederlands-Indië* (Amsterdam: KIT Publishers, 2010), 24.

80 Eye Filmmuseum, http://catalogus.eyefilm.nl/ce/Corporaties/Ministerie van Oorlog.

81 Eye Filmmuseum, http://catalogus.eyefilm.nl/ce/Corporaties/Ministerie van Koloniën.

82 G. Révész, J. F. Hazewinkel, 'Over de didactische waarde van de projectielantaarn en de bioscoop', *Paedagogische Studiën*, 4 (1923), 33–67; J. F. Hazewinkel, 'Over de didactische waarde van de projectielantaarn en de bioscoop', *Paedagogische Studiën*, 4 (1923), 169–84.

83 Frank N. Freeman, *The Handwriting Movement: a Study of the Motor Factors of Excellence in Penmanship, an Investigation Carried on with the Aid of a Subsidy by the General Education Board* (Chicago: University of Chicago Press, 1918).

focused on film and lantern projections, while the American experiments also included stereographs, photographs and other, both visual and non-visual presentation modes. And whereas the Dutch experiments measured didactic value by the degree of retention in questionnaires or free compositions, the American experiments also tested understanding through multiple choice, completion, Yes and No tests, and right-or-wrong tests. In fact, these experiments were of a wider scope, as they also included practical matters, such as their correlation with the curriculum or the handling of visual materials.[84] But what the two series of experiments did have in common was that, for the first time, methods of instruction with visual materials were systematically and comparatively tested in the place where they were used, the classroom, rather than in unsubstantiated polemics in newspapers, brochures or government reports — what Freeman dismissed as 'opinion'.[85]

The Dutch experiment was conducted at an Amsterdam secondary school among *c.* eighty students at the ages of thirteen to sixteen. They were tested on their memories of lantern and film projections about geographical topics; the materials came courtesy of the Colonial Institute. The fact that this institute had slides and films on the same topics made the experiment, according to the two psychologists, 'objective'. To guard this objectivity care was taken to adapt each slide projection to the length of its companion film.[86] As well the experiments were conducted without oral comments, because 'surely only then', they claimed, 'could the results of film and slide be determined purely and independently of the comments' subjective value'.[87] Test subjects were first and second graders, at the ages of thirteen to fourteen, for whom the topics — all about the Netherlands East-Indies — were new. A control group consisted of students in the third and fourth grades, who once had been instructed in some of these topics, albeit without visual aids. Over a period of four weeks all test subjects were shown two colonial topics. In the first week one class was shown a film, while a parallel class watched a slide projection about an identical topic; in the third week this procedure was repeated with another topic, while the visual aids were interchanged. In the week following each of these projections students were asked to write an essay about what they had seen, a method, the experimenters explained, to get a better sense of the visual teaching aids' after-effect, which they defined as stabilised memory. Their preference for this method was based on demonstrated minor differences

84 Révész, Hazewinkel (1923), 43; Frank N. Freeman, 'Introduction: problem and method of procedure', Freeman (ed.), *Visual Education: a Comparative Study of Motion Pictures and Other Methods of Instruction* (Chicago: University of Chicago Press, 1924), 13; F. Dean McClusky, 'Comparisons of Different Methods of Visual Instruction', Freeman (1924), 84.

85 Freeman (1924), 3; see also McClusky (1924), 86.

86 In the American experiments a similar procedure was followed in various comparative experiments with parallel groups by taking material contained in a film — the object most resistant to invasive measures — and adapt it to other modes of presentation: 'charts were taken directly from the film by projecting the film upon large cardboard sheets and tracing the outline of the picture upon it'; 'by copying and reproducing orally the titles and subtitles and then supplementing these by a few additional sentences' for a lecture; or to prepare material contained in a film 'in the form of slides and a mimeographed text illustrated by photographic prints' whenever permission was granted. Although reminiscent of Révész and Hazewinkel's 'objectivity', Freeman uses the more modest term 'identity'; Freeman (1924), 17, 29, 35, respectively.

87 Ibid., 44. This method was abandoned in the second experiment; see: Hazewinkel (1923). The original Dutch text reads: '*Alleen dan* kon immers het resultaat van film en diapositief geheel zuiver, onafhankelijk van de subjectieve waarde van die verduidelijking, bepaald worden.'

Fig. 3.4. Amsterdam Lyceum, site of the didactic experiments of Révész and Hazewinkel (Image via <https://archief.amsterdam/beeldbank/detail/abce87bb-8603-df67–64a8–54b9b91e50b0/media/703c42c4–94b9–9bf3-ffa3-c557ca946a5c?mode= detail&view= horizontal&q= Amsterdams%20Lyceum&rows= 1&page= 4>.

between answering questions and writing essays.[88] In a follow-up experiment ten months later, in November 1922, another but unannounced essay assignment was given.[89] A new experiment was also conducted that included, besides the projections, oral comments by a teacher. To minimize individual differences between teachers, only one person was asked to deliver the lecture in a largely identical way, i.e. in length and emphasis.[90]

'Our investigations have shown that the energetic propaganda made for the film on the strength of its alleged didactic importance is not well-founded. On the other hand, its educational importance has not been disproved. The opinion of those who hold that the lantern slide is to be preferred to the film is supported by our results in so far as they put beyond a doubt the great didactic value of the former for pupils up to the age of seventeen, i.e. for the majority of our school population.' That was the summary of their experiments' most important conclusions,[91] while the second experiment's conclusion was that 'it appears that the didactic value of film decreases more sharply than lantern

88 Ibid., 45–47.
89 Hazewinkel (1923), 169–70.
90 Ibid., 179.
91 G. Révész, J. F. Hazewinkel, 'The Didactic Value of Lantern Slides and Films', *British Journal of Psychology*, 15, #2 (1924), 197. This was a partial translation of the first experiments' research results (its review of the literature was left out). The second experiment's report was not translated.

slides among children beneath the age of twelve, *while neither method has at that point any didactic value without oral comments.*'[92]

Coming to an End

There is much to be said against these experiments. By initially omitting teachers' oral comments, for instance, comparison was enabled yet at the same time a setting was created that had little resemblance to and, therefore, probably less value for the very situation in which the experiment was conducted: a class. The ready-made lecture in the follow-up experiment also served 'objectivity' rather than naturalism. A serious omission, furthermore, is that the very setup of these comparative experiments precluded the realisation of the specific advantages of screening slides: simultaneous analysis and discussion. And an element that was not discussed at all was the presence of intertitles in the films, which at the time would be part of any complete print, most certainly nonfiction films. In contrast to slide projections, which lacked their common oral clarifications here, the films would have given pupils more information to understand or frame the images.[93] As there is no telling how this would have affected the results, caution is warranted. After all, the results focused on which and how many facts students were able to remember; it did not ask whether and to what extent they really understood what they had been watching.[94] As a matter of fact, the test materials' explicitness about temporal, spatial or ethnographic contexts left much to be desired. Finally, the experimenters' objective setup ignored contemporary, more child-centred educational approaches, such as the Montessori method or, indeed, the very location where they conducted their experiments, a secondary school that considered extra-curricular activities all-important for students' development, particularly in the lower grades.

What *is* an important fact, although a long time coming, is that these experiments were conducted at all. Praise is therefore due to Révész and Hazewinkel (as well as the American experimenters) for having put the debate regarding photographic teaching aids on a scientific footing, i.e. factually and verifiably rather than rhetorically and anecdotally. And that was less simple than it looks. Whereas the experimenters' discussion of the existing literature could rely on a small library, academic or polemic, on film, the section on the pros and cons of lantern projections mentioned no references at all — hence its more introspective character. It is all the more sad, then, that this first serious discussion of photographic teaching aids was for a long time also the last one in the Netherlands. A hopeful beginning turned out to be an unnoted ending, as the Psychological-Pedagogical Lab closed its doors not long after for lack of funding. In this country, as far as we know, a long silence in the field of experimental pedagogical media followed.

---

92 Hazewinkel (1923), 184. The original Dutch text reads: 'Het schijnt of dat de waarde voor het geheugen bij kinderen beneden 12 jaar bij het gebruik van film veel sterker afneemt, dan bij gebruik van diapositieven, *geen van beide methoden echter hebben dan nog didactische waarde zonder mondelinge toelichting.*'

93 Similar American experiments did not mention this difference between film and slide either. It seems unlikely that the Colonial Institute would have sent prints without titles, for the simple reasons that there weren't any; the making of title-less prints would certainly have been rejected as financially unwise or impractical.

94 For a more extensive evaluation of the experiments, see: Jan Elen, 'Beweging en stilstand: anders en toch weer niet', *Pedagogische Studiën*, 90 (2013), 27–30.

And with the adoption of the Cinema Law, in 1926, another thing came to an end. With its narrow focus on permits and censorship this piece of legislation clearly showed that guarding good morals was higher on the government's agenda than investing in education and heritage.[95] It left educational film largely to good intentions and the market. Insofar as school cinemas had not closed already before the end of the decade, the Great Depression would eventually finish them off. And a more structural problem, now that the government had clearly withdrawn from supporting the educational film sector, was that the Dutch film market became an essentially self-regulating affair. Soon it became the remit of an all-powerful organisation, the *Nederlandsche Bioscoop Bond* (NBB; Dutch Cinema League), which united cinema proprietors, distributors, and, since the early 1930s, producers that operated within Dutch territory. A particularly unfavourable measure for educational cinema was the stipulation, also adopted in 1926, that League members could only trade with each other, except when it concerned Dutch-made materials. But that, as we have seen, was just the problem.[96] While the Hague School Cinema continued the practice of acquiring its films abroad,[97] the Rotterdam school cinema was apparently able to continue showing a mix of self-made, donated or rented Dutch productions. Some relief came in 1936, when the League introduced its *List of no objection,* a registry 'of cultural, social or educational institutes that were allowed to screen films without NBB membership, but under strict conditions that prevented them from competing commercially with regular exhibitors.'[98] But by that time it was too little too late.

95 'Tekst der Wet. Wet van den 14en mei 1926 ter bestrijding van de zedelijke en maatschappelijke gevaren van den bioscoop', *De Bioscoopwet* (Alphen aan den Rijn: N. Samsom, 1927), 25–30.

96 The earliest official statement on this matter that we found was in a 1933 NBB brochure: *Statuten, reglementen, verhuur- en huurvoorwaarden en bedrijfsbesluiten* (Nederlandsche Bioscoop-Bond: Amsterdam, February 1, 1933), 52, Eye Filmmuseum, NBB Archive, 137. A1. folder 1.

97 'Voor de Haagsche Schoolbioscoop', *Kunst en Amusement*, January 23, 1926, p. 52.

98 Thunnis van Oort, 'Resurrection in Slow Motion: the Delayed Restoration of the Cinema Exhibition Industry in Post-war Rotterdam (1940–1965), *European Review of History/Revue européenne de l'Histoire*, 24 (2017), 10; *Nieuwe leden- en zakenbesluit* (Nederlandsche Bioscoop-Bond: Amsterdam, February 10, 1936), 16, Eye Filmmuseum, NBB Archive, 137. A1. folder 1.

Agents of Change

WOUTER EGELMEERS

'Deep and Lasting Traces'. How and Why Belgian Teachers Integrated the Optical Lantern in their Teaching (1895–1940)

Introduction

From 1895 onwards, a number of Belgian teachers began to devote themselves to the use of the optical lantern in their schools. Astonished by the results of lantern projections in evening classes for adults, a group of enthusiastic Brussels primary school headmasters and teachers decided to introduce the medium into their schools in 1894. Five years later, six Brussels public primary schools together already possessed over 5000 slides with which they organized 1313 illustrated lessons within a year.[1] In 1899, the City Alderman for Education set up a commission consisting of the most prominent Brussels teachers that made use of the optical lantern to draw up a report on the possible uses of the optical lantern as a teaching aid and the further expansion of its use in Brussels schools. The authors of the report stated that long after the projection sessions, pupils could still reproduce the details of the projected views with a fidelity that astonished staff members. 'The impressions produced by the projected, enlarged image stimulate their curiosity and thus leave deep and lasting traces in their minds while simultaneously producing sharp ideas and clear notions' of what they are taught.[2] The lantern was an outstanding teaching aid, the educators concluded, since it produced large, clear images by means of which teachers could take their classes on virtual field trips. Apart from that, the luminous

* I would like to thank the editors of this volume for their valuable feedback on earlier versions of this text. This work was supported by Research Foundation — Flanders (FWO) and the Fonds de la Recherche Scientifique (FNRS) under Grant of Excellence of Science (EOS) project number 30802346 (B-magic project 'The Magic Lantern and Its Cultural Impact as Visual Mass Medium in Belgium (1830–1940).

1 *Bulletin communal: Première partie. Compte rendu des séances* (Veuve Julien Baertsoen, 1897), II, p. 682; *Bulletin communal: Première partie. Compte rendu des séances* (Veuve Julien Baertsoen, 1899), II, p. 442.

2 'Les impressions produites par l'image projetée et agrandie, en excitant vivement la curiosité, ont laissé dans l'esprit des traces profondes et durables, ont fait naître des idées nettes et des notions claires.' *Bulletin communal: Première partie. Compte rendu des séances*, II, p. 441.

Wouter Egelmeers • KU Leuven, wouter.egelmeers@kuleuven.be

Learning with Light and Shadows: Educational Lantern and Film Projection, 1860-1990, ed. by Nelleke Teughels and Kaat Wils, TECHNE-MPH, 8 (Turnhout, 2022), pp. 101-121

10.1484/M.TECHNE-MPH-EB.5.131496

images captured pupils' attention and inspired them, while the analysis of the images in class greatly benefited their observation skills.[3]

Although the optical lantern was predominantly used for entertainment purposes during the first half of the nineteenth century, by the late nineteenth century, it had become a very widespread and prolific means of communication for propaganda and education. It was now mostly used in settings of a more serious nature, such as meetings organized by learned societies, social organizations, religious groups, adult education societies, and universities.[4] Research focusing on educational lantern use suggests that the city of Brussels was not alone in introducing the apparatus into primary and secondary schools: from the late nineteenth century onwards, schools in various parts of the world started adding it to their teaching collections.[5] How and why teachers made use of it, however, still remains unclear.

Notwithstanding the apparent relevance of studying the implications of the introduction of this new medium in schools, historians of education have not yet thoroughly engaged with the optical lantern's educational history. From the early 2000s onwards, historians of education have increasingly focused both on the visual material used in education, such as wall charts and textbook illustrations, and on how they can make use of visual sources like photographs or documentary films of historical teaching situations.[6] Although these publications did sometimes mention the optical lantern amongst a broad range of visual educational media, they seldomly critically engaged with the medium.[7]

---

3 'Ville de Bruxelles — Enseignement par l'aspect — Epreuve' (Brussels, 1897), Brussels City Archive, ASB IP II 1906-Enseignement par l'aspect (cinéma, etc.), 1894–1922.

4 Kessler, Frank, 'Researching the Lantern', in *A Million Pictures: Magic Lantern Slides in the History of Learning*, ed. by Sarah Dellmann and Frank Kessler (London: John Libbey Publishing, 2020), pp. 13–19 (pp. 17–18).

5 Gaulupeau, Yves, 'Une technologie nouvelle au service de l'enseignement', in *Lumineuses projections! La projection fixe éducative*, ed. by Anne Quillien (Rouen: Canopé éditions, 2016), pp. 25–37 (pp. 34–35); Wells, Kentwood D., 'The Magic Lantern in Russia', *Magic Lantern Gazette*, 21.1 (2009), 3–14 (pp. 5–6); López San Segundo, Carmen, Beatriz González de Garay Domínguez, Francisco Javier Frutos Esteban, and Manuela Carmona García, 'The projection of images in the Spanish secondary school classrooms in the first third of the 20th century', *Fonseca, Journal of Communication*, 16 (2018), 31–45 <https://doi.org/10.14201/fjc2018163145>; Hollman, V. C., 'Glass Lantern Slides and Visual Instruction for School Teachers in Early Twentieth-Century Argentina', *Early Popular Visual Culture*, 14.1 (2016), 1–15 <https://doi.org/10.1080/17460654.2015.1092390>; Egelmeers, Wouter, and Nelleke Teughels, '"A Thousand Times More Interesting": Introducing the Optical Lantern into the Belgian Classroom, 1880–1920', *History of Education*, 50.6 (2021), 784–801.

6 See, for example, Depaepe, Marc, and Bregt Henkens, 'The History of Education and the Challenge of the Visual', *Paedagogica Historica*, 36.1 (2000), 10–17 <https://doi.org/10.1080/0030923000360101>. Mietzner, Ulrike, Kevin Myers, and Nick Peim, eds, *Visual History: Images of Education* (Bern: Peter Lang, 2005); Comàs Rubì, Francisca, ed., 'Photography and the History of Education,' special issue of *Educació i Història. Revista d'Historia de l'Educacío* 15 (2010); Warmington, Paul, Angelo Van Gorp, and Ian Grosvenor, eds, 'Education in Motion: Uses of Documentary Film in Educational Research,' special issue of *Paedagogica Historica* 47.4 (2011). Recently, *Paedagogica Historica* published a special issue on the visual in histories of education which rightly addresses new and more systematic ways of exploiting the potential for visual resources to add new insights into the history of education. Dussel, Inés, and Karin Priem, eds, 'Images and Films as Objects to Think With: A Reappraisal of Visual Studies in Histories of Education', special issue, *Paedagogica Historica* 53.6 (2017). An important exception that does address the optical lantern is Catteeuw, Karl, 'Als de muren konden spreken… Schoolwandplaten en de geschiedenis van het Belgisch lager onderwijs' (unpublished doctoral thesis, KU Leuven, 2005).

7 Depaepe and Henkens, 'The History of Education', p. 13.

The relative unfamiliarity of historians of education with the historical uses of lantern slides might in part be due to the late development of academic interest in the medium in general.[8] However, since the late twentieth century, an interest in intermedial relations and the optical lantern as a medium in its own right has emerged, focussing on the broad range of its everyday uses.[9] Scholars addressing the role of the optical lantern in the dissemination of knowledge mainly focused their attention on how it was used in popular scientific lectures and adult education, travelogues, and academic teaching, while others have mainly engaged with the commercial producers of educational slide series.[10] On the day-to-day use of slides in primary and secondary education, only a modest number of publications has appeared. These works have provided valuable initial analyses of the use of lantern slides in schools exploring, for instance, the medium's position in specific national contexts or its role in the construction and consolidation of nationalist and colonialist worldviews.[11] Due to a lack of source material directly related to class practice however, these studies were predominantly based on quantitative analyses of large collections of slides and small numbers of normative texts written by authorities and slide producers. They thus mainly highlight the top-down implementation of the new medium by national authorities, slide lending services or school boards.[12] As Nelleke Teughels has recently asserted, the ways in which educators 'developed their own stance towards visual media technology, and how their individual interests shaped their own unique pedagogical practices of media use and production remains an underexplored area of study'.[13]

Previous research, then, has often not engaged with how teachers reflected on the medium and how they integrated it into their lessons. Nevertheless, teachers and schools have played a major role in shaping the worldviews of generations of schoolchildren through the ways in which they included images into their teaching practice.[14] As Inés

8 Kember, Joe, 'The Magic Lantern: Open Medium', *Early Popular Visual Culture*, 17.1 (2019), 1–8 (p. 2) <https://doi.org/10.1080/17460654.2019.1640605>.

9 Dellmann, Sarah, 'The Many Perspectives on A Million Pictures', in Dellmann and Kessler, eds, *A Million Pictures*, pp. 3–12 (p. 4); Kessler, p. 17.

10 See, for example: Smith, Lester, 'Entertainment and Amusement, Education and Instruction: Lectures at the Royal Polytechnic Institution', in *Realms of Light: Uses and Perceptions of the Magic Lantern from the 17th to the 21st Century*, ed. by Richard Crangle, Mervyn Heard, and Ine Van Dooren (London: Magic Lantern Society, 2005), pp. 138–45; Gaulupeau, 'Une technologie nouvelle'; Ruchatz, Jens, 'Travelling by Slide: How the Art of Projection Met the World of Travel', in Crangle and others, eds, *Realms of Light*, pp. 34–41; Nelson, Robert S., 'The Slide Lecture, or the Work of Art "History" in the Age of Mechanical Reproduction', *Critical Inquiry*, 26.3 (2000), 414–34; Kessler, Frank, and Sabine Lenk, 'Une "collection virtuelle": les plaques de projection pour l'enseignement de la société Ed. Liesegang en Allemagne avant 1914', *Transbordeur*, 1 (2017), 96–105.

11 Good, Katie Day, *Bring the World to the Child: Technologies of Global Citizenship in American Education* (Cambridge: The MIT Press, 2020); Quillien, Anne, ed., *Lumineuses projections! La projection fixe educative* (Rouen: Canopé éditions, 2016); López San Segundo and others, 'The projection of images'; Pitarch, Daniel, '"To Transform the Blackboard into a Blank Screen": Magic Lanterns and Phantasmagorias in Nineteenth-Century Spanish High Schools', *Fonseca, Journal of Communication*, 16 (2018), 81–102; Hollman, 'Glass Lantern Slides and Visual Instruction'; Hartrick, Elizabeth, *The Magic Lantern in Colonial Australia & New Zealand* (Melbourne: Australian Scholarly Publishing, 2017), pp. 90–109.

12 For a similar argument, see Chapter 6 in this volume.

13 Teughels, Nelleke, 'Expectation versus Reality: How Visual Media Use in Belgian Catholic Secondary Schools Was Envisioned, Encouraged and Put into Practice (*c.* 1900–1940)', *Paedagogica Historica* (published online ahead of print 2021), 1–18 (p. 4) <https://doi.org/10.1080/00309230.2020.185653>.

14 Dussel, Inés, 'Education and the Production of Global Imaginaries: A Reflection on Teachers' Visual Culture', *Yearbook of the National Society for the Study of Education*, 108.2 (2009), 89–110 (pp. 92–93).

Dussel puts it, teaching contexts have been 'crucial in the transformation of modern scopic regimes and in the production of particular kinds of visualities and visual imaginaries' ever since the late nineteenth century.[15] Indeed, various researchers have pointed out that the media-specific qualities of newly introduced optical instruments have often played an important role in the adoption of these technologies in educational circles. Victoria Cain, for example, has argued that turn-of-the-century American producers of photographic media like the stereoscope — a device that could be used to create immersive three-dimensional photographic experiences — promised to instill students with 'a coolly rational way of approaching a visually disorienting world', while Meredith Bak has demonstrated how US educators hoped the stereoscope would transform 'the child's unfocused look into [...] a productive educational exercise'.[16] In line with these indications, this contribution takes into account the ways in which Belgian teachers reflected on the specific characteristics of the optical lantern as a medium and how they hoped to use these for their own benefit.

Inspired by recent scholarship that focuses on everyday educational uses of teaching aids, this contribution does not focus on the arguments used by policymakers and commercial slide producers, but instead engages with those who actually made use of the medium in their day-to-day teaching.[17] It offers a case study providing a general image of primary and secondary school teachers' opinions on and use of the medium between 1895 and 1940, the period in which the lantern's educational use in Belgium was most widespread.[18] It focuses on both primary and secondary education, organized by both catholic and state schools.[19] Since source material on educational lantern use is relatively scarce, it is based on a broad range of sources that can roughly be divided into two groups: normative texts published in teachers' journals, brochures aimed at fellow teachers, the proceedings of teachers' conferences, and government reports on education on the one hand; and archival sources relating to educational optical lantern use scattered over various archives on the other.[20] The first group of sources mainly date from the late nineteenth century and are mostly authored by liberal teachers arguing in favour of the lantern's introduction. The sources relating to actual lantern use date from 1920–1940 and originate in two catholic

15 Dussel, 'Education', p. 91.

16 Cain, Victoria E. M., 'Seeing the World: Media and Vision in US Geography Classrooms, 1890–1930', *Early Popular Visual Culture*, 13.4 (2015), 276–92 (p. 287) <https://doi.org/10.1080/17460654.2015.1111591>; Bak, Meredith A., 'Democracy and Discipline: Object Lessons and the Stereoscope in American Education, 1870–1920', *Early Popular Visual Culture*, 10.2 (2012), 147–67 (p. 159) <https://doi.org/10.1080/17460654.2012.664746>.

17 Good, *Bring the World to the Child*; Teughels, 'Expectation versus Reality'.

18 From the 1940s onwards, glass lantern slides were gradually replaced by modernized alternatives to project still images, such as filmstrips and the 35 mm diapositive. On how various educational projection media succeeded each other, see: Petroski, Henry, 'Chapter 1. From Plato's Cave to PowerPoint', in *Success Through Failure: The Paradox of Design* (Princeton University Press, 2018), pp. 10–43.

19 Although nineteenth-century and early twentieth-century Belgium was a pillarized society in which free-thinking liberals and catholics lived their lives segregated along ideological lines, schools associated with both pillars eagerly adopted the optical lantern from the late nineteenth century onwards. On how both pillars made use of the optical lantern, see: Kessler, Frank, and Sabine Lenk, 'Fighting the Enemy with the Lantern: How French and Belgian Catholic Priests Lectured against Their Common Laic Enemies before 1914', *Early Popular Visual Culture*, 17.1 (2019), 89–111 <https://doi.org/10.1080/17460654.2019.1641971>.

20 I would like to thank Nelleke Teughels for exploring numerous editions of educational journals with me in search of traces of optical lantern use.

schools and the liberal Ghent school museum, which organized illustrated lessons for both public and catholic schools from Ghent and its surroundings.[21]

This contribution analyses both the reflections of primary and secondary school teachers on the use of the optical lantern in their teaching and the visual and textual narratives of a small number of preserved slide series that were used in schools. After a brief discussion of the introduction of the optical lantern to Belgian schools, I firstly engage with the way in which teachers argued that the medium captivated children's attention. Then I analyse how they believed lantern slides could depict practically everything imaginable and thus enabled them to take their pupils on virtual trips to a large range of places. Finally, I suggest that teachers were attracted to the fact that the medium could also be used to inform their class as to how images should be regarded, thus improving their observational skills.

The Introduction of the Optical Lantern

In the early 1880s, the Belgian government decided to counteract the focus on rote learning that had reportedly characterized Belgian teaching up to then. It instructed secondary school teachers to render their classes more intuitive by employing a method based on experimentation and the 'observation of natural objects, models, and wall charts, and by making use of schematic figures and diagrams'.[22] In line with this new emphasis on illustrative teaching, the Ministry of Education provided state secondary schools and teacher training colleges with an optical lantern that could be used in science lessons.[23] Simultaneously, it instructed geography teachers at teacher training colleges to include slide projections into their lessons in order to make geography teaching less dry and monotonous.[24] According to teachers looking back upon the campaign twenty years later, however, the government did not provide schools with the lantern slides that were needed in order to teach by means of projected images, which probably caused most of the provided lanterns to end up in physics laboratory cabinets.[25]

The government's devotion to intuitive and illustrative teaching was inspired by a number of contemporary pedagogical movements, the most prominent of which was the pressure group *Ligue de l'enseignement*, which argued that visually aided, more experiential forms of teaching would greatly benefit pupils' ability to memorize what they were taught. This new 'active-intuitive' method originated from a positivist epistemology that assumed

---

21 Unfortunately, the limited body of sources renders it difficult to determine the influence of the respective ideologies on how teachers assessed the optical lantern's qualities as a medium.

22 'D'observations portant sur les choses de la nature, sur des modèles, sur des planches murales, et de faire usage de figures schématiques, de diagrammes.' *Rapport triennal sur la situation de l'instruction primaire en Belgique. Années 1879–1880-1881* (Brussels: Fr. Gobbaerts, 1884), p. lxxv. See also pp. ci–cii.

23 Van Humbeéck, Pierre, 'Instruction moyenne. — Athénées royaux et écoles moyennes de l'État. — Acquisition d'objets pour l'Étude de la physique et de la chimie', *Bulletin du ministère de l'instruction publique*, 4 (1881), 35–47 (p. 39, 43). It is not clear how the provided lanterns were intended to be used. In state secondary schools, they possibly only served to illustrate lessons on optics and the properties of light. See: Egelmeers and Teughels, '"A Thousand Times More Interesting"', p. 790.

24 *Rapport triennal sur la situation de l'instruction primaire, 1879–1881*, p. cxxxiii–cxxxiv.

25 Dony, Emile, 'Les procédés intuitifs dans l'enseignement de l'histoire', *Revue de l'instruction publique en Belgique*, 45 (1902), 81–96 (p. 86).

that children would better remember their lessons when they actively engaged with objects by seeing, touching, smelling, and listening to them.[26] Notwithstanding the popularity of the new policy's emphasis on the active-intuitive teaching method amongst policymakers and pedagogues, teachers were not immediately won over by it. Late nineteenth-century pedagogical journals tried to reconcile government and pedagogical calls for more intuitive teaching with the poor financial situation of many schools and daily teaching practice which was traditionally oriented towards the verbal transmission of information.[27]

The introduction and use of the optical lantern was not discussed by teachers' journals until the final years of the nineteenth century, when a small number of liberal teachers appealed for the actual introduction of projection to Belgian classrooms after the first distribution of lanterns had not produced the desired results.[28] These progressive forerunners supposed that using the lantern in education ensured the formation of strong memories by captivating the pupils' attention while also enabling teachers to show almost everything they could think of. Additionally, the medium projected very clear images that could be used to teach their pupils how to meticulously observe the depicted objects and scenes. The lantern advocates managed to convince the government to renew its efforts to introduce the optical lantern to Belgian schools. From the turn of the twentieth century onwards, it provided subsidies for the acquisition of the required materials and recommended primary and secondary school teachers to make use of it.[29] Although some teachers expressed concerns about the supposed frivolity of the new medium, it seems that most of them were positive about it. Interestingly, the introduction of cinema, which was considered as particularly harmful for developing children, led many educators to view the optical lantern as a possible antidote that could help them maintain control over the images that children were exposed to.[30] By the 1920s, the use of the optical lantern started to generalize and by the 1930s it was an established visual teaching aid that was casually mentioned by pedagogical journals as a possible addition to lessons.[31]

'Like a Magnet Attracting Iron': Captivating Attention

An influential late nineteenth-century advocate for the use of the optical lantern in secondary education was the Liège-based science teacher Georges Kemna. In the published version

26 Grootaers, Dominique, 'Belgische schoolhervormingen in het licht van de "Éducation nouvelle" (1870–1970)', in *Reformpedagogiek in België en Nederland*, Jaarboek voor de geschiedenis van opvoeding en onderwijs 2001, 2002, pp. 9–33 (pp. 14–15).

27 Depaepe, Marc, *Order in Progress: Everyday Educational Practice in Primary Schools, Belgium, 1880–1970*, Studia Paedagogica, 29 (Leuven: University Press, 2000), p. 44; Egelmeers and Teughels, '"A Thousand Times More Interesting"', p. 792.

28 The implementation of the lantern was a long-winded process, which was partly due to the fact that the liberal government that insisted on intuitive teaching was succeeded by a catholic one that cut down subsidies for teaching aids. Additionally, the lanterns produced in the 1880s were still relatively costly and difficult to operate. See: Egelmeers and Teughels, '"A Thousand Times More Interesting"', pp. 792–93.

29 *Rapport triennal sur l'état de l'enseignement moyen en Belgique. Années 1897–1898-1899* (Brussels: J. Goemaere, 1900), pp. 294–95; *Rapport triennal sur la situation de l'instruction primaire en Belgique. Années 1897–1898-1899* (Brussels: J. Goemaere, 1900), pp. 97, 102.

30 Egelmeers and Teughels, '"A Thousand Times More Interesting"', p. 799.

31 Egelmeers and Teughels, '"A Thousand Times More Interesting"', p. 799–800.

of a lecture that he held for his colleagues at the state secondary school in Liège, he recalled how the impulse for more practical and intuitive teaching had led to the half-hearted and unsuccessful introduction of the optical lantern into Belgian secondary schools in the 1880s. He stated that the introduction of the medium was indeed long overdue in light of the strong movement for more intuitive teaching.[32] To Kemna, an important argument for its use was the fact that the optical lantern was very effective in captivating children's attention. In order to create lasting memories in their minds, he argued that 'the spirit has to be struck, the attention has to be aroused'.[33] Everything connected to the lecture — the announcement of the session, the preparations, the darkening of the room — contributed to the creation of strong impressions: 'You need staging, decorum, you need what is known in religious matters as ritual manifestations.'[34] In other words, teachers should take advantage of all resources offered to them by the child, and exploit children's natural appetite for images.

Other advocates for the educational use of the optical lantern agreed with Kemna on the power of projected images in school contexts. Véron De Deyne, a secondary school science teacher who published the first Belgian manual for the educational use of the optical lantern expressed it as follows: projecting slides in the classroom 'is simultaneously educational and recreational, useful and enjoyable'.[35] Ixelles primary school drawing teacher Pascal Mattot explained that the magic lantern that had fascinated children with its projections for generations had now become an advanced instrument for the dissemination of knowledge, even though it still made use of the universal principle that projected images attract the human eye 'like a magnet attracts iron'.[36] The images lighting up in the darkness made sure that the attention of the pupils was entirely focused on the picture. As a result, it had a great impact on their emotions, which ensured that the knowledge presented to them was 'engraved in their brains' and 'preserved without the slightest difficulty'.[37] Still, even illustrated lectures could become dull if the presentation only contained black-and-white pictures, according to De Deyne. Therefore, he advised his readers to break possible monotony by adding a piece of tinted glass to the slides. Even though the coloured views thus created were merely plainly coloured, this would still result in more captivating views.[38]

The sources preserved in school archives underscore the thesis that teachers found it important to capture the imagination of their pupils when making use of optical lantern

32 Kemna, Georges, *Les projections lumineuses dans l'enseignement* (Antwerp: Buschmann, 1895), pp. 5–6.
33 'Il faut que l'esprit soit frappé, que l'attention soit éveillée.' Kemna, *Les projections lumineuses*, p. 7.
34 'Il faut de la mise en scène, du décorum, il faut ce qu'on appelle en matière de religion le culte extérieur.' Kemna, *Les projections lumineuses*, p. 7.
35 'C'est à la fois un enseignement et une récréation, l'utile et l'agréable.' De Deyne, Véron, *La lanterne de projections à l'école: Propagation de l'enseignement scientifique par les projections lumineuses* (Brussels: J. Lebègue et Cie, 1897), p. 1.
36 Mattot, Pascal, *Les projections lumineuses dans l'enseignement primaire* (Brussels: J. Lebègue et Cie, 1897), pp. 7–8.
37 'Gravées dans les cerveaux [...] se retiennent sans la moindre difficulté.' Mattot, *Les projections lumineuses*, p. 8.
38 De Deyne, *La lanterne de projections*, p. 58; On the importance of colour in lantern slides, see: Moens, Bart, 'The Luminous Colours of the Magic Lantern: Shedding Light on the Palette of Life Model Slides', *Documenta*, 37.2 (2019), 13–43.

Fig. 4.1. Six brightly coloured slides from the series 'Relief'. Archive of Sint-Ursula Instituut, Onze-Lieve-Vrouw-Waver.

slides. Generally, substantial parts of surviving slide collections for school use contain large numbers of coloured images. The collections of the Sint-Ursula Instituut in Onze-Lieve-Vrouw-Waver and the Heilig Graf school in Turnhout, both catholic secondary schools for girls, for instance, hold large numbers of slides consisting of hand-coloured photographs. They were painted in extremely bright, almost unrealistic colours, and mostly used in lessons on the people and landscape types encountered in foreign countries. A series on geographic relief in the collection of the Saint Ursula Institute, for example, doubtlessly made a strong impression on pupils when it was projected (see Fig. 4.1). As a history teacher observed in a discussion of the slides created by De Deyne, projected images indeed struck the pupils in a powerful way: 'From this dry enumeration of slides one can only form an incomplete idea of the importance of these slides [...] one has to have seen them and moreover the faces, the eyes of the pupils!'[39] In a similar argument, the teacher's journal *De Opvoeder* concluded in a discussion of a new teaching method that included the projection of slides: 'the series of rich slides for projection make an impression on the children that will last for a lifetime.'[40]

39 'Uit deze droge opsomming der diapositieven kan men zich slechts een onvolledig gedacht vormen van het belang dat deze lichtbeelden hebben [...]. Men moet ze gezien hebben en daarbij de gezichten, de oogen der leerlingen!' Habets, A., 'Nieuwe banen in het geschiedenisonderwijs', *Dietsche Warande en Belfort* (1905), 19–36 (p. 31).

40 'Een reeks kwastige platen voor projectielantaarn die op de kinderen een indruk maken die levenslang zal bijblijven.' B. H. D., 'Nieuwe methode voor de behandeling der Belgische geschiedenis op de lagere school door Richard Wagner, *De Opvoeder*, 21 (1924), 333–34 (p. 334).

Fig. 4.2. 'Convertisseurs Bessemer et marteau pilon', slide 36 of the slide series 'Metaalnijverheid'. Archive of Heilig Graf secondary school, Turnhout.

The appearance of the slides, however, was not the only aspect of an illustrated lesson that could captivate the young audience's attention. As can be deduced from a number of lesson preparations that Heilig Graf teachers used from approximately the 1920s onwards, the narrative of the lesson also played an important role in keeping the audience captivated.[41] The surviving text and corresponding slides, in all probability used in a secondary school lesson on the production of steel and iron, provide an abundant amount of information on the production process — which could have caused the pupils' attention to wane after some time.[42] As a possible countermeasure, some of the slides that were projected towards the end of the session present very vividly coloured images (see Fig. 4.2). The colours are not necessarily realistic, but the strong contrast between the incandescent iron and its dark environment — an effect that was heightened by the fact that the projected image literally lit up a darkened classroom — did doubtlessly leave

41 On this source, see: Teughels, 'Expectation versus Reality', pp. 12–13.

42 De Deyne published a very similar collection of slides on the production of iron and steel for secondary school use from which a number of these slides were probably taken. See: De Deyne, Véron, *Excursions scolaires industrielles* (Menin: Impr. Émile Hoedt & sœurs, 1910), pp. 4–5.

a strong impression on the pupils. Interestingly, the text accompanying these slides was also clearly geared at striking the audience's attention, since it described the sounds and vivid scenery of the discussed stage in the production process in a highly imaginative fashion and with great detail:

> The activity of the converter is accompanied by a fierce roar — a rain of glistening stars is constantly escaping from its opening — after this initial period it begins to eject flames, first light flames (3 m long), then yellow ones, which continue to grow taller and thicker and soon form a brilliant bundle of about 10 m! The roaring constantly increases and is occasionally accompanied by explosions — after mixing with crude iron it acts like a raging monster, alternately throwing itself forward and backward like a living brew kettle.[43]

Still, the lantern's association with entertainment — its history as both a children's toy and its previous use in travelling phantasmagoria shows — made some teachers wary of its use in Belgian classrooms. Some of them considered projection sessions too frivolous for use alongside the traditional teaching aids and thought that it should only be put to use as a kind of educationally sound recreation that could be used in exceptional cases like school celebrations.[44] Even when used in this way, teachers had to beware of over-use: 'the school should by no means be transformed into a hall of spectacles'.[45] Others were even less accommodating to the new medium's pedagogical benefits. During a well-attended discussion on visual teaching aids at an international conference on secondary education, the teacher Arthur Mansion, for instance, strongly objected to its use. He turned around the argument that projected images ensure long-lasting memories by stating that instead, the fugitive nature of the projections created memories of an equally fleeting nature. Although the medium could be used to 'entertain the younger girls in secondary education or primary school children,' Mansion insisted that 'shadow play and phantasmagoria cannot be used as a basis for a scientific teaching in schools'.[46] Most of the conference participants, however, objected to Mansion's interpretation of the medium's effects, saying that a well-prepared teacher could indeed organize lantern sessions that were by no means too recreational, but in fact 'veritable lessons'.[47]

---

43 'De werking van den convertor gaat gepaard met een hevig gebrul — een regen v glinsterende sterren ontsnappen voortdurend uit zijn opening — na deze eerste periode begint hij vlammen uit te werpen, eerst lichte vlammen (3 tal m lang) daarna gele, die aanhoudend hooger en dikker worden en weldra een schitterende bundel vormen v een 10tal m.! Het gebrul neemt voortdurend toe en gaat nu en dan gepaard met ontploffingen — na de vermenging met ruwijzer gaat hij als een razend monster te werk, werpt zich beurtelings voor en achterover als een levende brouwketel.' Notebook 'IJzer staal', Archive of Heilig Graf secondary school, Turnhout: n.p.

44 Carrière, [Louis], 'La méthode catéchétique', in *Godsdienstige, liefdadige, sociale en economische werken: verslagen / Oeuvres religieuses, charitables, sociales et économiques: rapports* (Brussels: Goemaere, 1909), p. 148; Sluys, Alexis, 'Importance de la culture esthétique dans l'éducation générale de l'enfant', in *IIIe congrès international de l'art public: 1re Section* (Liège: L'Art Public, 1905), pp. 1–41 (p. 16).

45 'Geenszins de school herscheppen in een spectakelzaal.' 'Nota's over het gebruik van projecties met vaste en beweegbare beelden in de lagere scholen' [1928]. Archief Firmin D'Heygere 749, Liberas, Ghent.

46 'Pour amuser les petites filles des écoles moyennes ou les enfants des écoles primaires.' 'Il ne faut pas qu'à l'école, les ombres et la fantasmagorie servent de base à l'enseignement scientifique.' *Congrès international de l'enseignement moyen: Compte-rendu officiel* (Tournai: Decallonne-Liagre, 1901), pp. 227–28.

47 *Congrès international. Compte-rendu*, p. 231.

'Come On, Let's Go There!' The Possibility to Show Virtually Everything

Teachers making use of the optical lantern not only did so because the luminous images it projected captured their pupils' attention, they also did so because they believed that the medium enabled them to show nearly everything imaginable. Since any slide catalogue could contain up to 15,000 images to choose from, 'the photographs on glass that geography teachers can use to illustrate their lessons are innumerable', Kemna asserted.[48] In a 1902 article on intuitive history teaching, Emile Dony, history teacher at a state secondary school in Mons, provided his fellow history teachers with a detailed list of slide producers that they could choose from. Although he had to admit that there were not many Belgian manufacturers, he pointed out that there were large collections to choose from in neighbouring countries: Lévy et ses fils in Paris, for instance, could provide numerous slides on the monuments and ruins of Greece, Italy, Palestine, Egypt, artworks from Italy and the Louvre, a 'trip through Belgium' of at least 300 views, and a series of 700 slides on the history of France. Since the Berlin company A. W. Fuhrmann and the London Newton & Co. also provided numerous slide series, teachers 'should be able to easily find abroad a very satisfactory selection of slides to project on the screen'.[49] Even though not all of these commercial images were suitable for class use, Dony's colleague Kemna argued that the photographs that teachers needed did already exist, scattered among the collections of amateur and professional photographs — the only thing that was needed was a group of avid teachers that would collect them.[50]

While photography entailed the promise of enabling the reproduction and categorization of nearly everything visible, with a bit of creativity, drawings also enabled the representation of anything one wanted to show to a group of pupils — and to do so in a way that one thought best suited for class use.[51] When the city of Brussels introduced the optical lantern to its primary schools, the commission preparing its introduction discovered that commercial slides did not always meet their educational standards. A number of sympathetic colleagues then supplied the views that were needed by either drawing the required images or by themselves taking photographs of the 'monuments, sites, and engravings' that they wanted to show in class.[52] Therefore, the commission put their money on both modes of slide production and wrote a booklet in which they explained to their colleagues how to produce slides by either drawing or by using the photographic process.[53] Although Pascal Mattot, who was a drawing teacher, thought that the slides drawn by his Brussels

48 'Les photographies sur verre que le professeur de géographie peut utiliser pour illustrer ses cours sont innombrables.' Kemna, *Les projections lumineuses*, p. 12.

49 'Le professeur pourrait trouver aisément à l'étranger un choix très suffisant de clichés à faire défiler sur l'écran.' Dony, 'Les procédés intuitifs', pp. 87–88.

50 Kemna, *Les projections lumineuses*, pp. 16, 38; also see: De Deyne, *La lanterne de projections*, p. 3.

51 Ruchatz, Jens, *Licht und Wahrheit: Eine Mediumgeschichte der fotografischen Projektion* (Munich: Fink, 2003), p. 224.

52 De Kelper, D., and G. Van Deun, *L'enseignement par l'aspect. Rapport de la Commission* (Brussels: Commission instituée en vue d'étudier la question de l'enseignement par l'aspect, 17 July 1897), Brussels City Archive, ASB IP II 1906-Enseignement par l'aspect (cinéma, etc.), 1894–1922.

53 'Ville de Bruxelles—Enseignement par l'aspect', pp. 6–7.

colleagues were not always well executed, he still thought that drawn slides should play a predominant role in schools. After all, they could 'represent everything imagination can think of' and thus enabled teachers to show exactly what they wanted to highlight.[54]

Additionally, the relatively modest dimensions of lantern slides as compared to wall charts and other visual teaching aids enabled teachers to compile a collection of thousands of images in a relatively small place. Since 'no school could possibly acquire all specimens, models, maps and plates' that it needed, both educational slide lending services and the Belgian government insisted that projection slides could 'make the acquisition of large numbers of wall charts unnecessary'.[55] They could also serve as a way to go on virtual trips to places that normally could not be visited. Since both the Schools Inspectorate and many educators had judged geography teaching to be uninspiring and 'too dry' during the final decades of the nineteenth century, the possibility of enlivening this subject by means of projected views of faraway countries seemed an attractive antidote.[56] As Kemna explained, telescopic photographs of the moon could show its surface with such detail 'that one feels as if one were on a hot air balloon ride on a few kilometers above its surface'.[57] Teachers' journals indeed often referred to series of travel views in model geography lessons from the 1920s onwards, while archival records of various educational institutions, too, contain large amounts of slide sets relating to travel and geography, covering subjects from glaciers and volcanoes to the Belgian provinces, foreign countries, and entire continents.[58] The lessons on a specific country or region that were taught by means of these series were generally set up as virtual travels through the area.[59] A public primary school in Brussels, for instance, took its pupils on a 'boat trip from Antwerp to Constantinople', while a teacher working at Heilig Graf told her pupils during a discussion of the Spanish royal residence of El Escorial that 'if you have brought a guide from your hotel, the first thing you shall see is the man starting to dance in the middle of the choir in order to make you feel the 'elasticity of the wonder vault''.[60]

The teachers who illustrated their lessons by projecting lantern slides indeed aimed to transport their pupils to the places they showed and insisted that they did so almost literally. In

54 'Il peut représenter tout ce que l'imagination combine et propose.' Mattot, *Les projections lumineuses*, pp. 10–11.

55 'Zou geen enkel gesticht zich al de producten, modellen, kaarten en platen aanschaffen kunnen.' *Werk der lichtbeelden in't onderwijs: reglement; lijst der reeksen* (Antwerp: Werk der Lichtbeelden in't Onderwijs, 1925), p. 3. 'Despensera de l'achat d'un grand nombre de tableaux.' 'Ministère de l'Agriculture. — Ministère de l'Intérieur et de l'Instruction publique. — Sections professionelles primaires agricoles pour garçons', *Het Katholiek Onderwijs* 24 (1903): 416.

56 See, for instance: *Rapport triennal sur l'état de l'enseignement moyen, 1897–1899*, p. lxxxvi.

57 'Qu'on croirait planer en ballon à quelques kilomètres seulement au-dessus de la surface de l'astre.' Kemna, *Les projections lumineuses*, p. 10.

58 For model lessons using travel slides, see, for instance: De Brauwere, R., 'Practische beschouwingen over't onderwijs in de aardrijkskunde', *Nova et Vetera*, 4 (1921), 304–05; [Anon.], 'Tweede en derde graad: eenige grote reizen op den aardbol: Nansen's poolexpeditie', *De Opvoeder*, 22 (1925), 81–84.

59 Egelmeers, Wouter, 'Making Pupils See: The Use of Optical Lantern Slides in Geography Teaching in Belgian Catholic Schools', in *Faith in a Beam of Light: The Magic Lantern and Belief*, ed. by Natalija Majsova and Sabine Lenk (Brepols, forthcoming).

60 *L'enseignement par l'aspect: Sujets ayant été traités* (Brussels: Commission instituée en vue d'étudier la question de l'Enseignement par l'aspect, *c.* 1897), Brussels City Archive, ASB IP II 1906-Enseignement par l'aspect (cinéma, etc.), 1894–1922. 'Hebt ge van uw hotel een gids meegenomen, dan is het eerste wat ge te zien krijgt, dat de man midden in het koor begint te dansen, om "de elasticiteit van het wondergewelf te laten gevoelen".' Notebook 'Spanje IV', Archive of Heilig Graf secondary school, Turnhout: n.p.

Fig. 4.3. 'Slide 2'. Like the other drawn slides in this series, Wagner drew this slide of a 'moorland in the land of the Eburones' himself. He based it on a picture taken in the Belgian province of Limburg that was published in Massart, J., *Pour la Protection de la nature en Belgique*. The slide was reproduced in Wagner, Richard, *Nieuwe methode voor de behandeling der Belgische geschiedenis. 1ste tijdperk: Oud België. Platenalbum voor de leerlingen* (Antwerp: Krick, 1924), n.p.

a discussion of a new 'trail-blazing' history teaching method that combined a vivid story with the projection of lantern slides, the teacher reviewing the work stated that 'in imagination ... No: in reality [the pupils] have travelled back in time 2000 years. They are themselves Old Belgians living in Old Belgium before the arrival of Julius Caesar.'[61] The method, which was devised by the Antwerp public school teacher Richard Wagner, indeed took primary school pupils on an imaginary visit to the people who inhabited their country twenty centuries before them. It did so by providing them with a captivating narrative that was read aloud by the teacher, combined with the projection of photographs and drawings of the landscapes, archaeological finds, and the people that populated the story. After a general introduction on the prehistory of Belgium, the narrative proposed to go visit the ancient Belgians: 'Let us go back to those old days, to that wild land.'[62] While the optical lantern projected an image of a moorland (see Fig. 4.3), the teacher was to say: 'Here we are, in the vast sandy plains of the Campine, in the Limburg heath: sand, heather and a juniper tree here and there.'[63]

---

61 'De schoolkinderen [...] zijn in verbeelding... neen: in werkelijkheid 2000 jaren achteruit geplaatst. Zij zijn zelf Oude Belgen en leven in Oud-België voor de komst van Julius Cesar.' B. H. D., 'Nieuwe methode', p. 334. The new method was also greeted with approval by the Antwerp Paedology Association. See: 'Antwerpsch Nieuws-Algemeen Paedologisch Gezelschap', *Het Laatste Nieuws* (Brussels, 5 May 1924), p. 6.

62 'We zullen ons eens verplaatsen in dien ouden tijd, in dat wilde land.' Wagner, Richard, *Nieuwe methode voor de behandeling der Belgische geschiedenis: 1ste tijdperk: Oud België* (Antwerp: Krick, 1924), p. 9.

63 'Hier, we staan in de uitgestrekte zandvlakte van de Kempen, in de Limburgsche hei: zand, heidekruid en hier en daar een jeneverbes.' Wagner, *Nieuwe methode*, p. 9.

Wagner's text not only described what was to be seen on the slide, but also what it would feel like to actually be *in* the scenery depicted on the slide: 'We hear the humming of bees. [...] Perhaps we shall not find one single human being around here for hours.'[64] The narrative even invited the young audience to roam through the landscape in their imagination and warned them for the dangers that it might pose to them: 'Yonder, very far, that dark line, that is a pine grove. Come on, let's go there! But beware, don't walk into that bog. Along here, be careful.'[65] After they have made it to the forest, the children are asked to be extra quiet: 'Hush! I hear something.'[66] The teacher now projects a slide depicting a young boy and addresses him. In the conversation that ensues, the boy, whose name is Catal, tells the group about his life in ancient Belgium. Catal takes the group to his village, introduces them to his family, and tells the class about his rather exciting life.[67] The other lessons of the method, too, use Catal's life as a steppingstone to tell the pupils about life in prehistoric Belgium. The lessons often started by the teacher telling the pupils that they are going to visit the clan again and they all describe the people, surroundings, colours, sensations and smells as vividly as the introductory chapter.[68]

Wagner's progressive new teaching method was probably influenced by developments in Belgian history teaching in general. From the turn of the century onwards, the Schools Inspectorate had already suggested that Belgian history teachers should make their lessons more attractive by using illustrative teaching aids. It also hoped history teaching would become less dry if teachers would focus more on the history of civilizations instead of politics and wars.[69] Following the First World War this sentiment intensified, and many educators argued in favour of a still patriotic but less militaristic form of history education, with more attention for the history of cultures and civilizations.[70] Simultaneously, the calls for more intuitive and experience-based teaching combined with the development of new modes of transport during the interwar period had increasingly led school boards to take their pupils on school trips.[71] Both of these movements seem to have had a special significance to Wagner, who employed the optical lantern's immersive power to stimulate the imagination of his pupils and take them on a virtual field trip to experience the culture of their Belgian predecessors.

Schools not only used lantern slide series to virtually transport pupils to places in the past or in faraway countries, they also used them as an alternative to expensive, time-consuming school trips to places in Belgium, such as a match factory or the city of Bruges.[72]

---

64 'We hooren bijen gonzen. [...] Misschien vinden wij hier zelfs uren in den omtrek geen mensch.' Wagner, *Nieuwe methode*, p. 9.

65 'Ginds, heel ver, die donkere streep, dat is een dennenbosch. Komt, daarheen! Maar past op, loopt niet in dat moer. Langs hier, voorzichtig.' Wagner, *Nieuwe methode*, p. 9.

66 'Sst! Ik hoor iets.' Wagner, *Nieuwe methode*, p. 9.

67 Wagner, *Nieuwe methode*, pp. 9–12.

68 See: Wagner, *Nieuwe methode*, p. 13.

69 *Rapport triennal sur l'état de l'enseignement moyen, 1897–1899*, pp. lxxxv–lxxxvi.

70 Hens, Tine, Saartje Vanden Borre, and Kaat Wils, *Oorlog in tijden van vrede: de Eerste Wereldoorlog in de klas 1919–1940* (Kalmthout: Pelckmans, 2015), pp. 79–82.

71 Good, *Bring the World to the Child*, p. 105; Constandt, Marc, 'We reizen om te leren: schoolreizen in het interbellum', *Brood & Rozen*, 14.2 (2009), p. 66 <https://doi.org/10.21825/br.v14i2.3363>.

72 De Deyne, Véron, *La Belgique monumentale, historique et pittoresque: Série de vues de projections* (Bruges: A. Fockenier-Saelens, 1906), pp. 5–7; De Deyne, *Excursions scolaires industrielles*, pp. 12–13.

Further, they could use an illustrated lesson as a way to prepare or recapitulate trips. As Kemna argued, Belgian secondary education attached great value to the understanding of industrial production processes. These could perfectly be illustrated by using pictures taken in various Belgian factories. Although it would be difficult to capture the necessary photographs due to the unfavourable circumstances in these factories, 'the casting of iron in a blast furnace, [...] the production of steel in a Bessemer converter, the puddling, [and] the work with the drop hammer are exciting operations to follow because of the dazzling phenomena that accompany them'.[73] It seems that De Deyne took up the challenge, since, fifteen years later, he published a catalogue of slide series on Belgian industry containing seventy-two slides that were taken in the Belgian metalworks of John Cockerill & Cie. near the Belgian city of Liège which depict exactly the sights that Kemna had proposed.[74] De Deyne suggested that teachers should first discuss the factory that was to be visited during a lesson, then go on the field trip and subsequently deliver an illustrated lecture to wrap up what the children were supposed to have learnt.[75]

Even though the Heilig Graf illustrated lesson on the production of iron was clearly made up of slides provided by various manufacturers, judging by the similarities between a number of the slides in these series and their descriptions in the catalogue published by De Deyne, it is very likely that at least ten of them were acquired from him. The Turnhout slides follow roughly the same order as the slides produced by De Deyne, depicting the production of iron and steel from the crushing of ore to the production of train rails in a Belgian factory. The lesson notes corresponding with the series show very well how the illustrated class was in fact taught and that here too, the pupils were virtually transported to the factory. The feeling of being present is evoked by explanations that describe the production process as if it were actually happening in front of them: 'workmen are throwing the load into the blast-furnace' while 'the furnace is spitting out great flames, it roars and hisses so violently that ears and eyes perish'.[76] In combination with the clarity and great visibility of the luminous images projected onto the screen, these descriptions must have made a strong impression.

Concentrating Attention: Observing All Aspects of a Phenomenon

A final aspect of the optical lantern that was important to Belgian teachers was the clarity and visibility of the images that it projected. As a teacher at an Antwerp secondary school concluded from the contributions at a teachers' conference he attended, the use of lantern

73 'Une coulée de haut-fourneau, [...] la fabrication de l'acier au creuset Bessemer, le puddlage, [et] le travail au marteau pilon, sont des opérations émouvantes à suivre, en raison même des phénomènes éblouissants qui les accompagnent.' Kemna, *Les projections lumineuses*, pp. 22–23.

74 De Deyne, *Excursions scolaires industrielles*, pp. 4–6.

75 De Deyne, *Excursions scolaires industrielles*, p. 1. The practice of recapitulating school trips was common in Belgian schools. See: Hoegaerts, Josephine, 'Op 't bloedig oorlogsveld is ied're man een held. Hoe kinderen het slagveld verbeeldden en beleefden aan het eind van de negentiende eeuw', *Volkskunde*, 113.3 (2013), 306–24 (p. 311).

76 'Werklieden zijn bezig de vracht in den hoogoven te werpen.' 'Intusschen spuwt de oven geweldige vlammen, hij brult en sist met geweld dat ooren en zien [*sic*] vergaan.' Notebook 'IJzer staal', Archive of Heilig Graf secondary school, Turnhout: n.p.

slides should be recommended 'because they allow the teacher to provide explanations to the whole class at the very moment when he or she is showing the photographic reproductions to the pupils'.[77] By projecting these images, the entire group of pupils could simultaneously observe an image that was larger and clearer than a wall chart, while the teacher was also relieved of the time-consuming effort of passing on images through the class.[78] According to Mattot, wall charts were mostly badly drawn, with limited dimensions and exaggerated colours. Projected images, however, could 'represent nature in its finest details, reproducing all aspects of the earth with a perfect and striking reality'.[79] Due to their clarity, these projected photographs were even clearer than photographs on paper, since they were 'alive with reality' and kept their precision even though being magnified multiple times by the projector.[80] Kemna went a step further and supposed that it could be more effective to observe projected images than nature itself, since 'the art of the painter, of the sculptor, of the photographer even, exists precisely in not slavishly imitating nature, but in bringing out certain things, in camouflaging others,' and in making the observer see 'how the artist has grasped, has understood nature'.[81] All in all, it seems that to many teachers, the supposed objectivity of the projected images was indeed not as important as the medium's ability to create a large, clearly visible image of the exact phenomenon that the teacher wanted to show.[82]

By integrating the optical lantern into their lessons, teachers also responded to contemporary calls for more active teaching. By the late nineteenth century, the visually aided intuitive teaching style that educators saw as the remedy against declamatory teaching had led to the development of so-called object lessons. During these exercises, pupils had to observe physical objects that were placed in front of them — specimens of industrial products or a stuffed animal, for example — to gain an understanding of their place in the world. The main goal of these lessons was to teach pupils to carefully observe and then be able to communicate their observations, often by answering the questions of their teachers that were supposed to guide their thinking process. When concrete objects were not available, images could serve as stand-ins.[83] Of course, these images could also be projected — as long as they went hand in hand with serious thought and active observation, since a viewing session could by no means 'be a play time, it has to be a study'.[84]

---

77 'Il permet au professeur de donner des explications à toute la classe au moment même où il fait passer sous les yeux des élèves les reproductions.' *Congrès international de l'enseignement moyen. Rapports préliminaires* (Tournai: Decallonne-Liagre, 1901), p. 77.

78 Kemna, *Les projections lumineuses*, p. 6.

79 'Représentant la nature dans ses moindres détails, reproduisant avec une parfaite et saisissante réalité tous les aspects de la terre.' Mattot, *Les projections lumineuses*, p. 4.

80 'Vivantes de la réalité.' De Deyne, *La lanterne de projections*, p. 1.

81 'L'art du peintre, du sculpteur, du photographe même, consiste précisément à ne pas imiter servilement la nature, mais à faire ressortir certaines choses, à en gazer d'autres et à faire naître chez celui qui est en présence de l'oeuvre d'art le sentiment correspondant à la façon dont l'artiste a saisi, a compris la nature.' Kemna, *Les projections lumineuses*, p. 5.

82 For a discussion of the (un)importance of objectivity in lantern images to catholic Belgian teachers, see: Egelmeers, 'Making Pupils See'.

83 Depaepe, *Order in Progress*, p. 77; Bak, 'Democracy and Discipline', pp. 156–57.

84 'Het mag geen speeltijd, het moet een studie zijn.' Elebaers, Karel, 'De lichtbeelden in't onderwijs', in *Compte-rendu des travaux du Congrès national de l'Enseignement Moyen Libre de Belgique, tenu à Bonne-Espérance les 11, 12, 13 septembre 1911 / Verslag der werkzaamheden op den landdag van het Vrij Middelbaar Onderwijs van België,*

How exactly projected views could ensure a more active teaching method was explained by the Brussels intuitive teaching commission: 'The child does not *receive* the notion wrapped in a flood of words, they search for it, discover it by their own efforts, and pass from a state of passivity to a phase of internal activity.'[85] The commission argued that by means of this approach, the pupils were taught how to see well and how to read images. They suggested that the pupils' concentrated attention generated ineffaceable memories and gave rise to questions and feelings that they learnt to express more freely and clearly. Since the children had learnt how to reach conclusions by their own capacities, the commission's argument was that in turn, they would even compose better writing assignments.[86] Several years later, after the first introduction of educational film screenings to a number of Belgian schools, teachers continued to stress the importance of projecting still images. In line with the influential Belgian pedagogue Alexis Sluys, they argued that in geography teaching, for instance, one should only make use of film projections to show phenomena that were particularly full of movement, like the eruption of geysers or the rituals of people living in other parts of the world. Lantern slides, on the other hand, should be used when correct representations of shapes and details were required.[87] In line with this conviction, the Brussels primary school headmaster Léon Neyrinck had an additional optical lantern installed in his school's cinema classroom in 1923 so as to enable the teachers to also show still images, which he believed to be 'of the utmost necessity'.[88]

Of course, practice is often very different from theory. The sources documenting the use of slide projection in various lessons show that teachers often assumed that pupils did not automatically reach an active mode of seeing. Indeed, they often supported their pupils' observation by guiding their view through the image, thus informing them how and what to see. A number of geography slides used in secondary school geography lessons at the Saint-Ursula Institute, for example, present black-and-white photographs of geological formations in which the slide manufacturer had literally highlighted the earth layer that was of interest to the lesson in transparent pink ink (see Fig. 4.4). Preserved readings of illustrated lessons suggest that the teachers who wrote them had a similar approach. Wagner's history lessons on ancient Belgium, for example, often actively guided the pupil's gaze through the image that was projected by the optical lantern:

> If you follow the yellow line on the map that starts at the Scheldt, runs to the south, then to the east, encloses the Treveri and Mediomati and then runs to the west at the

gehouden te Bonne-Espérance den 11, 12, 13 september 1911 (Roeselare: De Meester, 1911), II, 338–51 (pp. 338–41); also see: Dony, 'Les procédés intuitifs', pp. 93–94.

85 'L'enfant ne *reçoit* pas la notion enveloppée dans un flot de paroles, il la cherche, la découvre par ses propres efforts, passe de la phase de passivité à une phase d'activité.' *Bulletin communal. Première partie. Compte rendu des séances*, II, p. 441. Italics in original.

86 *Bulletin communal. Première partie. Compte rendu des séances*, II, p. 441; also see: *Werk der lichtbeelden in het onderwijs*, p. 4.

87 [Anon.], 'De film als opvoedingsmiddel', *De Opvoeder*, 22 (1925), 160–62 (p. 160); A. V. C., 'De Kinematograaf in de School', *Ons Woord*, 30 (1923), 80–83 (p. 81). Also see: Sluys, Alexis, *Manuel de la cinématographie scolaire et éducative*, Union des villes et communes belges. Publications 19 (Brussels: Union des villes et communes belges), p. 60.

88 Letter from Léon Neyrinck to G. Van Deun, 25 March 1923, Brussels City Archive, ASB IP II 1909-Enseignement par l'aspect — Cinéma — Achat de matériel films, plaques, etc., 1920–1930,.

Fig. 4.4. Three slides from the series 'Relief' highlighting specific geological layers in pink ink. Archive of Sint-Ursula Instituut, Onze-Lieve-Vrouw-Waver.

> north side of the river Seine until it reaches the sea, then you have found the land that the Romans called Gallia Belgica.[89]

The notes used in lessons taught in Turnhout made use of a similar strategy, successively describing what was to be seen in various parts of the image, for instance by first describing the functioning of the Bessemer converter depicted in the foreground, after which the teacher briefly explained that in the background, the pupils could see a drop hammer, while the middle of the picture showed 'the transport of slag or molten steel'.[90]

An analysis of the images depicted in the slides of collections used in primary and secondary school education makes clear that lessons making use of the medium also strove to show a process or phenomenon in a way that enabled pupils to investigate a subject from all possible angles. They combined various kinds of images with each other in order to use the specific characteristics of each mode of visualization to their advantage and visualize the subject of their lesson as best as they could. This is exemplified by a collection of slides that were used in the Ghent *Onderwijsmuseum* or school museum, which was founded and led by the liberal primary school teacher Michel Thiery. Teachers of both public and catholic schools could take their class to the museum and use its materials for

89 'Als ge op de kaart de gele lijn volgt die begint aan de Schelde, naar't Zuiden loopt, dan naar't Oosten, om Trevieren en Mediomatrieken heengaat en dan ten Noorden van de Seine naar't Westen loopt tot aan de zee, dan hebt ge het land dat de Romeinen *Gallia Belgica* noemen.' Wagner, *Nieuwe methode. 1ste tijdperk*, p. 8.

90 Notebook 'IJzer staal', Archive of Heilig Graf secondary school, Turnhout: n.p.

their teaching.[91] Many of the slide series that were used in the museum indeed strove to show a phenomenon from all angles. A series on volcanism, for example, consisted of a mixture of photographs, reproductions of paintings, and home-made drawings. The drawings presented schematic cross-sections of various kinds of volcanoes while the reproduced paintings illustrated the origin of the word 'volcano'. The photographs, then, depicted a range of volcano types in various states and from various angles — even from the air. One slide combined two images that were taken ten minutes apart to demonstrate how slow lava flows.[92] Other slide series were similarly made up of a mixture of images from all kinds of sources: photographs, microphotographs, maps, schematics, botanical drawings, and even dried plant leaves.[93]

The collection of slides and lesson notes used at the Heilig Graf school in Turnhout displays a similar and constant urge to literally show phenomena as thoroughly as possible. Judging by the fact that many of the slide sets consist of slides provided by a broad range of producers and also include slides that were manufactured by the teachers themselves, Heilig Graf teachers clearly spent a considerable amount of time selecting the right images for their lessons.[94] A lesson on Spain, for example, is made up of a mixture of slides from six different slide series supplied by the publishing house Maison de la Bonne Presse, and complemented by home-made slides and slides produced by unknown manufacturers.[95] The depicted buildings are literally shown from various perspectives, both from the inside and the outside.[96]

The preserved notes describing what teachers were to say while projecting the images shed further light on their drive to show the subject of their lessons. They were working documents that were regularly updated and improved over the years and as a result, they display a proliferation of additions, insertions, changes, and even additional pieces of paper containing extra explanations. From these additions and modifications, one can deduce that the slides that were added to series often served to clarify aspects of the lesson that were not yet adequately illustrated in earlier versions of the lessons — possibly because the teacher had not been able to find the appropriate slides in the catalogues of slide producers. Initially, the lesson on the production of glass, for instance, mainly contained photographs depicting the working of the factory. These were at some point supplemented by more schematic, brightly coloured slides produced by the Paris publishing house of Mazo in order to further clarify what happened in the factory (see Fig. 4.5).[97] The same goes for the

---

91 Janssens, Suzanne, 'Het vroegere Gentse Schoolmuseum aan het Berouw', *Ghendtsche Tydinghen*, 12.3 (1983), 132–44 (p. 136).

92 Slide series *Vulkanen*. Inv. nr. 53076, Huis van Kind en Natuur, Ghent.

93 See, for instance, the slide series *Insecten IV*, *Bacteriën*, *Protozoën enz.*, *Jenner*, *Bijen II*, and *Plantkunde*. Inv. nrs. 53030, 53100, 53101, 53184, 53204, and 53236, Huis van Kind en Natuur, Ghent.

94 On the combination of slides from various producers in this collection, see: Egelmeers, 'Making Pupils See'.

95 The Maison de la Bonne Presse series from which slides were taken are 'Espagne du sud', 'à travers les deux Castilles', 'Capitales', 'Trois semaines en yacht', 'Pelerinage nord Espagne', and 'Pelerinage sud Espagne'. On this publishing house, see: Véronneau, Pierre, 'Le Fascinateur et la Bonne Presse: des médias catholiques pour publics francophones', *1895. Mille huit cent quatre-vingt-quinze. Revue de l'association française de recherche sur l'histoire du cinéma*, 40 (2003), 25–40 <https://doi.org/10.4000/1895.3282>.

96 Slide series *Spanje IV*, Archive of Heilig Graf secondary school, Turnhout.

97 On the slides produced by Mazo, see: Renonciat, Annie, 'Un média oublié d'enseignement Populaire: Les vues sur papier transparent pour projections lumineuses', in Quillien, ed., *Lumineuses projections!*, pp. 67–75.

Fig. 4.5. 'Four à bassin', slide 8 of a series on the production of glass published by Mazo. Slide 13 of the slide series 'Glasnijverheid'. Archive of Heilig Graf secondary school, Turnhout.

lesson on the production of steel and iron. Here, the notes on the lesson even suggest that the teacher has actively looked for slides that could illustrate parts of the lesson for which she had not yet found suitable images. In the margin of the part of the lesson discussing the puddling of iron, she first wrote 'no slide', which was later crossed out. In between the lines, she then wrote: '23. Mazo. Puddle furnace', referring to one of the inserted Mazo slides that depicted exactly the process happening in this piece of machinery.[98]

Conclusion

When engaging with the perspective of the first teachers that started to use the optical lantern in the late nineteenth and early twentieth century, one finds that they had many reasons to do so. The top-down incentives orchestrated by authorities and schoolboards and the arguments of teaching aid producers that have been the subject of previous

98 Slide series *Spanje IV*, Archive of Heilig Graf secondary school, Turnhout.

scholarship certainly played a role in the implementation of the medium, but it turns out that educators themselves also had valid reasons to include slide projection in their lessons. The medium's specific qualities played an important role in their reasoning about its use in class: the images it projected were not simply alternatives to wall charts, school trips, textbook illustrations, and drawings on a blackboard, but came with a wide range of implications that altered both teaching practices and the outcomes of the classes in which they were integrated. In the texts that teachers wrote about the use of the medium in instruction, they insisted that it enabled them to captivate their pupils' attention, give them the impression of travelling to faraway places, and helped them develop a more active method of observation — all of which reportedly led to a better retention of the lessons that they taught.

The scattered archival sources relating to the lessons that were taught by means of the optical lantern underline this conclusion. Teachers engaged with slide series in order to ensure that they would show the subjects of their lessons as best as they could. They combined slides from various producers and if they could not find suitable images, they added slides from a wide range of other sources. They either collected photographs made by amateurs or took them themselves. If they could not find the photographs that they needed, they drew images that did suit their needs. In the lessons taught by means of projection, they not only used vivid, colourful images, they also accompanied these with compelling stories and evocative descriptions of what could be seen on the slide. They aimed to stimulate their pupils' imaginations and observational skills and simultaneously help them to keep their mind focused on the lesson. By resorting to lantern slides, teachers were able to depict a plethora of phenomena and objects — from neighbouring cities to volcanoes or the life of the Ancient Belgians — from a wide array of perspectives and in a way that would not have been possible even if they could have witnessed them in real life.

Thus, in the eyes of Belgian teachers, the optical lantern was a powerful medium that helped their pupils engage with the world around them in a way that no other medium could. Arguably, the medium's specific qualities — the fact that it projected large, often colourful images that literally lit up darkened rooms while the teacher could verbally explain what could be seen — were responsible for this effect. The medium represented an extraordinary means for teachers to spark the imagination of their pupils and make them feel as if they were present *in* the projected images, roaming through them with their eyes. The medium allowed a kind of vision that was so immersive as to be almost sensorial. It is not at all surprising that many teachers believed this medium to be an ideal supplement to their teaching practices.

AUDREY HOSTETTLER

Progressive Education and Early Uses of Film in Swiss Schools

Introduction

In 1923 in Geneva, the neighbourhood newspaper *Le Plainpalistain* published an article criticising the use of so-called 'active methods' in local schools. The article claimed that these pedagogical methods failed to develop children's intelligence and gave excessive priority to sports and entertainment in day-to-day life of the school: 'The teaching staff neglects spelling while they organise walks, baths, showers, cinematographic recreations and intensive physical training.'[1] Primary schoolteacher and film enthusiast André Ehrler, who responded to this article shortly afterwards, did not comment on these activities, but underlined that active methods still remained marginal: 'For those who know how cautiously and even distrustfully the majority of our teaching staff try to apply the principles of "*Arbeitsschule*", such wild accusations remain unheeded.'[2] Beside the debate on the application of active methods, it is striking that the article's author seems to associate their use with the organisation of school film screenings, which were spreading in Switzerland at that time.

The active methods in question are now understood as 'progressive education' or 'new education'.[3] Scholars have underlined the multiplicity of reformist pedagogical theories developing internationally under various labels since the second half of the nineteenth century. Some propose to view progressive education as 'a nebula where philosophical

* This article is part of the Swiss National Science Foundation project 'Réformes scolaires et usages du film dans les écoles suisses durant l'entre-deux-guerres' (178348).

1 Author unknown, 'S. M. l'Orthographe', *Le Plainpalistain*, 1 August 1923. '*Le corps enseignant s'assied sur l'orthographe tandis qu'il organise promenades, bains, douches, réjouissances cinématographiques et culture physique intensive.*'

2 Ehrler, André, 'L'école à travers le monde', *Tribune de Genève*, 27 August 1923. '*Pour qui sait avec quelle prudence, voire quelque méfiance, la majorité de notre corps enseignant essaye d'appliquer les principes de l'"Arbeitsschule", d'aussi folles accusations restent lettre morte.*' As the original article in *Le Plainpalistain* could not be found, the quote is from Ehrler's article.

3 In French, '*école active*' or '*éducation nouvelle*' and in German, '*Arbeitsschule*' or '*Reformpädagogik*'. Hereafter, the English translation 'progressive education' will be used.

Audrey Hostettler • University of Lausanne, audrey.hostettler@gmail.com

Learning with Light and Shadows: Educational Lantern and Film Projection, 1860-1990, ed. by Nelleke Teughels and Kaat Wils, TECHNE-MPH, 8 (Turnhout, 2022), pp. 123-142
© BREPOLS PUBLISHERS 10.1484/M.TECHNE-MPH-EB.5.131497

theories, innovative practices, and political ideals coexist, and where conflicts abound'.[4] Others underline that it was never 'a static idea or coherent set of programmes'.[5] Despite its heterogeneity, its general characteristics can be sketched.

With its roots in the works of pedagogues like Jean-Jacques Rousseau, Johann Heinrich Pestalozzi, and John Dewey, and integrating a new understanding of the child brought by the development of psychology, progressive education aimed to break with the teaching practices of the time.[6] It refused the idea that knowledge had to be imposed upon students through language but instead encouraged engaging students' senses, especially their sight, and making them active in their learning. At its core, progressive education postulated that each child is driven by natural interests that need cultivating and automatically lead to spontaneous activity. This activity, be it intellectual or physical, was thought to be key for effective and long-lasting knowledge acquisition. Essentially, it was thought that the child learns best by doing. In this scenario, the teacher became a facilitator whose main task was not so much to transfer knowledge as it was to create the conditions where the student's natural interests and spontaneous activity could unfold.[7]

Even though progressive education's core characteristics such as child-centeredness or the use of visual teaching tools can be traced back to the nineteenth century and beyond,[8] the spread of experimental pedagogy in universities provoked a renewed reformist impetus in the first decades of the twentieth century.[9] In Switzerland, the creation in 1912 of the Jean-Jacques Rousseau Institute in Geneva, an educational sciences school, contributed to disseminating discourses on progressive education, leading to what has been called the 'roaring years of pedagogy'.[10] As the next sections will show, progressive education theories were a significant part of the pedagogical debates at that time, though their application, as André Ehrler rightly suggested, was not unquestioned by the teaching staff.

Around the same time, film began to be considered a potential teaching tool by teachers, school boards, and cantonal Departments of Public Instruction. While some school film screenings took place beginning in 1914, the first regular initiatives began in 1917 in Geneva, followed by the German-speaking part of the country a few years later (1921 in Bern and

4 Ohayon, Annick, Dominique Ottavi, and Antoine Savoye, 'Introduction', in *L'Éducation nouvelle, histoire, présence et devenir*, ed. by Annick Ohayon, Dominique Ottavi, and Antoine Savoye (Bern et al.: Peter Lang, 2004), pp. 1–7 (p. 4). '*En fait, nous avons affaire à une nébuleuse où voisinent théories philosophiques, pratiques novatrices, idéaux politiques, et où abondent les conflits.*'

5 Reese, William J., 'Progressive Education', in *The Routledge International Encyclopedia of Education*, ed. by Gary McCulloch, and David Crook (London, New York: Routledge, 2008), pp. 461–64 (p. 461).

6 Among other aspects, progressive education retains from experimental psychology the fact that the child is not an 'adult-to-be' but follows their own stages of development. See for instance Ferrière, Adolphe, *L'École active*, 2 vols (Neuchâtel, Genève: Éditions Forum, 1922), I, p. 9.

7 Reese, 'Progressive Education', pp. 461–64. See also Howlett, John, *Progressive Education: A Critical Introduction* (New York: Bloomsbury Academic, 2013).

8 Chung, Shunah, and Daniel J. Walsh, 'Unpacking Child-centredness: A History of Meanings', *Journal of Curriculum Studies*, 32.2 (2000), 215–34 <https://doi.org/10.1080/002202700182727>; Grunder, Hans-Ulrich, 'Seminarreform und Reformpädagogik in den Lehrerseminaren der Schweiz zwischen 1870 und 1930', *Jahrbuch für Historische Bildungsforschung*, 1 (1993), 109–31 <http://nbn-resolving.de/urn:nbn:de:0111-pedocs-158295> [accessed 2 June 2021].

9 Reese, 'Progressive Education', p. 462.

10 Durand, Gregory, Rita Hofstetter, and Georges Pasquier, eds, *Les bâtisseurs de l'école romande. 150 ans du Syndicat des enseignants romands et de l'Educateur* (Chêne-Bourg: Georg, 2015), p. 196. '*années folles de la pédagogie*'.

Basel). Since the article in *Le Plainpalistain* implied a particular affinity between film and 'active methods', one may question more generally how the contemporary pedagogical debates were related to the way film entered Swiss classrooms.

The introduction of film into schools has become a popular topic in film historical research following recent interest in nontheatrical screening practices, sometimes described as 'useful cinema'.[11] National and transnational studies have investigated how governments, school boards, and production companies worked to implement the use of film in teaching.[12] These institutional histories of school cinema rarely delve into the screening and teaching practices themselves, as their performative nature makes them particularly difficult to reconstruct.[13]

However, several scholars working on various geographical contexts have underlined how the investigation of teaching practices is essential to understand how film entered the classrooms. For example, Lucie Česálková has shown that previous research on early school film in Czechoslovakia, which focused on policy, did not account for the diversity of initiatives that were predominantly carried out by voluntary teachers outside of official educational and film production networks.[14] Similarly, Katie Day Good has underlined that classroom media use in the United States differed from the uses promoted by educational media industries or encouraged by school boards, as teachers preferred second-hand, do-it-yourself, and low-tech teaching tools over film or radio. Good called for a 'more user-centric and multimedia perspective', closer to teaching realities, where media was used in 'unpredictable and individualized ways according to the aims of the teacher in charge'.[15]

Therefore, in order to reconstruct how film entered the classrooms, it is necessary to question how the screenings were actually performed and what teaching methods were used. Given the particular pedagogical emulation at that time, it is probable that these methods were related to contemporary pedagogical debates. In one of the rare articles questioning the links between early school film screenings and reform pedagogy, Frank Kessler and Sabine Lenk have argued that in Germany, progressive education theories influenced and legitimised

---

11 Acland, Charles, Haidee Wasson, eds, *Useful Cinema* (Durham, London: Duke University Press, 2011). For Acland and Wasson, 'useful cinema' regroups films defined by their function rather than by their aesthetic qualities, such as promotional, touristic, medical, amateur or educational films.

12 See for instance Orgeron, Devin, Marsha Orgeron, and Dan Streible, eds, *Learning with the Lights Off. Educational Film in the United States* (New York: Oxford University Press, 2012); Dahlquist, Marina, and Joel Frykholm, eds, *The Institutionalization of Educational Cinema. North America and Europe in the 1910s and 1920s* (Bloomington: Indiana University Press, 2019); Gertiser, Anita, 'Schul- und Lehrfilme in der Schweiz', in *Schaufenster Schweiz. Dokumentarische Gebrauchsfilme 1896–1964*, ed. by Yvonne Zimmermann (Zurich: Limmat Verlag, 2011), pp. 384–471; Alovisio, Silvio, *La scuola dove si vede. Cinema ed educazione nell'Italia del primo Novecento* (Torino: Kaplan, 2016).

13 On the methodological difficulties underlying the reconstruction of teaching practices, see for example Alovisio, Silvio, and Paolo Bianchini, 'Introduzione', *Immagine. Note di storia del cinema*, 11 (2015), 7–16 (p. 12).

14 Česálková, Lucie, 'Cinema Outside Cinema: Czech Educational Cinema of the 1930s under the Control of Pedagogues, Scientists and Humanitarian Groups', *Studies in Eastern European Cinema*, 3.2 (2012), 175–92 <https://doi.org/10.1386/seec.3.2.175_1>.

15 Day Good, Katie, 'Making Do With Media: Teachers, Technology, and Tactics of Media Use in American Classrooms, 1919–1946', *Communication and Critical/Cultural Studies*, 13.1 (2016), 75–92 (p. 88) <https://doi.org/10.1080/14791420.2015.1092203>. On the need to investigate concrete screening practices, see also Curtis, Scott, *The Shape of Spectatorship. Art, Science, and Early Cinema in Germany* (New York: Columbia University Press, 2015), p. 2.

educational screenings made in '*Reformkinos*', specialised cinemas that served as a model for subsequent school film screenings.[16] According to them, the didactic principles that underlay the first educational film screenings had their origins in progressive education. While their analysis is useful in sketching theoretical affinities, postulating a temporal or causal relationship between theoretical texts and an established *dispositif* tends to eclipse the remediation processes that occur, the institutional constraints the new medium and methods are subject to, and the very individual and spontaneous ways teachers teach within this *dispositif*.[17]

The analysis of the links between pedagogical theories and teaching practices using film would therefore benefit from the more user-centric perspective that Katie Day Good advocates. Just as teachers' use of media was 'unpredictable and individualized', as she states, their understanding of pedagogical theories varied, and determined, alongside institutional and material constraints, their teaching methods in singular ways. By reviewing two case studies from French- and German-speaking Switzerland and focusing on primary and secondary teachers, this paper aims to further question the links between progressive education and the early uses of film in Swiss classrooms.

Pedagogic Theory and Teaching Practices

As historians have underlined, progressive education internationally was particularly influential on a theoretical level and only marginally determined educational practice.[18] While progressive education theories were very popular among Swiss teachers and cantonal school authorities, incorporating them into teaching methods was not straightforward. On the contrary, it resulted from concerted effort and discussions within the teaching profession. As they were facing different types of constraints, teachers needed to negotiate between pedagogical principles and the realities of the classroom.

Under Switzerland's federalist system, responsibility for public instruction falls to the twenty-five cantons. Each cantonal government is composed of five to seven ministers, one of whom is appointed head of the cantonal Department of Public Instruction (hereafter DPI). The Confederation has no saying on school policy, which is the sole responsibility of the twenty-five cantonal DPIs. Municipalities help to provide facilities and teaching aids.

Some of the cantonal curricula issued by the DPIs in the 1920s explicitly encouraged the use of so-called 'active methods'. The DPIs of Geneva, Bern, and Neuchâtel at least spoke in favour of progressive education, insisting on the need to start from the child, and to guarantee their active engagement.[19] In Zurich, however, it was underlined that as the

16 Kessler, Frank, and Sabine Lenk, '*Kinoreformbewegung* Revisited: Performing the Cinematograph as a Pedagogical Tool', in *Performing New Media, 1890–1915*, ed. by Kaveh Askari and others (New Barnett: John Libbey, 2014), pp. 163–74.

17 Kessler defines *dispositif* as the interrelation of a spectator, a technology and a representation. Kessler, Frank, 'La cinématographie comme dispositif (du) spectaculaire', *Cinémas*, 14.1 (2003), 21–34 (p. 31) <https://doi.org/10.7202/008956ar>.

18 Reese, 'Progressive education', p. 461.

19 *Programme de l'enseignement dans les écoles primaires* (Genève: Imprimerie Soullier, 1923), p. 4; *Plan d'études pour les écoles primaires françaises du canton de Berne: histoire, géographie et histoire naturelle* (Saignelégier: Al. Grimaitre, 1925), p. 13; *Programme d'enseignement pour les écoles enfantines et primaires du 14 janvier 1927* (Neuchâtel: Département de l'Instruction publique, 1927), pp. 3–6.

curriculum only determined the subject matter, it could not take a stand on progressive education and teaching methods in general, which were the teacher's responsibility.[20] In all cantons, it seems to have been widely acknowledged that the choice of teaching methods was not a matter of policy but was left to the teacher's expertise.

Accordingly, the school curricula did not refer to specific progressive education theoreticians. At that time, several Swiss scholars published extensively on the pedagogical renewal, and may have inspired the school field. In the French part of Switzerland, progressive education was more institutionalised thanks to the Jean-Jacques Rousseau Institute and its prominent members Adolphe Ferrière and Édouard Claparède, who both published seminal books on this question.[21] The Swiss-German part of the country was mainly influenced by German pedagogues such as Georg Kerschensteiner and Hugo Gaudig.[22]

Progressive education theorists showed affinities for the use of film in schools, even though they rarely provided concrete recommendations.[23] Because of its dynamism and realism, it was thought that film could bring life into the classroom. The new medium also embodied a certain ideal of modernity and cosmopolitanism that spoke to the leaders of the pedagogical movement internationally, who advocated for a 'new era'.[24] The rhetoric of modernity in progressive education aimed to highlight a radical break from earlier more intellectualist ways of teaching, proposing science-based conclusions insisting on the child's activity instead. At the core of progressive education's project was the desire to prepare the child for their future role in society. Therefore, schools had to be particularly attuned to the realities of the time, providing 'an education more suited to the needs and aspirations of modern life'.[25]

As progressive education maintained that interest led to active engagement and, ultimately, to knowledge, film was also considered useful because it appealed to children and could be the starting point for further research on a topic. This idea was defended by, among others, schoolteacher André Ehrler, who was explicitly in favour of progressive education and claimed that 'cinema's educational value lies in its appeal'.[26] As will be shown

---

20 Gassmann, Emil, *Der Lehrplan der Zukunftsschule* (Winterthur: A. Vogel, 1923), p. 21.

21 See for example Ferrière, Adolphe, *L'École active*, 2 vols (Neuchâtel, Genève: Éditions Forum, 1922); Claparède, Édouard, *L'éducation fonctionnelle* (Neuchâtel: Delachaux & Niestlé, 1931).

22 See Gassmann, *Der Lehrplan der Zukunftsschule*, p. 20.

23 Adolphe Ferrière often participated in discussions on school cinema and even attended the International Educational Film Congress in Rome in 1934. However, he was more interested in the use of film as a way to train teachers and to promote progressive education internationally than giving precise advice regarding the possible applications of film in schools. Ferrière, Adolphe, 'Formation professionnelle des éducateurs par la psychologie et la pédagogie scientifiques', *Revue internationale du cinéma éducateur*, 10 (October 1934), 806–10.

24 One of the progressive education movement's main publications in the 1920s was *Pour l'ère nouvelle/The New Era/Das Werdende Zeitalter*, a journal published in three languages whose French issue was edited by Ferrière. See Haenggeli-Jenni, Béatrice, 'Pour l'Ere Nouvelle: une revue-carrefour entre science et militance (1922–1940)' (unpublished doctoral thesis, University of Geneva, 2011).

25 *The New Era in Home and School*, 11.43 (1930), quoted in Haenggeli-Jenni, 'Pour l'Ere Nouvelle: une revue-carrefour entre science et militance (1922–1940)', p. 77.

26 Ehrler, André, undated manuscript, Geneva State Archives (Archives d'état de Genève, hereafter AEG) AP 95.2.2.4. *'La valeur éducative du cinéma réside dans son attrait'*. Progressive education theoretician Édouard Claparède also mentions in an article that a film about Robin Hood, for instance, could be the starting point of a lesson on the Crusades. Claparède, Édouard, 'Réflexions d'un psychologue', *Annuaire de l'instruction publique en Suisse*, 16 (1925), 7–60 (p. 23).

in the next section, the use of film in teaching was often criticised by more conservative teachers, who insisted on the need for children to learn to accomplish less appealing tasks in order to develop their discipline and capacity for effort.[27]

Even though their theoretical references differed, teachers in several places in the country noted that progressive education principles were not always clear. In the Swiss German part, *Arbeitsschule* was supposedly often mistaken for being the mere encouragement of manual work.[28] In the French part of the country, some teachers also expressed perplexity. In 1924, a leading teachers' union, the *Société Pédagogique Romande* (Pedagogic Society of West Switzerland), decided to hold its quadrennial congress on the adaptation of progressive education to primary school. As the congress's *rapporteur* underlined, the use of progressive education principles had been delayed by 'a clear indecision and lack of clarity in the conclusions and in the proposed application of active theories'.[29] He stressed that many teachers had read progressive education theories with enthusiasm but struggled to incorporate them into their teaching.[30] This is hardly surprising, knowing that several theoreticians insisted on not prescribing teaching methods. According to them, methods had to be designed according to the students' capacities and interests.[31]

Interestingly, this congress did not provide many details on how to apply progressive education theories. The proceedings ended with a list of several closing theses. The central thesis merely reasserted the theoreticians' statement that 'In order to adapt the active school to the primary school, the teacher should seek to introduce interest-based activities in all school work'.[32] Taking the students' needs and wishes into consideration, along with designing lessons that would make them active, seem to be the main principles that Swiss teachers retained from progressive education theories.

Historians have shown that this congress was a major symbolic step for progressive education that could be understood as its officialization by the Swiss school system, were it not for signs of dissent emerging between theorists and the teaching staff.[33] The congress *rapporteur* underlined how progressive education theories made sense in an ideal world but took no consideration of the specificities of the current teaching practicalities, the number of students, and the expectations and constraints of the institution.[34] Teachers were aware

27 This opposition between advocates of effort and interest was underlined by John Dewey, among others. In Switzerland, teachers defending the need to teach effort mainly expressed their views in conservative journals like *Pro Juventute* or *Schweizer Schule*. Dewey, John, *Interest and Effort in Education* (Boston: Houghton Mifflin, 1913) <https://archive.org/details/interestandeffooodeweuoft/>.

28 Gassmann, *Der Lehrplan der Zukunftsschule*, p. 20.

29 Richard, Albert, *XXIe Congrès de la Société Pédagogique de la Suisse romande* (Genève: Sonor, 1924), p. 11. '*Une indécision manifeste et [...] un manque de netteté dans les conclusions et dans l'application proposée des théories actives*'.

30 Richard, *XXIe Congrès de la Société Pédagogique*, p. 11. Richard mentions that Swiss teachers were particularly inspired by the works of the Rousseau Institute and the Belgian pedagogue Ovide Decroly.

31 Ferrière, *L'École active*, I, pp. 9–10; Claparède, Édouard, 'L'école et la psychologie expérimentale', *Annuaire de l'instruction publique en Suisse*, 7 (1916), 71–130 (p. 74).

32 Richard, *XXIe Congrès de la Société Pédagogique*, p. 65. '*Pour adapter l'école active à l'école primaire, le maître devra chercher à introduire dans tous les travaux scolaires l'activité basée sur l'intérêt*'.

33 Hofstetter, Rita, *Genève, creuset des sciences de l'éducation (fin du XIXe siècle-première moitié du XXe siècle)* (Genève: Librairie Droz, 2010) p. 322.

34 Richard, *XXIe Congrès de la Société Pédagogique de la Suisse romande*, p. 16.

of their role as intermediaries between the theoretical need to base their teaching entirely on the child and what the institution thought was desirable and possible: they seemed to be particularly attentive to their day-to-day reality of managing the classroom.

Even though progressive education was rather popular among Swiss teachers, its implementation into teaching methods needed to be discussed, considered, and negotiated together with other constraints. While there seemed to be affinities between the new pedagogic theories and film, the next sections will show that teaching staff also needed to navigate the arrival of this new media in ways that complicate these conceptual affinities.

Putting Pedagogy at the Heart of the School Film Debate

When film began to be considered a potential teaching tool, many different people participated in the debates concerning its possible uses. Teachers strove to preserve their authority on practices that they thought were, above all, pedagogical matters. The need to foster the students' active engagement and to design child-centred lessons were at the centre of their conceptualisation of film's usefulness for school instruction.

Political authorities were involved in school film activities to varying degrees. As cantons were responsible for public instruction, there was no federal policy concerning these screenings and therefore no substantial subvention until the 1940s.[35] In 1931, an intercantonal body regrouping the heads of all twenty-five Departments of Public Instruction (the Intercantonal Conference of the Heads of Departments of Public Instruction) showed symbolic support for school film activities, stating that 'classroom film, when used intelligently and carefully, can be an important enrichment of teaching. It is hoped that the Confederation and the cantons will give their sympathy and effective support to the efforts made in favour of classroom films'.[36]

However, this apparent openness to the use of film in schools only partially reflected actual support. On the cantonal level, several DPIs, such as those of Basel and Geneva, showed interest in school film activities very early on. They allocated them a budget and entrusted their organisation to ad hoc services. Other cantons tolerated the organization of screenings by teachers but did not provide consequent financial support; others still were rather opposed to the use of film.[37]

35 The first national strategy regarding cinema consisted in the creation in 1938 of a national Film Chamber that mainly addressed commercial film problematics and tried to protect the Swiss market as international exchanges became increasingly difficult. The Confederation's first significant support for school film screenings comes with its sponsorship of the newly founded Swiss Association of Educational Film Offices in 1948. Haver, Gianni, and Pierre-Emmanuel Jaques, *Le spectacle cinématographique en Suisse (1895–1945)* (Lausanne: Antipodes & Société d'Histoire de la Suisse romande, 2003), p. 73; Borel, Antoine, 'Le film au service de l'école', *Archiv für das schweizerische Unterrichtswesen*, 38 (1952), 1–28 (pp. 6–7).

36 Letter from the Intercantonal Conference of the Heads of Departments of Public Instruction, 23 May 1931, AEG 1985 va 5 5.3.260. '*Le film scolaire, lorsqu'il est employé intelligemment et prudemment, peut être un enrichissement important de l'enseignement. Il y a lieu de souhaiter que la Confédération et les cantons accordent leur sympathie et leur appui effectif aux efforts faits en faveur des films scolaires.*'

37 In the 1930s, the head of Lucerne's DPI was still reluctant to allow film screenings in schools. Letter from Lucerne's DPI, 24 November 1933, Lucerne State Archive, A 606/11.

In addition to the political authorities, from the beginning of the 1910s, many people from different backgrounds took a stance on the subject of children and film. These views ranged from concerns about morality, hygiene, and safety to enthusiasm for film's educational potential. Lawyers, health professionals, charities, church officials, parents' associations, and many others participated in what had become a public debate.[38]

Much of this debate revolved around film's alleged stronghold on the spectators' psyche. Many feared film's potential hypnotic effects, particularly on people who were perceived as weak — typically children, women or people with mental disabilities.[39] In addition, film was considered the most elaborate and effective teaching tool ever produced, to the extent that some people feared it could replace teachers altogether.[40] While this replacement evidently never happened, Frank Kessler and Sabine Lenk have shown that as opposed to the optical lantern, film tended to restrain the lecturer's or teacher's role. Using Philippe Marion's terminology, they described the lantern as 'heterochronic', meaning that slides could be shown in the order and rhythm that suited the lecturer and the audience. Film, however, is on the 'homochronic' side of the continuum, as the medium determines the duration of reception.[41]

Compared to other teaching tools such as the optical lantern or the textbook, film's form resulted in the transmission of fixed units of knowledge. They were industrially produced, rather standardized, and therefore little adaptable to the classroom's particularities. The new medium was not ideal for teachers that sought to encourage student activity, and many complained about its lack of flexibility. However, rather than rejecting it altogether, many teachers were quick to pivot the discussions to how film could be used in teaching, rather than spending time debating whether it should be used at all. Through numerous articles in the generalist and specialised press, they insisted on the need for a pedagogical framing of the screenings.

In the mid-1920s, for example, the conservative teacher Rosine Tissot argued in several conferences and publications that film had the dangerous tendency to make children's minds 'lazy [and] inattentive' and therefore prevent concentration and active learning.[42]

---

38 Meier-Kern, Paul, *Verbrecherschule oder Kulturfaktor? Kino und Film in Basel 1896–1916* (Bâle: Helbing & Lichtenhahn, 1992), pp. 102–06; Engel, Roland, *Gegen Festseuche und Sensationslust. Zürichs Kulturpolitik 1914–1930 im Zeichen der konservativen Erneuerung* (Zurich: Chronos, 1990), pp. 121–22.

39 Berton, Mireille, *Le corps nerveux des spectateurs. Cinéma et sciences du psychisme autour de 1900* (Lausanne: L'Âge d'Homme, 2015), pp. 413–14.

40 This fear was fuelled by, among others, enthusiastic statements by Thomas E. Edison. See Masson, Eef, *Watch and Learn. Rhetorical Devices in Classroom Films after 1940* (Amsterdam: Amsterdam University Press, Eye Film Institute Netherlands, 2012), pp. 29–31.

41 Kessler, Frank, and Sabine Lenk, 'The Voice of the Lecturer. Image-Word Relations in Optical Lantern and Early Film Performances', *International Film and Media Studies Conference*, 'Retuning the Screen. Sound Methods and the Aural Dimension of Film & Media History', Gorizia/UdineGorizia/Udine (online), 3 November 2020 <https://ff.2020.filmforumfestival.it/panel/sound-before-sound-aurality-in-early-and-silent-cinema/> [accessed 2 June 2021]; Marion, Philippe, 'Narratologie médiatique et médiagénie des récits', *Recherches en communication*, 7 (1997), pp. 61–88.

42 Tissot, Rosine, 'Ce que lisent nos enfants', *Pro Juventute*, 8 (August 1924), 405–12 (p. 407). '*Le film rend l'esprit paresseux, inattentif. [...] il tue le goût de la libre recherche, la possibilité de la concentration*'. / 'The film makes the mind lazy, inattentive. [...] it kills the desire for free research, the possibility of concentration'.

Similarly, Ernest Briod, a teacher who was relatively in favour of progressive education,[43] warned against the use of film:

> It can be a useful and enjoyable diversion; it can create new sources of interest, sketch out intuitions, give some knowledge, prepare somewhat for action. But it *is not* action, and it even offers the danger of making the mind lazy in the presence of painful and necessary schoolwork that it would be bad to avoid.[44]

Many other teachers refuted these claims, such as Ernest Savary, who published the first in-depth study of the school film in Switzerland in 1925. Quoting Tissot, Savary objected that the effect of the film depended primarily on how it was used. In his view, Tissot's criticism could only apply to hour-long screenings comprising several films and shown to several classes at a time, therefore not leaving enough space for the teacher's pedagogical action.[45]

This insistence on the need for a pedagogical framing of film was rarely accompanied by concrete propositions for what it could look like, for it was thought to be dependent on the context. Even in his extensive study on the subject, Savary framed the few practical didactic suggestions he made by stating that 'it is not possible to set precise rules on how to illustrate a lesson with animated projections. The makeup of the class, the degree of intellectual development of the students, their abilities and the subject matter determine all lessons'.[46] While it is true that Savary's words suggest that the film should be used at the end of the lesson to 'illustrate' it, he leaves the responsibility of designing the surrounding activities to the teachers, who know best what their students need.

By insisting early on that the success of an educational screening depended on the way in which the films were shown and used in class, teachers gradually turned the public debate into a pedagogical one, in which they were the experts. Putting pedagogical arguments at the centre of the debates was a way for teachers to secure their place as organizers of knowledge transmission. They also aimed to guarantee film's usefulness for their particular students. Even though not all teachers who supported progressive education equally supported the use of film, the need for the child's active engagement and the possibilities of child-centred lessons were central concerns in teachers' early conceptualisations of film use in schools.

The next sections will focus on the cantons of Geneva and Zurich to show that the respect of pedagogical interests did not always come naturally when film was implemented as a teaching tool. Teachers faced different obstacles when trying to defend their freedom to use the teaching methods they favoured.

---

43 In a study destined for local teachers, Briod encouraged the use of active methods, but also underlined that their systematisation would be problematic for the development of discipline and memorisation. Briod, Ernest, *L'école active et l'enseignement secondaire* (Lausanne: Payot, 1925), pp. 30–31.

44 Briod, Ernest, 'L'économie des forces au service du progrès scolaire et social', *l'Éducateur*, 19 (10 May 1919), 289–96 (p. 295). *'Il peut créer des sources nouvelles d'intérêt, ébaucher des intuitions, donner quelques connaissances, préparer quelque peu à l'action. Mais il n'est pas l'action, et il offre même le danger de rendre l'esprit paresseux en présence de devoirs pénibles et nécessaires qu'il serait mauvais de lui éviter.'*

45 Savary, Ernest, 'Le cinéma et l'école', *Annuaire de l'instruction publique en Suisse*, 15 (1924), 45–82 (p. 54).

46 Ibid., pp. 65–66. *'Il n'est pas possible de fixer des règles précises sur la manière d'illustrer une leçon avec des projections animées. La formation de la classe, le degré de développement intellectuel des élèves, leurs aptitudes et le sujet traité conditionnent toutes les leçons'.*

Opposing Centralisation

When cantonal school authorities started reflecting on the possible use of film for teaching, pedagogical concerns were addressed but did not dominate in the discussions. In 1910, in Geneva, a teachers' syndicate called on the cantonal Department of Public Instruction to express its concerns regarding film's potential negative effects on children.[47] The DPI subsequently issued regulations on children's cinema attendance, and decided to set up screenings of good films to educate their taste and keep them out of licentious screenings.[48] The canton had also been offered a large sum of money by a local patron specifically for developing school film screenings.[49] The Department of Public Instruction commissioned teacher Éloi Métraux to study the different possibilities regarding school film screenings. Métraux asked for progressive education theoretician Adolphe Ferrière's advice, who emphasized the necessity to frame the film with a variety of school exercises, if possible immediately after the screening. Following these conceptions, Métraux's study advocates the screening of short films of maximum twenty minutes, if possible in class.[50]

However, in the project's implementation, the practical arguments outweighed the pedagogical ones. The DPI chose to make use of a large hall that was situated in the city centre and could assemble up to eight hundred students, rather than following the recommendation of equipping schools with lighter projection facilities. At first, the service tolerated the parallel organization of classroom screenings by teachers, who found ways to buy small projectors and dealt directly with film distributors.[51] Over time, however, the service changed its strategy at the expense of these individual initiatives. In the early 1920s, school authorities appointed a new head of the school cinema service, teacher Emmanuel Duvillard, who was one of the organisers of the 1924 pedagogical congress. Despite Duvillard's personal commitment to progressive education, financial and logistic constraints led him to reorganise the service in a way that made it difficult for teachers to apply the new pedagogical principles. The school cinema service started mandating the use of its collection of films and working to improve efficiency in its organization.[52] Consequently, implementing the one-and-a-half-hour programmes offered by the service (using the films they had acquired) became compulsory. These films were either to be viewed in the central hall or rented for an in-school screening if the school was too far

47 'La littérature immorale', *Journal de Genève*, 14 February 1910, p. 2.

48 'Les cantons et l'instruction publique en 1916–1917', *Annuaire de l'Instruction publique en Suisse* (1918), 280–89 (p. 289). On cinema regulations in Geneva, see Frauenfelder, Consuelo, *Le Temps du mouvement. Le cinéma des attractions à Genève (1896–1917)* (Genève: Presses d'histoire suisse, 2005).

49 Letter from Éloi Métraux to the head of the DPI, 30 March 1914, AEG 1985 va 5.3.102. It is not clear whether the 2000 Swiss francs came with conditions as to what specific form the screenings should take.

50 Métraux, Éloi, 'Notes sur la création d'un cinématographe scolaire à Genève', 30 March 1914, AEG 1985 va 5.3.115.

51 The cantonal authorities allowed and paid for up to 1500 m. of film per month per school, which could be freely chosen. Gielly, J., 'Rapport sur le cinéma scolaire de l'école du Grutli, année 1924', AEG 1985 va 5.3.152.

52 Duvillard regularly invoked his service's limited budget when responding to critics regarding school film activities' centralisation. It is also probable that buying a certain number of films once and for all was considered cheaper and more efficient for the school cinema service than reimbursing individual schools or teachers for rentals they made directly from film distributors. See for instance Duvillard, Emmanuel, 'Rapport sur la marche du service du cinéma scolaire et des projections pendant l'année 1927', [1928], AEG 1985 va 5.3.213.

from the city centre. In 1929, most small projectors used in schools were banned by the cantonal Department of Hygiene for not meeting security requirements, putting an end to most in-class screenings in Geneva.[53]

With its pre-designed compulsory programmes, the new system provoked great discontent among teaching staff. Many criticised its lack of flexibility, and many regretted not being able to offer adequate pedagogical guidance in relation to using the films. As one teacher pointed out, the system where the schools organized their own screenings was 'more elastic' and allowed them to apply tried-and-tested didactic techniques, such as alternating films and slide projection:

> [The former system] allowed us to better adapt our screening sessions to the programme covered each year. Above all, it made it possible to organise our mixed screenings, which only used cinema insofar as it could be useful to show what slides cannot show.[54]

One of the most vocal opponents of the centralization of the screenings was André Ehrler, a schoolteacher, film critic and socialist politician who also explicitly defended progressive education. In the daily, socialist and cinematographic press, he published numerous articles criticizing the status quo and reasserting the expertise of the teacher. He stated, 'It is not enough to "show cinema" to our schoolchildren. It is necessary to know whether it is profitable for them. Only the teachers are qualified to say so'.[55] Ehrler benefitted from a certain visibility as a politician and did not hesitate to be outspoken in his criticism.[56] 'We don't have a "school" cinema', 'Who are we laughing at?', 'Educational Cinematography. A fiasco', 'We are sleeping!': his articles' titles demonstrate his indignation and his involvement in making teachers' voices and pedagogical arguments heard.[57]

André Ehrler defended the new pedagogical movements and was particularly inspired by the works of the Rousseau Institute, which he seemed to consider compatible with the use of film. Around 1928, he had started writing a book on classroom film, which was never published. Some parts of the manuscript that have been preserved contain extensive quotes of the Institute's founder Édouard Claparède on the concepts of interest and activity.[58] As was shown in the introduction, Ehrler did not hesitate to defend progressive education in the local press, and regretted that it was not better accepted by the teaching staff.

---

53 AEG 1985 va 5.3.243.

54 Gielly, J., 'Rapport sur le musée et la bibliothèque scolaires. École du Grutli', January 1924, AEG 1985 va 5.3.152. '*[L'ancien système] permettait de mieux adapter nos séances de projections au programme parcouru dans chaque année. Il permettait surtout d'organiser nos séances mixtes qui ne faisaient appel au cinéma que dans la mesure où cela pouvait être utile pour montrer ce que les vues fixes ne peuvent pas montrer.*'

55 Ehrler, André, 'Pour une organisation rationnelle de la cinématographie éducative. I. Principes généraux', *Le Travail*, 7 January 1929, [page unknown]. '*Il ne suffit pas de "montrer du cinéma" à nos écoliers. Il faut savoir si cela leur est profitable. Les maîtres sont seuls qualifiés pour le dire.*'

56 Ehrler was an influential member of the Geneva Socialist Party, of which he became president in the early 1930s. During this decade, he was elected to several legislative positions on a cantonal and federal level. Heimberg, Charles, 'Ehrler, André', in *Dictionnaire historique de la Suisse* <https://hls-dhs-dss.ch> [accessed 2 June 2021].

57 'Nous n'avons pas de film "scolaire"', 'De qui se moque-t-on?', 'Cinématographie éducative. Un fiasco', 'Nous dormons!'. These articles were published in *Tribune de Genève* (August 1923), *Cinéma Suisse*, (10 August 1928, pp. 3–6) and *Le Travail*, (22 April 1929 and 17 December 1929), respectively.

58 AEG AP 95 2.2.4.

In January 1930, Ehrler refused to attend a compulsory screening of a film about Finland organised by the school cinema service, a decision he subsequently had to answer for. He wrote a letter of explanation, and the reasons he gave for his absence deserve to be quoted in full, as they encapsulate the opposition between a vision of school screenings aligned with progressive education principles and the realities of school cinema at the time:

> A. The film, *Finland*, which I have seen, did not seem to me, as a whole, to help my teaching. For several months now, I have been giving my students, who read and study it at their leisure, a booklet with many illustrations. The film would not give them any new knowledge on the subject.
> B. A delay of more than a month in the implementation of my programme (sixth B) obliges me to avoid any waste of time.
> C. The way in which the films are presented harms teaching more than benefits it.
> D. The dangers of crowding several hundred children in a room, even a large one, are not outweighed by the interest that a schoolchild or educator may find in such events.
> E. The link between the school curriculum and the film material is not established; this breaks the continuity of teaching; the teacher forces himself into unfortunate comments from which confusion and disorder arise.
> F. The teacher has the right to choose from the material proposed to him; he is the sole judge of the didactic elements that are suitable for his teaching.[59]

In addition to criticising the central screenings as dangerous and pedagogically doubtful, Ehrler's letter confronted his superiors with the specificities of his teaching. He reminded them of his responsibility for designing and maintaining his programme and for choosing materials based on the needs of his class. His refusal to attend the screening was the rejection of a standardised system that left little space for his agency or that of his students.

In the same letter, Ehrler claimed that long screenings prevented any pedagogical work, made the children passive and were a 'negation of the active school'.[60] While it is clear that Ehrler's criticism of the organisation of the screenings in Geneva was linked to a personal political agenda, it can also be seen as a consequence of his commitment to progressive education as a movement that encouraged flexibility and made the teacher responsible for creating the best possible learning environment. The school institution, as it was subject to economic and logistical constraints, was an obstacle to the realisation of these ideals.

---

59 Ehrler, André, 'Rapport à M. le Directeur Mingard', 14 February 1930, AEG AP 95 2.2.7. 'A. *Le film, la Finlande, que j'ai vu, ne m'a pas paru, dans son ensemble, de nature à aider mon enseignement. Depuis plusieurs mois, j'ai remis à mes élèves, qui la lisent et l'étudient à loisir, une plaquette largement illustrée. Le film ne leur apporterait aucune connaissance nouvelle sur le sujet.*
B. *Un retard de plus d'un mois dans l'application de mon programme (6e B) m'oblige à éviter toute perte de temps.*
C. *La manière dont sont présentés les films nuit à l'enseignement plus qu'elle ne lui profite.*
D. *Les dangers qu'offre l'entassement de plusieurs centaines d'enfants dans une salle, même vaste, ne sont point balancés par l'intérêt qu'un écolier ou qu'un éducateur peut trouver à ces manifestations.*
E. *La liaison entre le programme scolaire et les documents cinématographiques n'est pas établie; cela rompt la continuité de l'enseignement; Le maître se contraint à des incises fâcheuses d'où naissent confusion et désordre.*
F. *Le maître a droit de choisir dans le matériel qui lui est proposé; il est seul juge des éléments didactiques qui conviennent à son enseignement.*'

60 Ibid., '*impossibilité de "travailler" pédagogiquement le film, les enfants demeurent passifs, négation de l'école active [...]*'.

Classroom Films for Classroom Use

One of the central concerns among Swiss teaching staff was that the available educational films were not adapted to their specific teaching context. As they were often produced in other countries or for different types of screening venues, the films themselves were an obstacle to child-centred teaching. Consequently, in the German part of the country, some teachers assembled in order to establish educational film production by teachers for teachers, which was thought to enable better adaptation of the films to the specifics of the Swiss school setting.

In 1929, a group of teachers from the cantons of Zurich and Basel created a cooperative, the *Schweizerische Arbeitsgemeinschaft für Unterrichtskinematographie* (Swiss Working Committee for Classroom Film, hereafter SAFU). They aimed to produce their own films and set up distribution facilities specifically for schools. They wanted films that were adapted or adaptable to the different realities of Swiss classrooms. In Zurich, where the SAFU was most active, the DPI was long reluctant to support classroom film initiatives. Unlike Geneva's school cinema service, the SAFU always remained an independent teachers' association. There was no central organisation of film screenings in this canton.

The creation of the SAFU can be understood as a reaction to the activities of a private company that also produced and distributed films, the *Schweizer Schul- und Volkskino* (Swiss School and People's Cinema, hereafter SSVK). The SSVK was founded in Bern in 1921 by a Swiss businessman, M. R. Hartmann. It was the leading nontheatrical film company in the country, and it organized screenings for schools and popular education institutions.[61] The SSVK travelled the country with portable projectors and screened the same programme for schoolchildren in the afternoon and adults in the evening. It was popular among teachers, as it had a generous catalogue and offered turnkey solutions. In addition, for several years, it had little local competition. However, other teachers were opposed to the use of SSVK films, which were predominantly *Kulturfilme*, or cultural films; they deemed them too general and sometimes too entertaining for school purposes. This opinion was at the core of the SAFU's creation.[62] Its members thought that if a film could be shown to an adult audience and still be understood and enjoyed, it did not belong in a classroom. Ernst Rüst, one of the initiators of the SAFU, drew a clear line between cultural films and classroom films, saying:

> our school is not an entertainment institute but a working school [*Arbeitsschule*]. We demand active participation from the students. We want to develop their abilities and guide them to build on their existing knowledge. Therefore, the educational film must presuppose a certain knowledge and ability, i.e. it must be adapted to a certain school level.[63]

---

61 Gertiser, 'Schul- und Lehrfilme in der Schweiz', pp. 447–48.

62 Schweizerische Arbeitsgemeinschaft für Unterrichtskinematographie, '10 Jahre SAFU', offprint from *Schweizer. Zeitschrift für Gemeinnützigkeit*, 78.11, (December 1939), 5–6.

63 Rüst, Ernst, 'Kulturfilm und Lehrfilm', *Der Bildwart*, 3 (March 1931), 114–16 (p. 115). '*Unsere Schule ist aber kein Unterhaltungsinstitut, sondern eine Arbeitsschule. Wir verlangen vom Schüler tätige Mitarbeit, wir wollen seine Fähigkeiten entwickeln, ihn anleiten, auf vorhandenen Kenntnissen weiter zu bauen. Daher muß der Unterrichtsfilm ein bestimmtes Wissen und Können voraussetzen, d. h. er muß einer bestimmten Schulstufe angepaßt sein.*'

According to Rüst, only by adapting the film to school levels can the active participation promoted by progressive education be enabled, as the content will then be built on existing knowledge. More generally, Rüst thought that 'It is essential that the educational film takes into account the requirements of the school (age of the children, type of other lessons, etc.). The films must be created from the pedagogical circumstances'.[64] Producing about two short films a year beginning in 1929, the SAFU could not adapt them to regional or local needs: the films had to circulate widely to be economically viable. Nevertheless, the cooperative organised itself in a way that enabled a certain decentralization, so that it could stay close to what the field required.

The SAFU was made up of several local groups of teachers who volunteered for the development of classroom films. These autonomous groups were responsible for establishing *Filmbedarfspläne*, or plans of film needs. Teachers in several regions were encouraged to highlight specific topics for which they thought a film would be useful, and the local groups then discussed the possibility of taking the idea into production. Even though filming was done by professional filmmakers, the teacher volunteers remained involved throughout the production process (Fig. 5.1). Members of the association published numerous articles in the pedagogic press, some of which describe the long and animated discussions that took place during the development of the screenplays.

It is evident that the SAFU members drew extensively on their classroom experience to envision and create the films that they needed. For instance, some primary school teachers had informed their local SAFU group that they would benefit from a film about crafts and suggested 'how to build a table' as a topic. The local group then wrote a script with the help of a carpenter. Since making table legs seemed too difficult for the young schoolchildren the film was designed for, they decided to skip this step and worked with pre-cut legs instead. After reviewing the first draft of the script, the group felt the film was too dry. One teacher suggested integrating a child into the story, who asks the carpenter questions and draws attention to certain details. Towards the end of the process, the teachers decided to add conversational interjections, such as 'Hans, that's the way to do it!' or 'Ouch, how rough!', predicting that the students could read them in unison during the screening.[65]

To guarantee that their films could be used in the most autonomous way, preferably in class, the SAFU also set up training programmes for teachers to learn how to use the projectors. They published multiple articles and brochures, held conferences and did trial lessons with the public that helped illustrate the many ways their films could be used in class. The teachers who organised themselves in the SAFU cooperative did so to keep a certain control over a teaching tool that was otherwise standardized and sometimes not destined for school use at all. By being present at the production and

---

64 Rüst, Ernst, quoted in 'Protokoll der Konferenz zur Abklärung des gegenseitigen Verhältnisses der schweizerischen Lehrfilmorganisationen', 23 February 1938, Swiss Federal Archive, E3001A#1000/729#225*. '*Beim Unterrichtsfilm muss unbedingt auf die Erfordernisse der Schule (Alter der Kinder, Art des übrigen Unterrichts, etc.) Rücksicht genommen werden. Die Filme müssen aus den pädagogischen Gegebenheiten heraus geschaffen werden.*'

65 Rüst, Ernst, 'Wie Unterrichtsfilme geschaffen werden', Sonderdruck aus *Schweizer Erziehungs-Rundschau*, 10 (January 1939), [pages unknown]. '*Hans, das macht man so!*', '*Au, wie rauh!*'. On SAFU's production of classroom films, see also Rüst, Ernst, 'La production des films d'enseignement', *Revue internationale du cinéma éducateur*, 6 (June 1934), 492–504.

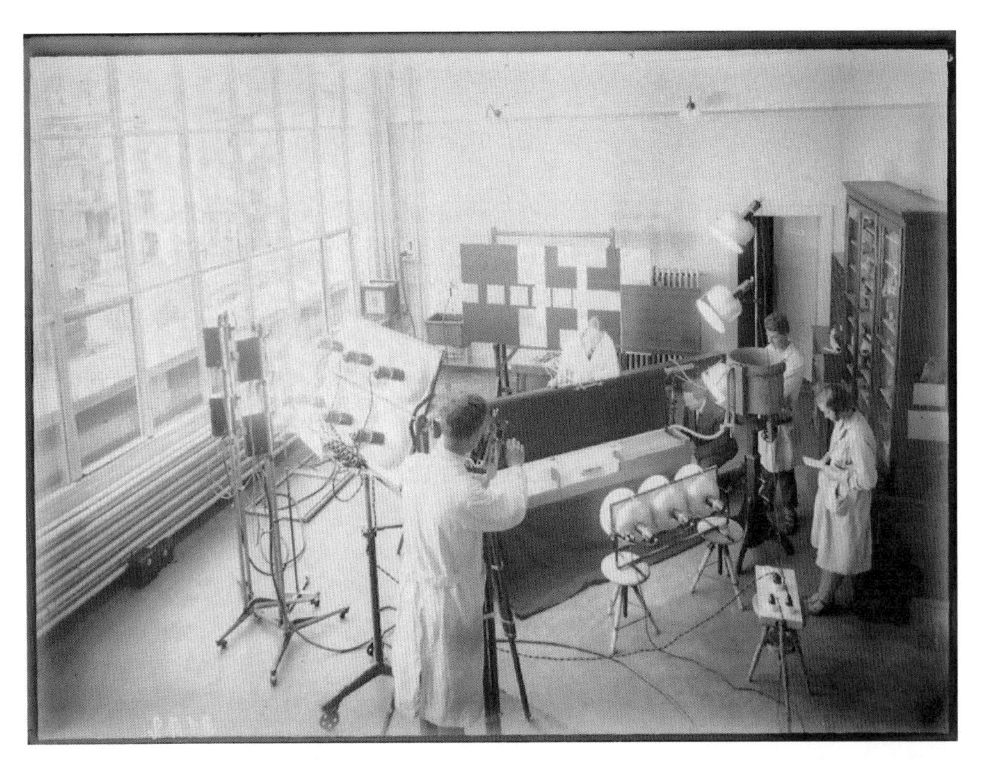

Fig. 5.1. On the set of the SAFU film *Schleuse* (*Lock*, 1931). ETH-Bibliothek Zürich, Bildarchiv / Photographisches Institut der ETH Zürich / PI_31-A-0002 / CC BY-SA 4.0.

distribution stages, communicating extensively, and training other teachers to use films in class, they found a way to bring the responsibility of school screenings into the hands of teachers themselves.

Even though teachers in Geneva and those from the German part of Switzerland faced distinct challenges, they had similar objectives. By refusing to attend a group screening or creating an association that tried its best to listen to the needs of those in the field, teachers generated strategies to defend their expertise and agency when using films in schools. This defence was a crucial step in remaining free to apply the methods they favoured and able to design their lessons to best suit their students.

Designing Lessons Around Screenings in Cinemas

Not only did teachers help guarantee adapted film screenings by defending their authority in the face of external obstacles, they also tried to teach with film in a way that was flexible enough to be easily adapted to different scenarios. Despite facing distinct constraints, Ehrler and the SAFU teachers developed teaching practices that left space for student expression and activity. Ehrler's solution was to carefully develop lessons in relation to the films and to focus on the work in class, as that was the space where student expression was possible.

Fig. 5.2. Poster made by André Ehrler about the film *The Yellow Cruise*, 1939. AEG Archives privées 95 2.5.18.

Two of Ehrler's agendas and approximately fifty posters he made with his students in the late 1930s have been preserved and help to document his teachings. At that time, the school cinema service was no longer operational and the central hall was unused, having been declared too dangerous. However, the school authorities had agreements with local cinemas, and it is in this context that André Ehrler occasionally watched films with his class.[66] The films were feature-length documentaries or works of fiction with an instructive potential, typically for geography lessons. Over time, Ehrler and his classes saw *Nanook of the North* (Robert J. Flaherty, 1922), *The Yellow Cruise* (Léon Poirier, André Sauvage, 1934), *Elephant Boy* (Robert J. Flaherty, Zoltan Korda, 1937), *The Story of Louis Pasteur* (William Dieterle, 1936), and *Tundra* (Norman Dawn, 1936), among others. As these long screenings in cinemas were similar to a frontal mode of teaching, Ehrler tried to foster the children's active involvement by planning a multiplicity of exercises to be made in class, which the posters document.

These posters made in conjunction with the screenings seem to have been both a support for learning and a demonstration of the results. They contain rich documentation assembled and annotated by the teacher (photographs, maps, stamps, etc.) and also display the students' creations: compositions, drawings, and grammar and math exercises. *The Yellow Cruise*, for example, was part of an extensive lesson on the different countries

66 AEG 1985 va 5.3.333.

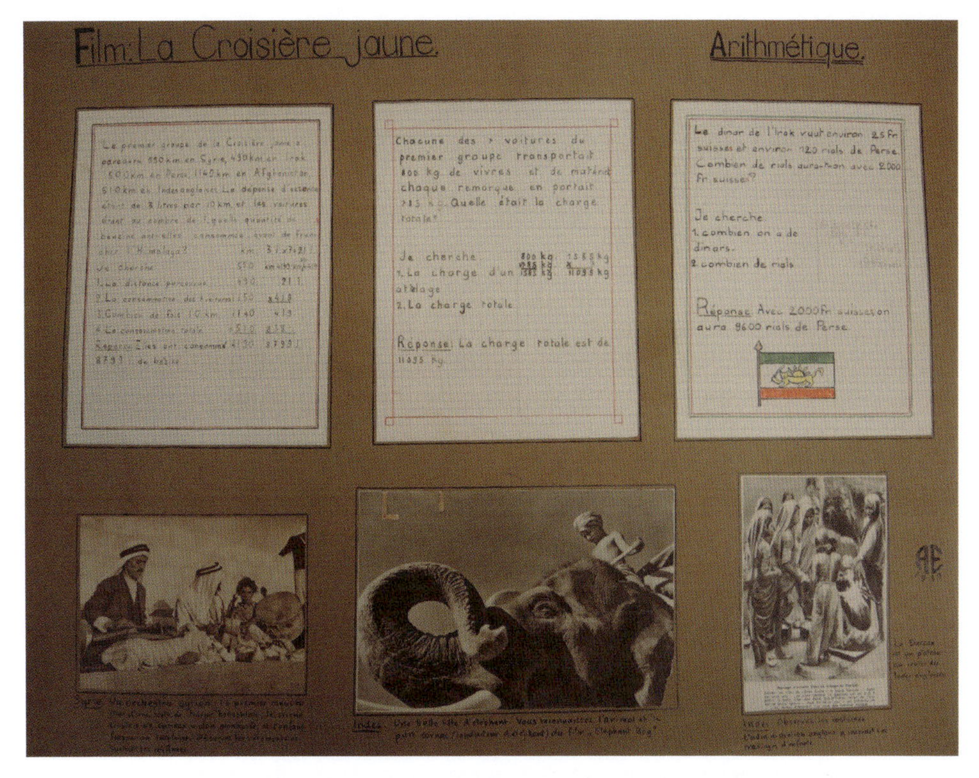

Fig. 5.3. Poster made by André Ehrler about the film *The Yellow Cruise*, 1939. AEG Archives privées 95 2.5.18.

crossed by the Citroën expedition described in the film. Through twenty-five posters, Ehrler worked with his students on the geography of the Asian continent with the help of maps, photographs, and press articles (Fig. 5.2–5.3). The children were led to do math exercises related to the topic and were also encouraged to draw sketches. Reaching across disciplinary lines, Ehrler centred his teaching on one subject that was of potential interest to his students, a film on an expedition through Asia, and hoped to thereby enable their active learning. As he underlined, it was critical that the films be close to the school programmes.[67]

This type of lesson was following one of the only practical recommendations progressive education theorists had provided. In *L'École active*, Ferrière explained that according to their age, children manifest interests of different natures and complexities and that these interests could be the starting point for any type of exercise.[68] During the 1924 pedagogical congress, teachers agreed that to encourage activity, one should facilitate 'the execution of works radiating from a *centre of interest*', regardless of the discipline.[69] Ehrler seems to

67 Undated manuscript, AEG AP 95.2.2.4. 'It is in the close cohesion between film projection and the subject matter of school curricula that the real method lies.' / '*C'est dans l'étroite cohésion entre la projection cinématographique et la matière des programmes scolaires que réside la véritable méthode.*'

68 Ferrière, *L'École active*, pp. 330–31.

69 Richard, *XXIe Congrès de la Société Pédagogique de la Suisse romande*, p. 47. 'L'exécution de travaux rayonnant d'un *centre d'intérêt*.'

have followed this recommendation, as his teaching strongly capitalized on the children's interest in the cinematic medium and organized the transmission of different kinds of knowledge around it (mathematics, grammar, drawing, etc.). Ehrler's agendas suggest he also made regular use of other media like the epidiascope (a projector for still opaque images) or the radio. While film remained central, the way he designed his lessons around the students' interests, and his alternate use of several media suggests that he tried to create a dynamic space for student expression.

Leaving Space for Individuals in the Classroom

In the German part of the country, the SAFU acted on the medium itself and recommended the use of film in ways that would make space for the people in the classroom. The cooperative was committed to allowing flexible uses of its films and did not want to standardize any fixed application of the films. However, they also had a very precise vision of what their films should be. For them, classroom films needed to be short, silent and focus on what the medium was specifically good for: showing movement. This distinctive form almost automatically called for specific uses.

In a brochure containing typical lessons edited by one local SAFU group, we find examples of how five of their early films could be used.[70] The example of *Die Lachmöwe* (*The Black-Headed Gull*, 1930), a film they considered emblematic of their vision, shows how their conception differed from large group screenings in cinemas.[71] The SAFU recommended first having a discussion of what the students already knew about gulls. The class could then go for an excursion to the lake, where the teacher would direct their attention to certain details: what the gulls ate, how they behaved, how they flew. The children could also try to imitate the bird's call. After this excursion, several activities could be done in class, starting with a comparison of the students' observations, a composition on a topic given by the teacher, or drawings and cuttings of birds. Then, migration and the importance of nature preservation could be discussed. It was also recommended to show the slides that the SAFU rented with the film that illustrated certain details like the eggs or the nest. The children could comment verbally on these images. After this, the film was screened once, discussed, and then shown a second time to clarify remaining misconceptions. Finally, another composition could be written on a subject of the students' choice.

Just as with André Ehrler's methods, the SAFU provided a profusion of tools and activities that were combined within a lesson. Film was but one of these tools, and it was only used complementarily to show movement and provide a closer view than what is possible in nature (Fig. 5.4). This enumeration of activities was primarily meant to present all the possible ways to pedagogically frame a film and was most likely never meant to be applied integrally. However, this shows that with a silent, thematically adapted fifteen-minute film, teachers had many possibilities and students had space to express themselves, follow their interests, be active, and learn.

---

70 Arbeitsgemeinschaft für Filmunterricht, *Fünf Lehrproben aus der Praxis der Filmverwertung* (Zurich, May 1931).

71 Ibid., pp. 10–18.

Fig. 5.4. Still from the SAFU film *Die Lachmöwe*. Lichtspiel Kinemathek Bern.

One of the SAFU's founding members, schoolteacher Gottlieb Imhof, insisted on student participation in their own knowledge acquisition, which he thought was a distinctive feature of progressive education. According to him, teacher and students needed to collaborate in order to use the film in a way that would serve their goals:

> Even the best educational film is only raw material in the hands of the creative teacher. He should always be the artist who, together with the young people entrusted to him, brings out the best in the film. Film education is a working school [*Arbeitsschule*] in the best sense of the word.[72]

The SAFU encouraged teachers to tailor films to fit their needs by discussing the films, pausing them to comment on the details, screening them twice or leaving them for the end of the lesson. These actions suggest that the medium could be customized to become more 'heterochronic', allowing those in the classroom to adapt the film to their needs and curiosities.

---

72 Imhof, Gottlieb, 'Die Bedeutung des Schmalfilmes für die Schulkinematorgaphie', *Der Bildwart*, 12 (December 1929), 769–78 (p. 772). '*Auch der beste Lehrfilm ist nur Rohmaterial in der Hand des schöpferischen Lehrers. Dieser soll immer der Künstler sein, der mit der ihm anvertrauten Jugend aus dem Film in gemeinsamem Erleben und Erarbeiten das Beste erst herausholt. Filmunterricht ist Arbeitsschule im besten Sinne des Wortes.*'

Conclusion

In this chapter, I have argued that analysing the links between progressive education and early school film screenings in terms of theoretic filiation, as Kessler and Lenk do, neglects the major role of teachers, who had to negotiate material and institutional constraints in order to make their pedagogic arguments heard.

Even though the Swiss context, devoid of national educational or cinematographic policies, is undoubtedly specific, my analysis has shown the necessity of taking into consideration the various obstacles that complicate the conceptual correspondences between pedagogical theories and teaching *dispositifs*. The work done by teachers becomes central, as they navigate the contemporary pedagogical debates, the arrival of a new teaching aid, and the constraints linked to their institutional affiliation.

Teachers' agency on a pedagogical level tended to be limited by the DPIs, for whom financial and logistic aspects prevailed, by the film industry which largely disregarded school programmes, and finally by the films themselves, as industrial and heavily standardised tools. By vocally opposing these systems, being involved in classroom film production, and developing in their lessons screening strategies that preserved considerable space for their students' expression, teachers acted to develop uses of film in class that allowed them to apply their favoured teaching methods.

It will probably remain impossible to determine to what extent teachers thought their teaching practices were examples of progressive education. The last sections have shown the work done by some teachers to design lessons with film that would foster their students' active engagement. It seems clear at least that the teachers who were invested in the development of school cinema set up a system and practices that were flexible enough to allow the principal tendencies of progressive education to unfold: a child-led teaching practice guaranteeing the student's active engagement.

Objects and Spaces of Change

NELLEKE TEUGHELS

Teachers' Agency and the Introduction of New Materialities of Schooling: The Projection Lantern and Classroom Transformations in Antwerp Municipal Schools, *c.* 1900–1940

Introduction

'Are we still in time to talk about lantern slide projection?', Antwerp teacher Hendrik van Tichelen asked his fellow members of the Antwerp teachers' association 'Diesterweg' in the summer of 1912.[1] His question was prompted by the growing popularity of both commercial cinema and educational film, an evolution he observed with interest, but also with concern. Although he was convinced that the projection of still images using an optical lantern would leave children with a lasting impression and thus raise educational effectiveness, van Tichelen, like many educators at the time, argued that moving images could have a detrimental effect on children's eyes and nerves.[2] Moreover, he feared that film projection would take the place of the optical lantern in the classroom even before the latter had been successfully incorporated into teachers' classroom routine.

Yet the educational lantern projection had had a thirty-year head start on educational film. Attempts to introduce the optical lantern in Belgian education had begun in the final quarter of the nineteenth century. In 1875, the Belgian liberal *Ligue de l'Enseignement*,[3] at

* This chapter was written as part of the research project 'B-magic. The Magic Lantern and Its Cultural Impact as Visual Mass Medium in Belgium (1830–1940)'. 'B-magic' is an Excellence of Science project (EOS-contract 30802346, www.b-magic.eu), supported by the Research Foundation Flanders (FWO) and the Fonds de la Recherche Scientifique (FNRS). I would like to thank my colleagues of the Research Group Cultural History since 1750 at KU Leuven for providing insightful feedback and valuable suggestions on earlier versions of this chapter.

1 Van Tichelen, Hendrik, 'De goede weg op', *Ons Woord*, 8–9 (1912), 246–52 (p. 246).

2 'De goede weg op', pp. 246–47; Lefebvre, Thierry, 'Une "maladie" au tournant du sciècle; la "cinématophtalmie"', *Revue d'histoire de la pharmacie*, 81, no. 297 (1993), 225–30. See also: Deckx, H., 'De kinema-plaag', *Het Katholiek Onderwijs*, 33 (1911–12), 545–47; V. R., 'Het kinema-gevaar', *Het Katholiek Onderwijs*, 35 (1913–14), 332–36; A. V. C., 'De Kinematograaf in de School', *Ons Woord* 30, no. 4 (1923), 80–83.

3 The *Ligue de l'Enseignement* was created in 1864 by progressive liberals, and the Brussels freemasons in particular, to discuss and foster the secularization and expansion of Belgian public education. Alexis Sluys, *Mémoires d'un pédagogue* (Bruxelles: Éditions de la Ligue de l'Enseignement, 1939), pp. 55–86; Tyssens, Jeffrey, *Om de*

Nelleke Teughels • Geheugen Collectief, nelleke@geheugencollectief.be

Learning with Light and Shadows: Educational Lantern and Film Projection, 1860-1990, ed. by Nelleke Teughels and Kaat Wils, TECHNE-MPH, 8 (Turnhout, 2022), pp. 145-167

10.1484/M.TECHNE-MPH-EB.5.131498

the instigation of the liberal-thinking politician Charles Buls, had established a secular model school for primary education in Brussels. The model school served as a testing ground for new pedagogical principles, the so-called 'active-intuitive method'. Whereas older pedagogical principles had favoured an exclusively textual mediation of knowledge through books and lectures, the members of the *Ligue de l'Enseignement* advocated a more immediate and sensory style of learning.[4] The premise was that knowledge was transferred more effectively when children were stimulated to observe and study natural phenomena, processes and objects using all of their senses.[5] When direct observation was impossible, models, illustrations and projected images were considered good alternatives.[6] The school architecture reflected these pedagogical and didactic principles and included a large projection hall and a music hall which could also be used for illustrated lectures.[7] In the final decades of the nineteenth century, the more visually oriented didactic of the active-intuitive method established itself in Belgium as an antidote to the passivity and inefficiency that supposedly characterized text-oriented teaching methods.[8] Policymakers and a growing number of teachers joined educational reformers in their quest to transform the classroom from a language-oriented to a multisensorial learning environment. However, despite repeated promotional efforts by the Education Ministry, ardent pleas by educational reformers,[9] and a great willingness among his fellow Antwerp teachers, in 1912 van Tichelen regretfully had to conclude that the educational use of the optical lantern still fell short of expectations.

As this article will demonstrate, the incorporation of lantern slide projection into teachers' routines would not be made a reality simply by providing schools with a lantern, or by the willingness of teachers to change classroom practice alone. It required a physical transformation of the classroom, in the sense of installing a projection screen, blackout curtains, a projector stand and electricity (Fig. 6.1), or, alternatively, necessitated the construction of a designated room for light projection. And whereas many of the ideas propagated by Buls' model school had become widely accepted by the beginning of the twentieth century, the inert nature of school infrastructure was at odds with faster innovation cycles in teaching technology.

The efforts of policymakers, educators and commercial media producers to introduce new, high-profile educational technology such as film, radio, television, and computers

---

schone ziel van't kind… Het onderwijsconflict als een breuklijn in de Belgische politiek (Ghent: Provinciebestuur Oost-Vlaanderen, 1998), pp. 55–79.

4 *Mémoires d'un pédagogue*, pp. 55–86; Grootaers, Dominique, 'Belgische schoolhervormingen in het licht van de 'Éducation Nouvelle'(1870–1970)', *Jaarboek voor de geschiedenis van opvoeding en onderwijs* (2001), 9–33 (pp. 9–15).

5 Grootaers, p. 14.

6 *Mémoires d'un pédagogue*, pp. 55–86; Grootaers, p. 14; Depaepe, Marc and others, *Order in Progress: Everyday Educational Practice in Primary Schools, Belgium, 1880–1970*, Studia Paedagogica, 29 (Leuven: University Press, 2000), p. 55.

7 Ligue de l'Enseignement, *Notice sur les travaux de la Ligue de l'Enseignement et sur l'école modèle fondée par cette association* (Bruxelles: Alliance Typographique, 1878), p. 21; Jurion-de Waha, Françoise, 'Architecture scolaire à Bruxelles', *Bruxelles Patrimoines*, 1 (2011), 7–23 (p. 13).

8 Grootaers, pp. 9–15.

9 Egelmeers, Wouter, and Nelleke Teughels, 'A Thousand Times More Interesting. Introducing the Optical Lantern into the Belgian Classroom, 1880–1920', *Journal for the History of Education* (2021) <https://doi.org/10.1080/0046760X.2021.1918271> [accessed 7 August 2021].

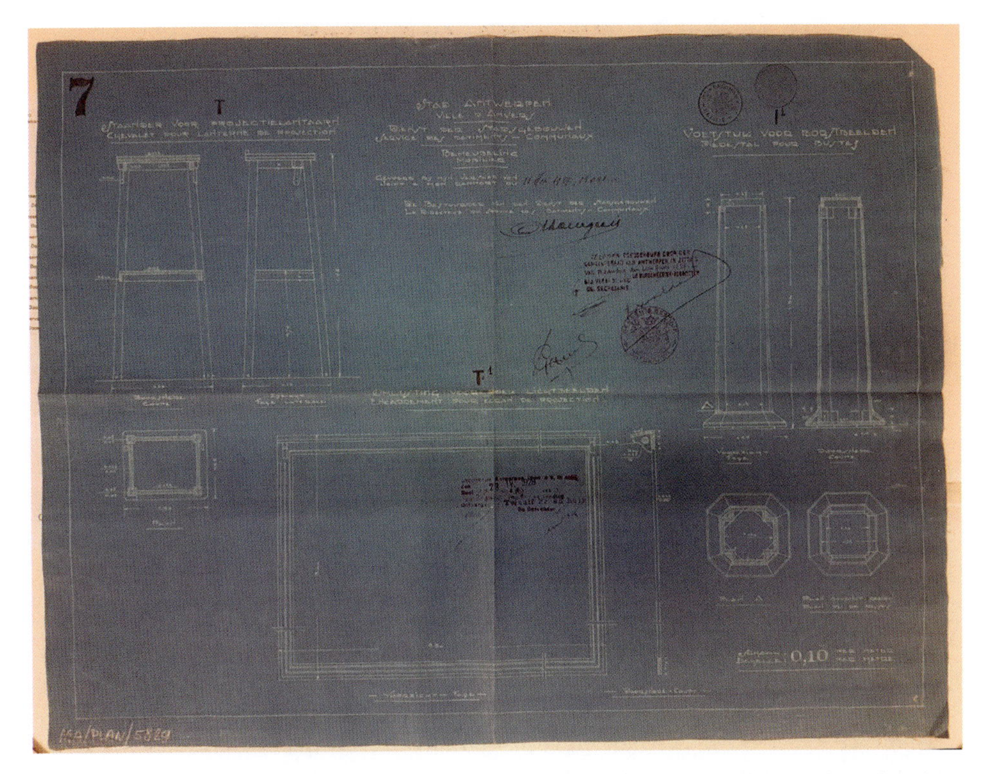

Fig. 6.1. Blueprint for the construction of a projection screen, a projector stand and a column for a bust for the boys' school in Napelsstraat, Antwerp. Design by the Antwerp city buildings service, 1927. Felix-archief Antwerpen (FA) 697 # 5829.

in Western education have been studied by several media historians and historians of education.[10] Some of these studies specifically looked at how popular media technologies, initially associated with entertainment and leisure, were legitimized for instructional purposes.[11] Most of the existing historical scholarship has focused on the discourse and priorities of the industry, policymakers and/or educational reformers.[12] However, in

10 I.a. Bianchi, William, *Schools of the Air: A History of Instructional Programs on Radio in the United States* (Jefferson, N. C.: McFarland & Co, 2008); Cuban, Larry, *Teachers and Machines: The Use of Classroom Technology Since 1920* (New York: Teachers College Press, 1986); Cunningham, Peter, 'Moving Images: Propaganda Film and British Education 1940–1945', *Paedagogica Historica*, 36.1 (2000), 389–406; Orgeron, Devin, Marsha Orgeron, and Dan Streible, *Learning with the Lights Off: Educational Film in the United States* (New York: Oxford University Press, 2012); Peterson, Jennifer, '"The Five-Cent University": Educational Films and the Drive to Uplift the Cinema', in *Education in the School of Dreams: Travelogues and Early Nonfiction Films* (Durham: Duke University Press, 2013), 101–36; Quillien, Anne, *Lumineuses projections. La projection fixe éducative* (Chasseneuil du Poitou: Réseau Canopé, 2016); and Taggart, Robert J., 'The Promise and Failure of Educational Television in a Statewide System: Delaware, 1964–1971', *American Educational History Journal*, 34.1–2 (2007), 111–22.

11 Cain, Victoria E. M., 'Seeing the world. Media and vision in US geography rooms, 1890–1930', *Early Popular Visual Culture*, 13.4 (2105), 276–92; Day Good, Katie, 'Sight-Seeing in School: Visual Technology, Virtual Experience, and World Citizenship in American Education, 1900–1930', *Technology and Culture*, 60.1 (2019), 98–131.

12 e.g. Bak, Meredith, 'Democracy and Discipline: Object Lessons and the Stereoscope in American Education, 1870–1920', *Early Popular Visual Culture*, 10.2 (2012), 147–67; Fuchs, Ekhard, Anne Bruch, and Michael Annegarn-Gläss, 'Educational Films: A Historical Review of Media Innovation in Schools', *Journal of Education Media*,

recent years, the materiality of schooling has received increasing attention from historians. Whereas up until the early 2000s the artefacts associated with the classroom were mostly treated as subsidiary, there is now growing interest in the various ways in which objects have altered classroom routines and necessitated changes to the physical learning space. In their edited work from 2005, Martin Lawn and Ian Grosvenor focused on the question of 'how objects arrive in school, how they exist there and what happens to them' in order to shed light on less visible aspects of the history of education.[13] The contributions in their volume explore the ways in which classroom artefacts are given meaning, how they are used and their role in closely connected networks of people, objects and practices. They pay attention to how shifting demands on schooling from the middle of the nineteenth century onwards resulted in new artefacts and specialist places to use and/or store them. The important book by Estrid Sørensen, *The Materiality of Learning*, adopts a posthumanist stance to investigate how the material culture of schooling uses humans and influences educational practice.[14] Likewise, the volume edited by Marc Depaepe and Paul Smeyers aims to further explore this relationship between individuals and the material culture of education to gain an insight into the hitherto mostly hidden aspects of the history of schooling.[15] In 2012, Henning Schmidgen studied the architectures and technologies of the lantern projections by the German physiologist and pharmacologist Carl Jacobj (1857–1944) in order to broaden our understanding of how the means of producing images transformed both the teaching space and the viewing experience.[16]

Nevertheless, the materiality of educational technology, how this materiality was linked to changing views on educational purposes, and the various ways in which it necessitated changes to the spaces of education have remained underexplored areas of study. Moreover, very few studies have addressed the question of teachers' role in the implementation of innovative teaching aids and the introduction of the new materialities of schooling. This paper seeks to shed light on some of the imbrications between projection equipment, school infrastructure, pedagogical considerations, government policy and teachers' agency that influenced the educational use of optical lantern projection in Belgian secular education. In order to do this, it analyses the discourses and practicalities surrounding the introduction and use of the optical lantern in Antwerp municipal schools during the first four decades of the twentieth century. This study takes into account the normative discourse of educational media advocates and governmental actors. It investigates the Antwerp city council's views on lantern projection in municipal schools and its actions

---

8.1 (2016), 1–13; Vignaux, Valérie, 'Le film fixe Pathéorama (1921) ou généalogie d'une invention', *Trema*, 41 (2014), <https://journals.openedition.org/trema/3128> [Accessed 3 January 2020]; Willis, Artemis, 'Between Nonfiction Screen Practice and Nonfiction Peep Practice: The Keystone "600 set" and the Geographical Mode of Representation', *Early Popular Visual Culture*, 13.4 (2015), 293–312.

13 Lawn, Martin, and Ian Grosvenor, eds, *Materialities of Schooling: Design, Technology, Objects, Routines* (Oxford: Symposium Books, 2005), p. 7.

14 Sørensen, Estrid, *The Materiality of Learning. Technology and Knowledge in Educational Practice* (Cambridge: Cambridge University Press, 2009).

15 *Educational Research: Material Culture and Its Representation*, ed. by Marc Depaepe, and Paul Smeyers (Cham: Springer, 2014).

16 Schmidgen, Henning, 'Cinematography without Film: Architectures and Technologies of Visual Instructon in Biology around 1900', in *The Educated Eye: Visual Culture and Pedagogy in the Life Sciences*, ed. by Nancy Anderson, and Michael R. Dietrich (Hanover: Dartmouth College Press, 2012), pp. 94–120.

to encourage the uptake of this instructional technology, but also the discourse of school directors and teachers on the problems associated with, opportunities offered by and further implications of incorporating lantern projection into daily classroom practice. This case study was chosen on the one hand because it is exceptionally rich in source materials. On the other hand, it is of particular interest because it can offer an insight into the extent to which Belgium's culture war (see below) between liberals and Catholics, in which education was the crucial area of disagreement, impacted on the introduction of these new materialities of schooling. At a national level, the Catholic political party held an absolute majority between 1884 and 1918 and would continue to largely dominate Belgian politics during the interwar years. In Antwerp, by contrast, until the First World War, education policy was shaped by the liberals.

Theory and Methodology

This paper agrees with Marc Depaepe, Frank Simon and Pieter Verstraete that the artefacts and spaces of schooling — like other historical objects and constructions — should be considered as mediating agencies that take on meaning from their organizational, social and cultural relationships and that they moreover have an effect on those relationships.[17] In addition, this study follows the assumption put forward by Sørensen that 'new as well as already established technologies take part in and contribute to forming school practices [...]'.[18] This kind of approach takes into account the agency of material culture and of the people using, modifying or working around it. This is an important advantage over other frameworks, such as Foucault's disciplining concept, which tend to present the introduction of new artefacts in education as a top-down intervention that was imposed on passive teachers and their pupils. In addition, the concept of 'the grammar of schooling', as posited by David Tyack, William Tobin and Larry Cuban, also occupies a central place in this research.[19] Those authors demonstrated how the paradox between the long-lasting, unchanging core elements of schooling — the internal dynamics of classroom practice — and the external development of innovative ideas and social transformation resulted in new pedagogical theories inevitably being absorbed into existing and dominant practices of teaching.

In order to take into account artefacts' and teachers' agency, as well as the difficulties teachers encountered in implementation, the findings presented in this paper are based on a wide variety of archival sources. Information on Antwerp municipal teachers' stance on visual media use in general and lantern slide projection in particular was found in *Ons Woord* (1894–1961), the monthly periodical published by the Antwerp municipal schoolteachers'

17 Depaepe, Marc, Frank Simon, and Pieter Verstraete, 'Valorising the Cultural Heritage of the School Deks through Historical Research', in *Educational Research: Material Culture and Its Representation*, ed. by Paul Smeyers, and Marc Depaepe (Cham: Springer, 2014), 13–30 (p. 18).

18 Sørensen, p. 3.

19 Tyack, David B., and William Tobin, 'The "Grammar" of Schooling: Why Has it Been so Hard to Change?', *American Educational Research Journal*, 31.3 (1994), 453–79; Tyack, David B., and Larry Cuban, *Tinkering Toward Utopia. A Century of Public School Reform* (Cambridge: Harvard University Press, 1995).

association *Diesterweg*.[20] The association, founded by free-thinking liberals, not only represented teachers' interests, but also strove to improve the wellbeing of pupils. It sought to actively engage in the battle for free, neutral (secular), scientific and compulsory education. One of its key achievements was the establishment of an 'Improvement Council' in 1907, which was charged with the task of investigating the pedagogical and practical problems faced by municipal schoolteachers. It advised the *schepen* (alderman) with responsibility for education on how to improve the municipal schools in terms of teaching methods, teaching aids, infrastructure, etc.[21] The minutes of the Antwerp Improvement Council are kept in the Antwerp city archive (Felixarchief). In addition, we can learn more about what Antwerp municipal school principals and teachers considered to be their schools' best material assets, their most serious defects and what was on their 'wish list', from the detailed responses to a survey conducted by the city council in 1923 and follow-up surveys in the 1930s, which are also kept in Felixarchief, as well as numerous municipal school construction and transformation plans. The correspondence between municipal school principals or teachers and the Antwerp city departments and council offers an insight into educators' views on and concerns about teaching aids and school infrastructure. The writings of Hendrik van Tichelen, a member of *Diesterweg* and fierce campaigner for the use of more modern visual teaching aids, are also an important source of information, as he closely followed and documented all developments in the availability of and problems concerning visual media technology in Antwerp municipal schools. We compare the findings from these sources with the normative discourse of Belgian educational media advocates, government recommendations in triennial inspection reports, catalogues of recommended schoolbooks, teaching aids and materials, and the views and actions of the Antwerp city council regarding the use of visual teaching aids in general, and the optical lantern in particular.

Classroom Culture War

In 1830, Belgium was born out of a compromise between a new liberal industrial bourgeoisie on the one hand and a Catholic party, backed by the landed nobility and a large part of the rural population, on the other. However, their opposing views on the principles of social order soon resulted in a democratic system that was based on the principle of 'pillarization'.[22] Both Catholics and liberals wished to see their visions imposed on the wider population and to bind the Belgian state to their own project. Education was one

20 'Diesterweg' was founded in 1892 and named in honour of the German pedagogue Friedrich Adolph Wilhelm Diesterweg (1790–1866), who campaigned for the secularization of schools and who was convinced that schools could be a major force for social change. De Vroede, Maurits, and others, *Bijdragen tot de geschiedenis van het pedagogisch leven in België in de 19de en 20ste eeuw. Deel II: De periodieken 1878–1895* (Leuven: KU Leuven, 1974), pp. 100; 565–96.

21 *Bijdragen, Deel II: De periodieken 1878–1895*, pp. 565–96.

22 Belgian society was pillarized in the sense of being segregated in every domain along ideological lines. There were two main 'pillars' in nineteenth and early twentieth-century Belgium: Catholics and free-thinking liberals, who had separate political parties, published and read separate newspapers, had their own education systems, etc.

of the main battlegrounds in this culture war. From the middle of the nineteenth century onwards, progressive liberals started a campaign to reconstitute education on the basis of their own rationalist beliefs in science as a driving force of individual and societal progress. Public secular education, they argued, should accord an important place to science in the curriculum and be accessible to all. The direct observation of the active-intuitive method would contribute to a valorisation of objectivity and the adoption of common positions. As a result, there would emerge a new, collective consciousness: reason, it was believed, was the basis of universal morality.[23]

With these ideas and goals in mind, the progressive liberals created the *Ligue de l'Enseignement* in 1864. It was to become a highly influential pressure group, greatly shaping the liberal education policy. In 1871 it presented its *Projet d'Organisation d'Enseignement Populaire*, advocating free and secular instruction that would introduce the general public to the scientific method with the aim of social emancipation. When the liberal party won the elections in 1878, the Minister of Education, Pierre Van Humbeéck, set out to modernize and secularize primary and secondary education following the principles formulated in the *Projet* and put into practice in the model school. Members of the *Ligue* teaching in municipal schools in Brussels were quick to adopt the method, which was also propagated by the Brussels normal school.[24] The liberal Antwerp city council followed suit and strove to expand, modernize and democratize Antwerp municipal education, taking their inspiration from Brussels.[25]

In 1881, to facilitate the more visually oriented and experiential method of learning, the liberal Government provided all state normal and secondary schools with a wide selection of educational media and instruments deemed necessary for teaching natural history, chemistry and physics in an active and intuitive way. This included an optical lantern, using a kerosene lamp or limelight as a light source. Lantern slides, however, were not included. From the sources it remains unclear whether the lantern was intended for use as a teaching aid or rather as an object of study in itself.[26]

Soon after, in 1884, the Catholic party swept back into power with a landslide victory and would remain in power until 1914. In the new curriculum that was introduced as soon as the Catholic government took office, physics was made an optional subject. In addition, the Catholics made deep cuts in the generous subsidies for teaching aids that had

---

23 Grootaers, pp. 11; 17–20; Witte, pp. 118–19.

24 Grootaers, p. 14. A *normal school* is an institution for the training of teachers by educating them in the norms of pedagogy and curriculum. For further reading on normal schools, see for example: Dhondt, Pieter, 'Teacher Training Inside or Outside the University: the Belgian Compromise (1815–1890),' *Paedagogica Historica*, 44.5 (2008), 587–605.

25 Van Daele, Henk, *150 jaar stedelijk onderwijs te Antwerpen 1819–1969* (Antwerpen: Stad Antwerpen, 1969), pp. 24–25; 57.

26 Van Humbeéck, Pierre, 'Instruction moyenne. — Athénées royaux et écoles moyennes de l'État. — Acquisition d'objets pour l'Étude de l'histoire naturelle', *Bulletin du ministère de l'instruction publique* 4 (1881), 30–47; De Burlet, J., 'Programme de l'enseignement à donner dans les écoles normales et les sections normales primaires de l'Etat', *Het Katholiek Onderwijs*, 14 (1892–93), 81–96; [Anon.], *Rapport triennal sur la situation de l'instruction primaire en Belgique. Quinzième période triennale: 1885–1886-1887* (Bruxelles: Imprimerie Gobbaerts, 1889), p. 87; [Anon.], *Rapport triennal sur l'état de l'enseignement moyen en Belgique présenté aux chambres législatives le 21 novembre 1900. Seizième période triennale, 1897–1898-1899* (Bruxelles: J. Goemaere, 1900), pp. cxxviii; 295; Verschaffelt, Eduard, 'L'enseignement intuitif à l'aide des projections lumineuses', *L'Abeille*, 27, no. 11 (1882), 469–75 (p. 469).

previously been granted to state-supported schools.[27] In line with this more frugal policy, the Education Ministry refrained from promoting the use of costly high-tech equipment such as the projection lantern during the last two decades of the nineteenth century.[28]

The liberal Antwerp city council (1872–1921) tried to compensate for these government retrenchments by increasing the budget for education every year. However, the new law on education issued by the Catholic government seriously hampered the further expansion of state-supported and municipal schools. Antwerp could no longer count on government support for building new schools, and new school construction plans were seldom approved.[29] As I will argue in more detail further in this text, it is especially this inertia in school infrastructure that might help explain why, on the eve of the First World War, optical lantern use in Antwerp municipal schools was still, as van Tichelen regretfully observed, the exception rather than the rule. This was despite the fact that, around the turn of the nineteenth century, even Catholic circles had come to favour a more sensorial and 'intuitive' approach to teaching and had recognised the optical lantern as a tool to modernize and optimize teaching.

This change of heart seems to have been induced by the writings and lectures of a handful of fervent lantern advocates, who also happened to be teachers at state-supported secular schools.[30] In November 1899, Education and Interior Minister Jules De Trooz issued a circular urging secondary schools to make use of the optical lantern in their lessons in history, geography and natural sciences.[31] Moreover, he commissioned a survey among state secondary schools to establish to what extent they already possessed the necessary equipment for lantern projection. Based on the resultant report, in 1901 the government decided to offer subsidies to schools to enable them to purchase the required materials.[32]

'What One Wants Is Not Necessarily Within Easy Reach'

Ons Woord had begun singing the lantern's praises in 1894. The publications by lantern advocates Véron De Deyne and the Brussels primary school teacher Pascal Mattot received brief but positive reviews in *Ons Woord*, in 1897 and 1899, respectively.[33] Initially, lantern projection was promoted as a useful tool for adult education, a key element in the liberal quest for public instruction as an instrument of emancipation. By 1901, however, the

27 [Anon.], *Rapport triennal sur l'état de l'enseignement moyen en Belgique présenté aux chambres législatives le 7 août 1889. Années 1885–1886-1887* (Bruxelles: Fr. Gobbaerts, 1890), pp. lxv; xcv.

28 In 1888, a conference session did discuss the use of the lantern in primary education but repeatedly stressed that it had to be cheap in order to be acceptable. [Anon.], *Rapport triennal sur la situation de l'instruction primaire en Belgique présenté aux chambres législatives le 1er avril 1892. Années 1888–1889-1890* (Bruxelles: J. Goemaere, 1892), pp. 446–47.

29 Van Daele, p. 25.

30 Egelmeers and Teughels; Dewilde, Jan, and Frederik Vandewiere, *Véron De Deyne 1861–1920* (Ieper: Stedelijke Musea, 2012), pp. 7–8.

31 *Rapport triennal sur l'état de l'enseignement moyen en Belgique 1897–1898-1899*, p. lxxxvi.

32 [Anon.], *Rapport triennial sur l'état de l'enseignement moyen en Belgique présenté aux chambres législatives le 13 avril 1904. Dix-septième période triennale. Années 1900–1901-1902* (Bruxelles: J. Goemaere, 1904), pp. xcii–xcvi.

33 De Deyne, Véron, *La lanterne de projections à l'école. Propagation de l'enseignement scientifique par les projections lumineuses* (Bruxelles: J. Lebègue & Cie, 1896); Mattot, Pascal, *Les projections lumineuses dans l'enseignement primaire* (Bruxelles: J. Lebègue et Cie, 1897).

association had become convinced that lantern slide projection would also be beneficial for primary education and expressed the hope that it would soon become a regular part of Antwerp municipal school teaching practice.[34] In that same year, the city council made a small contribution to the introduction of lantern projection in Antwerp municipal schools: it started organizing a yearly lantern show for all senior pupils of municipal primary schools. At first, these shows were held in the *Nederlandsche Schouwburg*, a large theatre in the city centre. From 1903 the city council, assisted by the *Kring voor Photographische en Wetenschappelijke Studiën* (Circle for Photographic and Scientific Studies), invited the senior pupils once a year on a Sunday morning to a school gymnasium or kindergarten playroom turned into a projection hall, where they were taken on a virtual trip around Belgium or abroad using lantern slides. This initiative was commented on with enthusiasm by *Ons Woord*, which argued that light images could breathe life into otherwise dull, abstract subject matter and help pupils see the world in a more nuanced and meaningful way:

> The geographical names from their atlas take on shape, colour and life. London, for example, is no longer a six-letter word, but a thing that can strike the eye in a hundred different ways. [...] The pupils become acquainted with the natural beauty and art forms of other countries. As a result of this, the widely spread problem, which makes people see every stranger as a barbarian, whose country can be nothing but an unsightly wilderness, whose life can be nothing but a succession of atrocities, finally disappears.[35]

It was therefore all the more unfortunate that none of the schools had the necessary infrastructure to organize lantern projection on a more regular basis. Despite de Trooz's 1901 initiative to provide subsidies to all schools that had the ambition to introduce the new technology into their classrooms, Antwerp's municipal schools remained deprived of modern projection technology. From *Ons Woord* it is clear that the Antwerp teachers' association harboured a lot of resentment towards de Trooz, accusing him of systematically over-financing Catholic schools to the detriment of secular educational institutions. Many articles in pedagogical journals, including *Ons Woord*, addressed the fundamental problem: that all the talk about increasing teachers' efficiency through the use of modern technology would yield no results unless the government were to finally provide them with the necessary financial means to organize such intuitive teaching.

This was a defensive reaction to the frequently heard accusation that teachers were simply unwilling to change their ways. Even in the late 1920s and early 1930s, government reports and teachers' associations still sometimes complained about the verbalism and passivity that supposedly characterized many lessons.[36] But although teachers were often made the scapegoat in such matters, van Tichelen asserted that they were not responsible. In 1912, writing about his fellow Antwerp municipal schoolteachers, he said: 'Each one of us fully appreciates the highly enjoyable and lively nature of lectures with light projection

34 [Anon.], 'Een kwestie van waardigheid', *Ons Woord*, 8.1 (1901), 34.

35 [Anon.], 'Onderwijs bij middel van lichtbeelden', *Ons Woord*, 10.6 (1903), 216–18 (pp. 216–17).

36 Vauthier, Maurice, *Rapport sur la situation de l'instruction primaire en Belgique présenté aux chambres législatives le 14 mars 1934. Vingt-neuvième période triennal 1927–1928-1929* (Bruxelles: Ministère des Sciences et des Arts, 1934), p. lxxxiii; De Keyzer, J., 'De actieve Methode bij het Onderwijs in de Geschiedenis', *Nova et Vetera*, 18.2 (1936), 210–23 (p. 210).

and also realizes how much our lessons would gain in attractiveness and staying power if every school possessed a magic lantern.'[37]

In 1907, the teachers' association *Diesterweg* had rejoiced when, after years of pushing for its creation, the city council established an Improvement Council for Antwerp's municipal schools. This would provide them with an ally, so they thought, to promote the use of modern teaching aids. The ultimate goal, according to *Ons Woord*, was to fight the one-sided focus on scientific facts of 'traditional' instruction. This, they argued, had to be complemented by an aesthetic and therefore also moral education, otherwise children would grow up possessing a great deal of factual knowledge, and yet still lack the ability to think and question. The optical lantern offered teachers many possibilities to cultivate pupils' aesthetic sensibilities, *Ons Woord* argued. Consequently, the fitting out of projection rooms in as many schools as possible was a matter to be raised urgently with the Improvement Council.[38]

But, as van Tichelen rightly pointed out a few years later, 'what one wants is not necessarily within easy reach'.[39] The Improvement Council would not debate the use of lantern slide projection until over four years later, in April 1912. And it would take yet another year before the alderman for education, Victor Desguin, followed up on the Improvement Council's (very brief) report on the matter.[40] Referring to the recommendations made by the Improvement Council, Desguin addressed a letter to the municipal school principals urging them to consider the use of lantern slide projection as a means of making instruction more intuitive and attractive and awakening and nurturing their pupils' aesthetic sensibility. Furthermore, he declared that:

> [the schools] can purchase such a lantern from their budget for equipment; the black curtains and the conduction of light will then be paid for from the budget for the school buildings.[41]

As a result, in September 1912 van Tichelen expressed optimism and announced that, as far as light projection for instruction was concerned, his hometown was 'on the right path'.[42] Two primary schools for boys had recently started to incorporate slide projection into the daily classroom routine, by turning the gymnasium into a projection hall. At a school for 'learning-disabled' boys in Zwartzustersstraat, the school's pupils and pupils of the school for 'learning-disabled' girls, located over a kilometre away in Bogaardenstraat, were instructed using light projection every two weeks from 1906 onwards. Here too, the gymnasium was darkened to double as a projection room. In addition, six more municipal

37 van Tichelen, Hendrik, 'De kinema als onderwijsmiddel', *Ons Woord*, 19.3 (1912), 79–82 (p. 80).

38 [Anon.], 'Ter bevordering van het schoonheidsgevoel in het onderwijs', *Ons Woord*, 15.4 (1908), 129–39.

39 'De goede weg op', p. 248.

40 Victor Desguin (1838–1919) was a Belgian liberal politician and alderman with responsibility for education in the city of Antwerp between 1892 and 1918. He founded nearly twenty primary schools and kindergartens and supported a series of socio-medical initiatives which were in line with the liberal idea of raising the working class by physically strengthening children at school. In addition, he was the first to open up schools to mentally challenged children. Vandendriessche, Joris, *Geneeskunde en politiek. De Antwerpse carrière van Victor Desguin (1838–1919)* (unpublished Master's thesis, KU Leuven, 2009), pp. 83–106.

41 Alderman for Education Victor Desguin to Antwerp school principals, 29 April 1913, Felixarchief Antwerp (FA) 480#3435.

42 Ibid.

primary schools had acquired projection lanterns by 1912, although it is not clear how often they actually used them. When three municipal normal schools were founded in 1911, lantern projection was commonly used to instruct the new teachers-to-be.[43] Lantern projection was also used on a more irregular basis, organized by school teachers who were lantern enthusiasts for the pupils of several classes or even from several municipal schools at the same time.[44] The lantern slides previously used for the annual lantern show for municipal school pupils — from 1907 replaced by film showings — could now be borrowed by schools that possessed a lantern but not the necessary lantern slides. But the teachers' association Diesterweg also supported van Tichelen's proposal to create a 'municipal school and educational museum', offering teachers a more extensive collection of lantern slides, projectors and other teaching aids to use during their classes.[45]

Of Empty Words and Empty Pockets

However, implementing optical lantern projection into regular municipal instruction not only required schools to acquire or borrow the projection material, but also to make the necessary changes to their infrastructure if they really wished to encourage the actual use of the projectors. The Brussels model school, with its own projection hall, had served as an example to the teachers' association for over thirty-five years now. Liberal educators in Antwerp had looked with admiration and envy at the developments in Brussels, where between 1875 and 1920 over fifty-five schools would be built following the architectural plan of the model school and where lantern projection was apparently well established. Yet it seems that, prior to 1910, none of the Antwerp municipal schools had been purposely built or rebuilt to facilitate teaching with projection technology.

Despite these issues, Belgian primary school buildings were considered to be exemplary by other countries. In 1878, in his publication on public schools in Belgium and the Netherlands, the Paris city architect Félix Narjoux highlighted Belgium's leading role in terms of school buildings.[46] And at home, between 1878 and 1914, the percentage of Belgian municipal school buildings rated 'decent' by the government education inspectorate rose to 93.5%.[47] This was no doubt a result of the law of 14 August 1873, which set aside a budget of 20 million Belgian francs for municipalities to fund the construction and equipping of school buildings, though with the proviso that government funding would only be made available if the construction plans were in line with the government building programme. This programme, dating from 1852 and updated in 1874 to include forty-five model plans, building cost estimates and building specifications, was published by Lambert Blandot,

43 'De goede weg op', p. 247.

44 Survey of municipal schools in Antwerp 1923. Schoolstraat, Jongensgemeenteschool no. 10. FA595#264.

45 'Ter bevordering van het schoonheidsgevoel', 129–39; Van Daele, pp. 267–68.

46 Narjoux, Félix, *Les écoles publiques. Construction et installation en Belgique et Hollande. Documents officiels services intérieurs et extérieurs — Bâtiments scolaires — Mobiliers scolaires — Services Annexes* (Paris: Ve A. Morel et Cie, 1878).

47 D'hoker, Mark, 'De lagere-schoolgebouwen in België in de 19de eeuw: een kwantitatieve, kwalitatieve en architectonische benadering', in *Liber amicorum Karel De Clerck*, ed. by Frank Simon (Gent: CSHP, 2000), pp. 43–58 (p. 52).

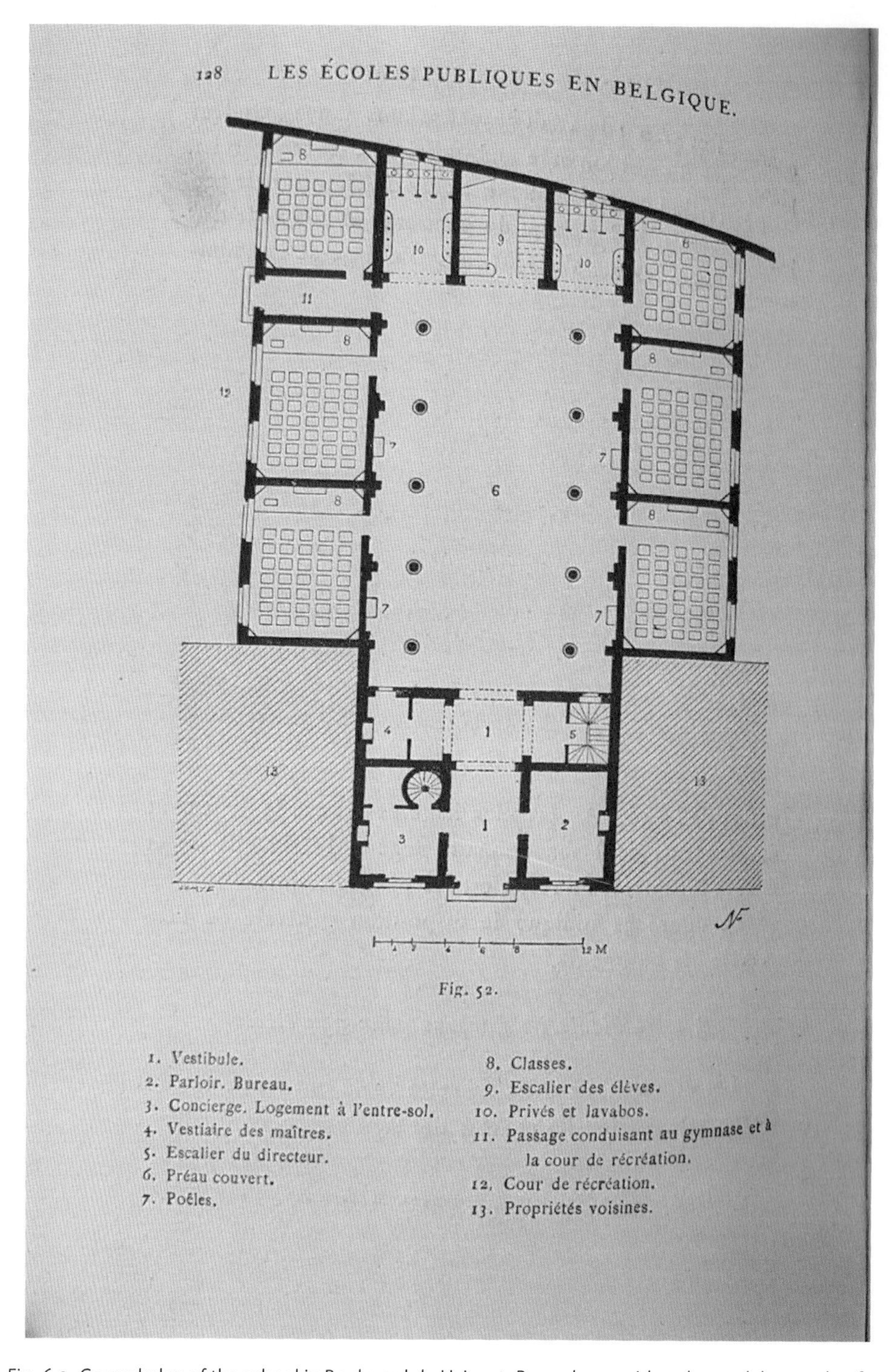
128 LES ÉCOLES PUBLIQUES EN BELGIQUE.

Fig. 52.

1. Vestibule.
2. Parloir. Bureau.
3. Concierge. Logement à l'entre-sol.
4. Vestiaire des maîtres.
5. Escalier du directeur.
6. Préau couvert.
7. Poêles.
8. Classes.
9. Escalier des élèves.
10. Privés et lavabos.
11. Passage conduisant au gymnase et à la cour de récréation.
12. Cour de récréation.
13. Propriétés voisines.

Fig. 6.2. Ground plan of the school in Boulevard de Hainaut, Brussels, considered a model example of an urban school by Felix Narjoux. Narjoux, Félix, *Les écoles publiques. Construction et installation en Belgique et Hollande. Documents officiels services intérieurs et extérieurs — Bâtiments scolaires — Mobiliers scolaires — Services Annexes* (Paris: Ve A. Morel et Cie, 1878), p. 128.

a private architect from the town of Huy.[48] Most attention in the book was given to the pedagogical shift away from the Lancasterian method to frontal, teacher-led lessons. The Lancasterian method, also known as the monitorial or mutual system, had been designed by the British educator Joseph Lancaster as a cheap way to make primary education more inclusive. The brightest children were trained by the teacher to be class monitors, whose task was to deliver simple lessons to their peers and maintain order. In this way one teacher could school hundreds of children in one, large classroom.[49] The teacher's desk was placed on a platform at the far end, while long benches and desks for the pupils occupied the middle of the room. Each desk contained one class, supervised by a monitor.[50] The Belgian Education Law of 1842 introduced a shift to the frontal, teacher-led lessons that are still prevalent today. This necessitated the construction of several smaller classrooms and the introduction of new educational tools and furniture, such as the blackboard.[51] Initially, the existing schools were divided into smaller spaces using partitions, but following the first Belgian government building programme of 1852, new public primary school buildings were built consisting of several smaller classrooms.[52] The updated programme of 1874 further reduced the number of children per classroom from seventy to between thirty and fifty, and devoted more attention to functional design and hygienic aspects, such as heating, ventilation, and sanitary installations.[53] Hygienic conditions and the demands resulting from a large and increasing school population were the most pressing issues dominating Antwerp school building design in the thirty years leading up until the First World War.[54]

The Antwerp schools that were built in the second half of the nineteenth century are characterized by a rather uniform design, largely corresponding to the school pictured in Figure 6.2, which was considered a model example of an urban school by Narjoux.[55] In general, schools had a symmetrical ground plan and consisted of six to twelve classes, each for thirty to fifty pupils. A central entranceway opened up into a large hall, which could be used as a vestibule or as a covered playground. Classrooms were located on either side of the central hall. These wings were preferably positioned perpendicular to the building line in order to reduce street noise. Plenty of natural light and good ventilation were considered vital for children's health, and the building programme therefore stipulated that easy-to-open windows should where possible be placed along the sides of the classroom. If this proved to be impossible, they should at least be placed on the left side of the students.[56] Like elsewhere in Belgium, Antwerp classrooms dating from this period had several large windows, placed around 0.80 cm apart, almost two meters wide and almost reaching the

48 Bertels, Inge, *Building the city. Antwerp 1819–1880* (unpublished doctoral thesis, Vrije Universiteit Brussel, 2008), pp. 297–99; D'hoker, pp. 48–49; 53–54.

49 Simon, Frank, Christian Vreugde, and Marc Depaepe, 'Lancasteronderwijs, meer dan een voetnoot in de Belgische pedagogische historiografie. Lager onderwijs in Brussel (1815–1875)', *Cahiers Bruxellois-Brusselse Cahiers*, 1 (2015), 30–55 (p. 39).

50 Bertels, p. 281.

51 Van Daele, pp. 46–47.

52 Bertels, pp. 293.

53 Bertels, pp. 293; 297; 305–07; D'hoker, p. 54.

54 Bertels, p. 305.

55 Bertels, p. 307.

56 Blandot, Lambert, *Instructions concernant la construction et l'ameublement des maisons d'école suivies de plans et de devis types* (Huy: Degrâce, 1875).

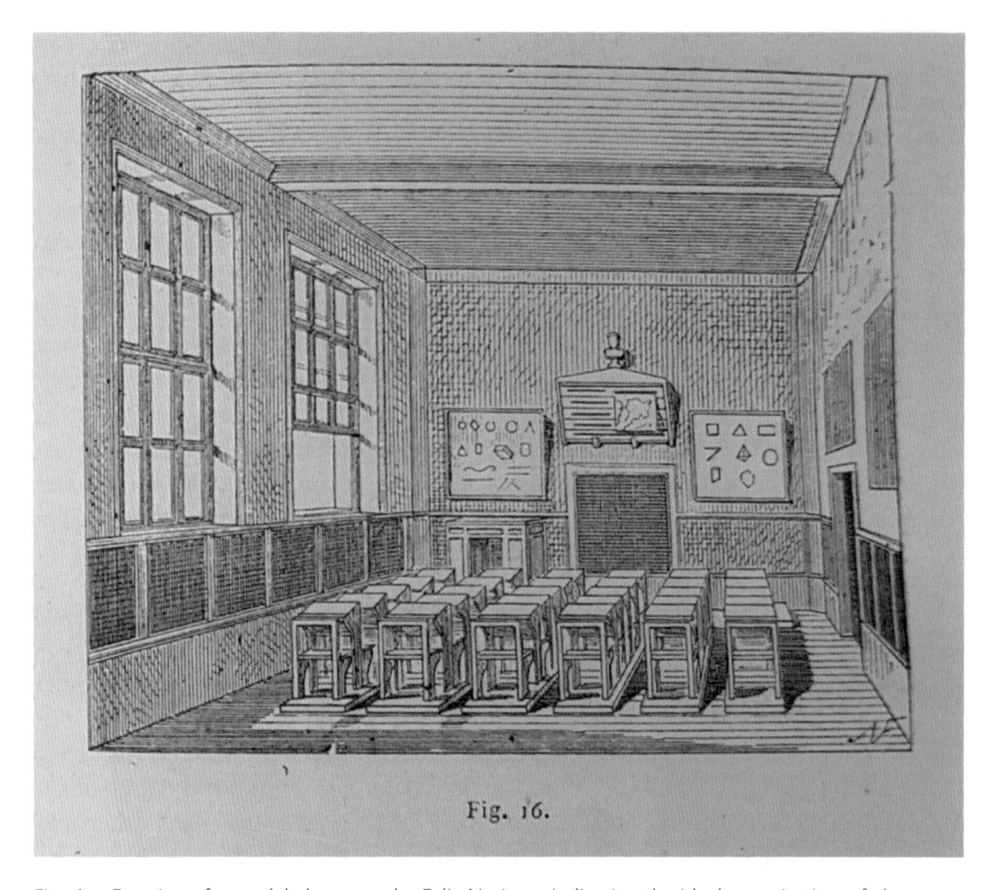

Fig. 6.3. Drawing of a model classroom by Felix Narjoux, indicating the ideal organisation of classroom space and the ideal dimensions of windows, furniture and the spaces in between. Narjoux, Félix, *Les écoles publiques. Construction et installation en Belgique et Hollande. Documents officiels services intérieurs et extérieurs — Bâtiments scolaires — Mobiliers scolaires — Services Annexes* (Paris: Ve A. Morel et Cie, 1878), p. 69.

4.5 metre-high ceiling (see Figure 6.3).[57] Bottom-up roller blinds in light grey cotton could be added to protect children from the hot summer sun without completely blocking the incoming light.[58] The growing attention for hygiene also led to new schools being built with a gymnasium. Alternatively, an exercise hall was erected in the immediate vicinity of existing schools.[59]

This preoccupation with hygienic conditions was clearly at odds with the material settings required for lantern projection. Moreover, despite the government's financial support, the budgetary implications of the new building programme for local authorities were still significant, and with limited building space available, a lecture hall specifically designed to allow lantern projection was seldom considered when municipal school construction plans were drawn. It is therefore no surprise that a commonly cited problem by Antwerp

57 Narjoux, pp. 41; 66–79.
58 Blandot, pp. 13; 58.
59 Bertels, pp. 313–14.

school teachers was the difficulty they encountered in trying to darken the classroom sufficiently to allow a clear projection of the images. This necessitated the installation of blackout curtains. In April 1913 J. Jaeckx, principal of the primary school in Antwerp's Zwartzustersstraat, raised the matter in a letter to the city administration and the alderman for education, Desguin.[60] Desguin asked the *Stadshoofdbouwmeester* (chief city architect) Alexis Van Mechelen to take action or offer advice on the issue. In his response dated 13 May 1913, however, Van Mechelen highlighted another issue that might need resolving, for he had encountered it elsewhere as well:

> [...] I have the honour of asking you whether the white roller blinds on which the images are to be displayed also have to be provided by my department. Until now, there has not been any mention of the acquisition of this white fabric; this is also the reason why in the two schools located in Indischestraat everything has been ready for several months now, but in the absence of the projection screen the lantern images cannot be shown.[61]

The alderman's response would moreover prove to be irrelevant, since one month later the chief city architect informed the city council that, contrary to what Desguin had communicated to the school principals earlier that spring, he did not have the financial resources to carry out the necessary transformations. At an estimated cost of three hundred Belgian francs per school, and even assuming that only about twenty-five of municipal schools would ask to have a classroom fitted with blackout curtains, white screens and other necessary equipment, the budget needed for these adaptations was 7500 francs. Moreover, this was a conservative estimate, for he was still awaiting the advice of the city engineer, Mr Kinart, regarding the cost of installing electricity.[62] The chief city architect decided he would ask for the sum of 15,000 francs to be included in the following year's budget in order to be able to carry out the necessary works in all the school buildings. However, a year later, he found that this request had not been granted.[63] All of this indicates that, even leaving aside the financial challenges, there was another common problem hindering the smooth implementation of lantern technology in Antwerp municipal education: the poor communication between the city council and the operational municipal services.

In the meantime, five other school principals had responded with enthusiasm and gratitude to Desguin's call and had written to the city council to ask for adaptations to be made to their school infrastructure from the school buildings budget. In June 1913, still unaware of the financial reality and other problems that would impede Desguin's initiative, *Ons Woord* also expressed great joy about the letter that had been received by the school principals. The (anonymous) author was delighted that the city council had decided to support municipal schools in introducing the optical lantern into the classroom, for it was a highly efficient teaching aid, more capable of fixing knowledge permanently in the young minds of pupils than any form of verbal instruction. An additional reason to rejoice was that the optical lantern offered the perfect antidote to the corruption of morals by films of dubious quality,

60 Principal J. Jaeckx of boys' school no. 21 to the city council, 23 April 1913. FA 480#3435.
61 Stadshoofdbouwmeester Alexis Van Mechelen to the city council, 13 May 1913. FA 480#3435.
62 Alexis Van Mechelen to the city council, 5 June 1913. FA 480#3435.
63 Alexis Van Mechelen to the city council, 10 June 1914. FA 480#3435.

a powerful weapon in the war on immoral motion pictures.[64] In his eagerness to 'turn theory into practice', the author estimated that at least fifty classrooms would need to be refurbished.[65]

Despite the mood of optimism, the author also expressed his concerns about the practical organization, which did not yet seem to be in place. His fears proved to be justified: due to miscommunication and a lack of budget, Desguin's letter was destined to remain an empty promise and lantern projection infrastructure continued to be the exception rather than the rule in schools for at least another decade.

This is not to say that no progress was being made. Van Mechelen seems to have taken to heart the growing demand for projection infrastructure from schools. When constructing new school buildings or renovating existing schools, he aimed to include a projection hall. The project to transform the Hotel Della Faille de Leverghem into a municipal normal school in 1911, for instance, included the construction of a new rear annex, which housed a large, 10.70 by 9 metre projection hall on the first floor (see Figure 6. 4). However, in 1923, the school principal complained that, although they had a designated hall for lantern projection available to them, they still did not own an epidiascope,[66] nor the necessary diapositive slides.[67]

Schools' Wish Lists

Unfortunately, due to a gap in the archival sources, we have no information about how the situation evolved during the First World War and the years immediately after. What we do know is that on 19 May 1914, the Belgian Parliament passed a law on schools that made primary education compulsory up to age fourteen. This law led to the creation of a fourth grade, or the seventh and eighth year of primary education. After the fourth grade, children could choose to start working full-time, to learn a profession at a technical school at secondary level, or follow a teacher training programme at a normal school. As a consequence, a number of schools had to be remodelled in order to accommodate this fourth grade education. The Minister of Arts and Sciences, Jules Destrée, published revised government instructions regarding the construction and furnishing of school buildings.[68] This document offered suggestions for the layout of fourth grade schools, and one of those suggestions was to include a classroom for light projection.

From a 1923 survey of Antwerp's municipal schools, commissioned by the city council with the aim of identifying their most pressing needs, we can conclude that by then, at least fourteen of the nearly eighty municipal schools possessed an optical lantern. Five of those schools reported that they had a lecture hall — or a gymnasium that could double as one — that was suitable for light projection. For example, when in 1922 the

64 [Anon.], 'Kijkjes links en rechts', *Ons Woord*, 20.6 (1913), 214.

65 Ibid., 215.

66 An epidiascope is an optical projection system which can easily be altered to project either transparent lantern slides or flat opaque objects such as postcards, textbook illustrations, drawings, etc.

67 Survey of Antwerp municipal schools 1923. Lange Gasthuisstraat 31, aangenomen Stedelijke Normaalschool voor Onderwijzeressen. FA595#311; Agentschap Onroerend Erfgoed 2021: Hotel Della Faille de Leverghem <https://id.erfgoed.net/erfgoedobjecten/5331> [Accessed 12 March 2021].

68 Destrée, Jules, 'Verordening op het oprichten, het inrichten en het meubelen van schoolgebouwen bestemd voor het lager onderwijs en het bewaarschoolonderwijs', in *Het Belgisch Staatsblad* (17 March 1920), 2094–2103.

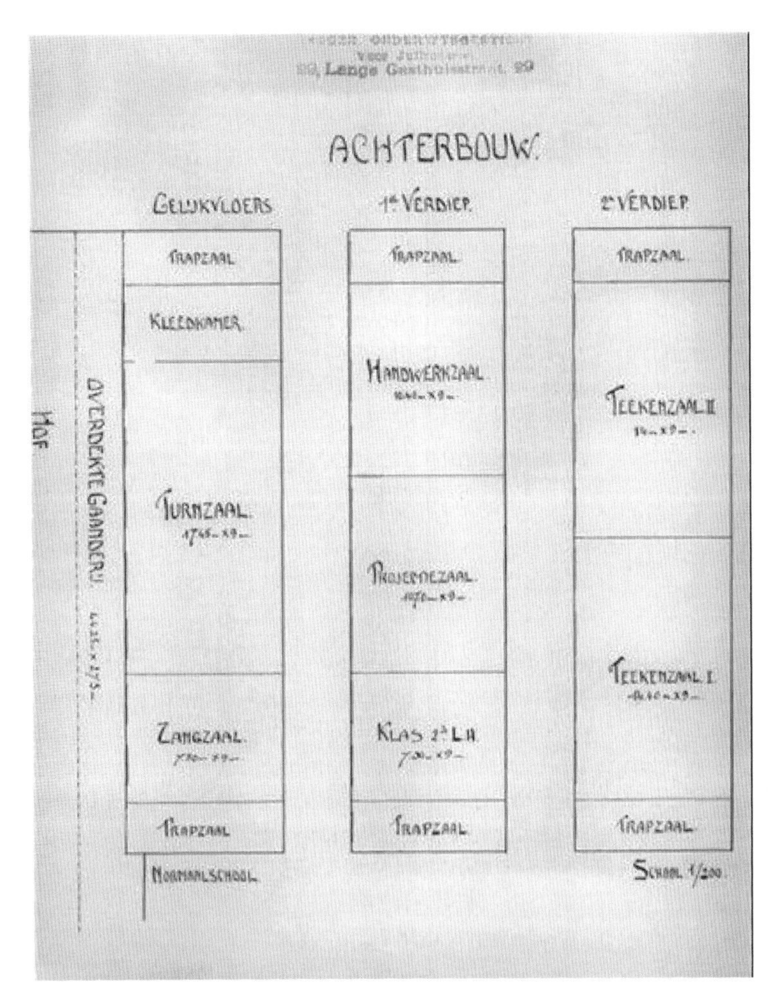

Fig. 6.4. Plans of the new rear annex, Institute for Higher Education for Girls, Lange Gasthuisstraat, Antwerp, including a projection room on the first floor (1911). FA 595#311.

girls' primary school located at Oever was transformed into an institution offering 'fourth grade' education[69] for girls, the freed-up space made it possible to include a projection hall, where geography and history were taught with the aid of a mirroscope.[70] When asked in the survey what they thought was still lacking in their schools, seven school principals

69 D'hoker, Mark, and Bregt Henkens, 'Van segmentering naar convergentie. Structuur en karakter van het secundair onderwijs in België in de 20ste eeuw', in *Paradoxen van pedagogisering. Handboek pedagogische historiografie*, ed. by Marc Depaepe, Frank Simon, and Angelo Van Gorp (Leuven: Acco, 2005), 159–76 (p. 164).

70 Survey of Antwerp municipal schools 1923. Oever, Lagere Hoofdschool voor meisjes no. 2. FA595#293. A mirroscope was an American type of episcope, which is the generic name for an opaque projector. Two electric light bulbs would illuminate the object to be projected on a wall or screen. The advantage of this technology was that it could be used to project any printed matter and even objects, whereas the optical lantern could only be used to project transparent lantern slides. Museums Victoria Collections <https://collections.museumsvictoria.com.au/items/721025> [Accessed 15 March 2021].

Fig. 6.5. The film and slide projectors exhibited at the Antwerp Exposition of modern teaching aids in 1922. FA FOTO-FO-#2316.

reported that they needed an optical lantern or epi(dia)scope, and six more expressed a desire to have a proper projection hall added to their school infrastructure.

A few months prior to the survey, in September 1922, van Tichelen, with the support of the city council, had organized an exhibition of modern teaching aids. This had no doubt served as a renewed incentive for municipal school principals to ask for the long-awaited projection technology and necessary adaptations to the school buildings. Despite his discouraging experiences before the War, van Tichelen had never ceased his efforts to introduce lantern projection into Antwerp's municipal schools. Since Desguin's plan to equip every school with projection technology had fallen through, mainly due to financial constraints, van Tichelen had started arguing more fervently for the establishment of a municipal school and educational museum.[71] And, as an integral part of that museum, for the creation of a 'lending service for learning and visual teaching aids'.[72] After publishing *Een school- en onderwijsmuzeum te Antwerpen* (A School and Educational Museum in Antwerp) in 1919, in which he painted a picture of what such a museum should offer to teachers, the school population and the broader public, and after addressing a formal letter to the city council

71 Van Daele, p. 268.

72 van Tichelen, Hendrik, *Een school- en onderwijsmuzeum te Antwerpen. Een gedachte en korte studie met schets* (Antwerpen: Van Resseler, 1919).

requesting the establishment of the museum, he finally convinced the municipal authorities: in the final months of 1920, the city council approved the creation of 'a municipal school museum' and appointed van Tichelen as its curator. Although the specificities and aims of the museum were clearly defined from the outset, its definitive location would remain under discussion for years. As a result, the museum collection was housed in two separate buildings, neither very suited to the proper exhibition and demonstration of teaching aids.[73] That the issue was never resolved in van Tichelen's seventeen years as museum curator is indicative of the project's neglectful treatment by the city council. Nevertheless, shortly after the new city council took office in July 1921, the socialist alderman for education, Camille Huysmans,[74] assigned van Tichelen the task of organizing a temporary exhibition of teaching aids, intended to familiarize teachers and the wider public with the enormous variety and potential of teaching equipment available.[75] For van Tichelen, it served another goal: to drive his point home that it was simply impossible to provide every school with all the teaching tools deemed useful for efficient and visual instruction, and that it was therefore necessary to install a lending service for such aids.[76] 'Projection equipment and accessories' formed a separate section in the exhibition and included, among other things, a mirroscope and four more types of epi(dia)scopes, accessories to allow the projection of natural science experiments, at least three film projectors, three screens, and a very small selection of lantern slides and films. Although van Tichelen was disappointed that budgetary constraints meant he could only display such a limited selection, he nonetheless seemed to be convinced that it offered a good starting point for further discussion among teachers and for the school museum collection.[77] Four days after its opening, alderman Huysmans decided to extend the exhibition's duration by another week due to its great success.[78] But apart from a renewed interest in projection equipment, the exhibition also sparked resentment and criticism. In a review of the exhibition in *Ons Woord*, Louis Louette, newly appointed president of *Diesterweg*, did not conceal his disappointment in both the city council and the exhibition:

> Alderman Huysmans outlined the purpose and essence of the exhibition in a few words. As a result of the War, he said, there was a lack of buildings and a crisis in school materials. It is a pity that he did not add that, even before the War, it was very difficult to get the city treasury to open the coffers when asking for improvements in education. [...] Had one been able to ask what teachers themselves considered useful, the exhibition would have taken a very different turn.[79]

73 Van Daele, 268; van Tichelen, Hendrik, *In vollen groei. Een woord tot afscheid van het stedelijk schoolmuseum te Antwerpen* (Antwerpen: Fantasio, 1938), pp. 19–28.

74 Camille Huysmans (1871–1968) was a Belgian socialist politician who was alderman for education (1921–1933) and later mayor of Antwerp (1933–1940 and 1944–1946). Van Daele, p. 27–30.

75 Van Daele, p. 268.

76 van Tichelen, Hendrik, 'Een ontleencentrale van leer- en aanschouwingsmiddelen in de schoot van het School- en Onderwijsmuzeum', *Ons Woord*, 29.1 (1922), 10–16 (p. 13).

77 van Tichelen, Hendrik, *Van woord tot daad. Het stedelijk schoolmuseum te Antwerpen. Een werkelijkheid* (Antwerpen: Viktor Resseler) pp. 19; 32–33.

78 Alderman for education Camille Huysmans, 'Tentoonstelling van schoolmateriaal', 14 September 1922. FA480#5447; alderman for education Camille Huysmans, 'Tentoonstelling van schoolmateriaal', 18 September 1922. FA 480#5447.

79 Louette, L., 'Tentoonstelling van leer- en aanschouwingsmiddelen', *Ons Woord*, 29.10 (1922), 238–41 (p. 238).

It seems that Huysmans at least took some of this to heart, since a couple of months later he commissioned the aforementioned survey, to be held among Antwerp's municipal schools, asking them to describe the school history, the school buildings, list the school's specific characteristics and identify the issues that needed to be addressed most urgently.

Despite the growing number of schools owning a lantern or planning to acquire one in the near future, many school principals saw their ambitions to efficiently organize instruction using light projection thwarted by inadequate school infrastructure. The separate boys' and girls' schools located in Grote Hondstraat, and built in 1902 as 'model schools' as regards the hygiene standards of the time, shared a hall that was used for light projection. Due to a lack of space to house the rapidly increasing school population, however, it also served as a drawing class and gymnasium.[80]

The girls' teaching practice school in Hopland, built in 1914, had inaugurated a 'mirroscope class' in October 1915. After the War, however, the number of pupils had risen so rapidly that the music room and principal's bedroom had to be turned into extra classrooms, while the projection room now served as a dressing room. In her response to the city council's survey, principal Van Kessel expressed the explicit wish to have a well-equipped projection room on the school grounds again.[81]

In 1913, chief city architect Alexis Van Mechelen had completed the transformation of the François Dhanis hotel in Belgiëlei into boys' school no. 3.[82] He had included a large lecture and projection hall measuring 12.41 by 5.43 metres on the third floor and conveniently located in the middle of the building, without any windows, so that it could be made completely dark in just a few seconds. Unfortunately, as the teacher René de Nave reported in the 1923 survey, the city's technical department had concluded that the latticework between the floor joists of the projection hall was not strong enough to support the weight of a large audience. Therefore, lessons with the school's mirroscope could not be held there until the hall had been completely renovated and declared safe for use. This was identified by de Nave as a priority, since the school was very proud to possess a wonderful collection of postcards for projection during geography and history instruction that now lay idle in a school cupboard.[83]

Four schools, including the municipal normal school mentioned earlier and the boys' primary school no. 1 in Louizastraat, housed a projection hall that was fully compliant with modern standards, but still lacked a projector.[84] The school in Violetstraat, which was one of the two primary schools for boys referenced by van Tichelen in 1912 as having just started to make regular use of slide projection during classes, finally bought a lantern in 1924. Most likely, they were previously using a lantern owned by an individual teacher.[85]

80 Survey of Antwerp municipal schools 1923. Grotehondstraat, Meisjesschool no. 16. FA 595#287.

81 Survey of Antwerp municipal schools 1923. Hopland-Oudaan, Meisjesoefenschool. FA 595#289.

82 Agentschap Onroerend Erfgoed 2021: Stedelijk Onderwijs Gesticht 3 voor Jongens [online] https://id.erfgoed.net/erfgoedobjecten/6491 [Accessed on 11 March 2021].

83 Survey of Antwerp municipal schools 1923. Belgiëlei, Jongensschool no. 3. FA 595#228.

84 Survey of Antwerp municipal schools 1923. Lange Gasthuisstraat 31, aangenomen Stedelijke Normaalschool voor Onderwijzeressen. FA595#311; Survey of Antwerp municipal schools. Louizastraat 17, Onderwijsgesticht voor jongens. FA595#223.

85 Survey of Antwerp municipal schools 1923. Violetstraat, Lagere Jongensschool no. 14. FA 595#238.

And there were also municipal schools that still had no dedicated projection room at all, even though some of them had owned a projector for a long time. A good example of this was the primary school for boys no. 7 in Markgravelei. In his letter of 1913 to encourage schools to start using lantern projection, Desguin had referred to the school's use of a lantern as an example of good practice.[86] However, ten years later, a lecture hall was still an item on the school's wish list.[87]

By then, the merits of educational films for primary and secondary education were no longer questioned by *Ons Woord*.[88] Since it was undeniable that cinematographic projection would join lantern slide and episcopic projection in the classroom, the journal warned that, when constructing new classrooms in the future, this should be taken into account. Preferably, it argued, a hall should be reserved specifically for lantern projection and film screenings. It would be wise, the article went on

> [...] to choose the darkest [room] and try to keep the number of windows to a minimum so that it can be made dark as if by magic and no time is lost or disorder caused, because lessons using projected images require a good deal of effort, preparation and careful documentation, and the difficulties should not be increased unnecessarily.[89]

Although the previous chief city architect, Alexis Van Mechelen, had already tried to apply these principles in his school renovation projects from the years just before the First World War, his successor, Emiel Van Averbeke, had other fish to fry. For what is particularly clear form the 1923 survey is how many municipal schools still lacked the most basic facilities and equipment, such as central heating, electric lighting, flush toilets, a gymnasium, decent blackboards of sufficient size, and, the most frequently heard request, simply more space to accommodate the growing school population. Moreover, many teachers were still having to make do with outdated low-tech media or without even the simplest of teaching aids, so that the incorporation of high-end media such as projection technology into their lessons must have seemed a pipe dream to them.

Dedicated Spaces for Teaching with the Lantern

As is apparent from repeated surveys of Antwerp's municipal schools, by the mid-1930s the use of film and lantern projectors had become more widespread. This was a general trend in Belgian schools, including Catholic education institutions.[90] Van Tichelen's wish for projectors to become established visual teaching aids, casually referenced by educational journals when discussing teaching methods,[91] had finally come to fruition. Moreover, this

---

86 Alderman for education Victor Desguin to Antwerp school principals, 29 April 1913, FA 480#3435.

87 Survey of Antwerp municipal schools 1923. Markgravelei, Jongensschool no. 7. FA 595#252.

88 'De Kinematograaf in de School', p. 80.

89 'De Kinematograaf in de School', pp. 82–83.

90 Teughels, Nelleke, 'Expectation versus Reality: How Visual Media Use in Belgian Catholic Secondary Schools Was Envisioned, Encouraged and Put into Practice (*c.* 1900–1940)', *Paedagogica Historica*, 0.0 (2021), 1–18<https://doi.org/10.1080/00309230.2020.185653> [accessed 7 July 2021]; Egelmeers and Teughels.

91 'Dierenbescherming en opvoeding', *Ons Woord*, 39 (1932): 251; 'Natuurwetenschappen en landbouw in den vierden graad', *De Opvoeder*, 29 (1932), 75–79.

trend was backed and applauded by central government, which now regularly reviewed new projectors and advised teachers to use one or more types of projectors, such as optical lanterns, episcopes and film projectors.[92]

In 1932 the Belgian government, after consulting the provincial chief architects, published a ministerial decree laying down the basic principles for school architecture. These facilitated the regular instructional use of film, slide and episcope projection, by strongly recommending the inclusion of 'a music and lecture hall, with a device for the projection of still or moving images'.[93] Two years later, the modernist Belgian architect and theorist Gaston Brunfaut reflected on 'the modern school' in Belgium in a special issue on school buildings of the periodical *Batir. Revue mensuelle illustrée d'architecture, d'art et de décoration.* According to Brunfaut, it was self-evident that all renewal in school architecture should closely follow the developments in pedagogical theory and methods. Since the ideas of the Brussels doctor Ovide Decroly held sway at the time in Belgium, Brunfaut argued, modern school architecture should be built according to his views.[94] Ovide Decroly belonged to the international 'New Education' movement, which positioned itself as a child-centred, progressive alternative to the purportedly passive, rigid, teacher-centred and text-based methods of traditional schooling. This may sound familiar: this 'new education' was in reality not radically different from the 'active-intuitive method' dating from the end of the nineteenth century.[95] Since this 'New Education' defined children as active beings, capable of observing and learning from life in a much more efficient way than from books, Brunfaut proposed that classrooms should be conceived as *ateliers*, that is, as workshops where materials and tools were brought together. They should be built to facilitate the use of equipment, collections, screens and film projection equipment and be large enough to allow the pupils to group freely around tables and work collaboratively.[96] The special issue still focused mainly on the hygiene-related aspects of school architecture, such as lighting, ventilation and ergonomic and easy-to-clean school furniture, but as the interview with Brunfaut shows, it had become accepted practice that new school buildings should also include a room equipped for projection. The advertisements for Belgian projection technology included in the special issue also testify to this.

That does not of course mean that all the problems had been solved overnight. In 1934, the principal of a boys' school wrote to the city council:

> Last year, electricity was brought into our boys' school n° 12. It was also planned that a classroom would be set up for light projection. Due to a lack of the necessary funds, the completion of the intended works was halted. Is there now not a way of finishing

92 *Rapport sur la situation de l'instruction primaire en Belgique. Années 1924–1925-1926* (Bruxelles: Th. Dewarichet, 1928), xlvii. See also *Suppléments* 7, 8 and 9 to the *Catalogue des ouvrages classiques* (Bruxelles: Imprimerie du Moniteur belge, 1926–1927, 1928, 1930).

93 Ministère des Sciences et des Arts, *Constructions scolaires. Aménagement et ameublement des locaux scolaires affectés à l'enseignement primaire, gardien et moyen. Arrêté ministériel du 25 mai 1932* (Bruxelles: Ministère des Sciences et des Arts, 1937), pp. 22–23.

94 Flouqet, Pierre Louis, 'L'école moderne. Interview de l'architecte Gaston Brunfaut', *Batir. Revue mensuelle illustrée d'architecture, d'art et de décoration*, 16 (1934), 589–96 (p. 589).

95 Depaepe, Marc, Frank Simon, and Angelo Van Gorp, 'The canonization of Ovide Decroly as a "Saint" of the New Education', *History of Education Quarterly*, 43.2 (2007), 224–28.

96 L'école moderne', p. 590.

> the room for light projection? A teacher at our school has a projection lantern so that we could use this room regularly. It would make teaching more efficient and a lot of time would be saved.[97]

And so it seems that, for a number of schools at least, very little had changed in practice. When *Batir* revisited the theme of school architecture in 1936, provincial chief architect Henri T. van Hall confirmed that a lot of work remained to be done in Belgium to bring schools up to modern standards.[98] However, the periodical's review of ten recently built 'model' schools in Belgium showed that a hall specifically equipped for the projection of still or moving images was now considered a standard part of school architecture. Projection technology had finally found a physical place in Belgian education and was there to stay, up until today.

Conclusion

For a long time, the introduction of new teaching technology has been studied as mainly the result of efforts by producers, pedagogues and policymakers to modernize instruction. As a result, teachers' role in classroom innovations has often been reduced to that of a consumer, either willing to use or rejecting instructional aids. Moreover, the various ways in which those objects themselves have necessitated modifications to the physical space for learning have been largely overlooked. In Belgium, ideologically charged educational policies have certainly played a major role in shaping the education system. As this article has demonstrated, however, their impact on pedagogical reality is certainly less significant than everyday classroom behaviour, which was in turn largely shaped by teachers, teaching spaces and material culture. From this case study, it became apparent that teachers were the driving force behind the spread of the use of lantern projection in municipal schools in Antwerp. Rather than simply trying to comply with government instructions, many of them took the initiative to try and facilitate and promote lantern use in schools. Moreover, they actively searched for alternative ways to achieve this goal when challenged by the inadequate, inert nature of school infrastructure and budgetary constraints which, it seems, were the real (f)actors responsible for the slow pace of incorporation of projection technology into Belgian education.

This illustrates that it is of paramount importance to take into account the agency of both teachers and material culture if we wish to gain more insight into the principles of the grammar of schooling and the tension between the theoretical frameworks from which pedagogical and didactic innovations emerged and the seemingly very conservative context in which they were intended to be used. Token compliance or wholehearted adoption, the various ways in which certain types of classroom media have been ignored or their use has been modified, cannot be fully understood unless we recognize the active role played by both teachers and school infrastructure in shaping classroom practices.

---

97 School principal Van Loock to Antwerp Alderman for education Willem Eekelers, 8 March 1934. FA 480#1520.

98 Flouqet, Pierre Louis, 'L'école nouvelle en Belgique. Interview de M. Henri T. van Hall, architecte provincial en chef du Brabant', *Bâtir: Revue mensuelle illustrée d'architecture, d'art et de decoration*, 45 (1936), 801–06.

MICHAEL MARKERT

Casting Long Shadows on the Teaching of Experimental Physics: The Projection Techniques of Physicist Robert Wichard Pohl (1884–1976)

Introduction: The New Age of Shadow Projection in Physics Lectures

Flickering shadows thrown on trees and rocks around a campfire were likely the very first human experience of artificial projection.[1] In harsh contrast to the parallel projection shadows caused by the sun or the moon, these 'new' shadows resulting from central projection could be very easily manipulated through the movement of opaque objects between the light source and the projection surface. The shadows are sharp and silhouette-like with the objects close to this end of the 'projection apparatus', while they are gigantic, distorted, and blurred with the objects close to the light source, which stimulates a sense of fantasy. Although projected shadows are, in essence, insubstantial, flat, artificial phenomena representing objects of the material world, their inherent pictorial qualities have elevated them to a distinctly dynamic and performative medium.[2]

In the shadow theatre of nineteenth-century Europe — undoubtedly the most elaborate and successful art form using shadows — the relationship between light and enlighted object is hidden behind a curtain.[3] Most modern forms of projection hide the techniques used to generate images, as in the cinema, where the film projector is placed in a separate room behind the audience. Accordingly, primordial types of projection, where the light source, illuminated object, and projected image are clearly visible to the viewers, are more or less unusual today. In contrast to this emphasis on the physical relationship between

1 Reust, Hans Rudolf, 'Animierte Gespenster', in *Ich sehe was, was du nicht siehst! — Sehmaschinen und Bilderwelten: Die Sammlung Werner Nekes*, ed. by Werner Nekes, and Bodo von Dewitz (Göttingen: Steidl, 2002), pp. 10–15 (p. 10).

2 Roth, Tim Otto, *Körper. Projektion. Bild: Eine Kulturgeschichte der Schattenbilder* (Paderborn: Wilhelm Fink, 2015), p. 528 (pp. 11–12).

3 Forgione, Nancy, '"The Shadow Only": Shadow and Silhouette in Late Nineteenth-Century Paris', *The Art Bulletin*, 81.3 (1999), 490–512 <https://doi.org/10.1080/00043079.1999.10786899>.

Michael Markert • Thüringer Universitäts- und Landesbibliothek, michael.markert@uni-jena.de

Learning with Light and Shadows: Educational Lantern and Film Projection, 1860-1990, ed. by Nelleke Teughels and Kaat Wils, TECHNE-MPH, 8 (Turnhout, 2022), pp. 169-195

10.1484/M.TECHNE-MPH-EB.5.131499

these three elements, modern media tends to be immersive, as Sophie Ernst pointed out in her Ph.D. thesis on 'The magic of projection: Augmentation and immersion in media art'.[4] As she stated, '[b]efore projection became an immersive (cinematic) experience it augmented reality',[5] which added something to the experienced world instead of overwriting it with virtual images. With augmented, that is 'reality expanding', projection, the observers thus 'become aware of that which is imagined or remembered in the context of that what is present'.[6]

The situation described by Plato in his allegory of the cave draws on this concept of projection. The spectators are fixated on the wall (projection screen) and thus are not able to notice the 'real' objects nor the light source that is present. The observers, on the other hand, can clearly 'see' the relationship between all elements. The imagined augmented projection in Plato's text is a pedagogical device to mediate knowledge about the relationship between the physical world and the world of ideas. Indeed, augmented projection was not only used in this educational way within fictional situations but also within the 'real' physical world as well. The projection of experimental devices and other material (often opaque, shadow-evoking objects) for educational purposes can be traced back to the eighteenth century, when it would go on to become a very successful teaching method over the next two centuries, especially in the field of natural sciences.[7] However, while the practices of research, including the performance of experiments,[8] have been widely researched by historians of science, the practices of teaching still are at the periphery of the discipline's interest.[9] Although there are some works on the role of textbooks,[10] research on teaching aids like experimental setups as well as media like lantern slides, wallcharts, or models have been widely absent in the literature.[11] Following David Kaiser's call for 'moving the pedagogy from the periphery to the center'[12] within the history of science, this

4 Ernst, Sophie J. G., 'The Magic of Projection: Augmentation and Immersion in Media Art' (unpublished Dissertation, Leiden University, 2016).

5 Ernst, p. 115.

6 Ernst, p. 49.

7 Hackmann, Willem, 'The Magic Lantern for Scientific Enlightment and Entertainment', in *Learning by Doing. Experiments and Instruments in the History of Science Teaching*, ed. by Peter Heering, and Roland Wittje (Stuttgart: Steiner, 2011), pp. 113–39.

8 *Experimentelle Wissenschaftsgeschichte*, ed. by Olaf Breidbach and others (München: Wilhelm Fink, 2010).

9 Mody, Cyrus C. M., and David Kaiser, 'Scientific Training and the Creation of Scientific Knowledge', *Handbook of Science and Technology Studies*, 2008, 377–402.

10 Warwick, Andrew, and David Kaiser, 'Kuhn, Focault, and the Power of Pedagogy', in *Teaching the History of Science*, ed. by M. Shortland, and A. Warick (Oxford: British Society for the History of Science and Basil Blackwell, 1989), pp. 393–409; Simon, Josep, 'Textbooks', in *A Companion to the History of Science* (John Wiley & Sons, Ltd, 2016), pp. 400–13 <https://doi.org/10.1002/9781118620762.ch28>; Badino, Massimiliano, and Jaume Navarro, 'Research and Pedagogy', 2013 <https://www.mprl-series.mpg.de/studies/2/index.html> [accessed 16 April 2020].

11 For some exceptions, see e.g., Bucchi, Massimiano, 'Images of Science in the Classroom. Wallcharts and Science Education 1850–1920', *British Journal for the History of Science*, 31 (1998), 161–84; *Learning by Doing. Experiments and Instruments in the History of Science Teaching*, ed. by Peter Heering and Roland Wittje (Stuttgart: Steiner, 2011); Markert, Michael, 'Embryonale Pluripotenz — Ein Lehrmodell zwischen Forschung, Ökonomie und Unterrichtung', in *Spiegel der Wirklichkeit — Anatomische und dermatologische Modelle in Heidelberg*, ed. by Sara Doll, and Navena Widulin (Heidelberg: Springer, 2019), pp. 73–85.

12 Kaiser, David, 'Moving Pedagogy from the Periphery to the Center', in *Pedagogy and the Practice of Science. Historical and Contemporary Perspectives*, ed. by David Kaiser (Cambridge: MIT Press, 2005), pp. 1–8.

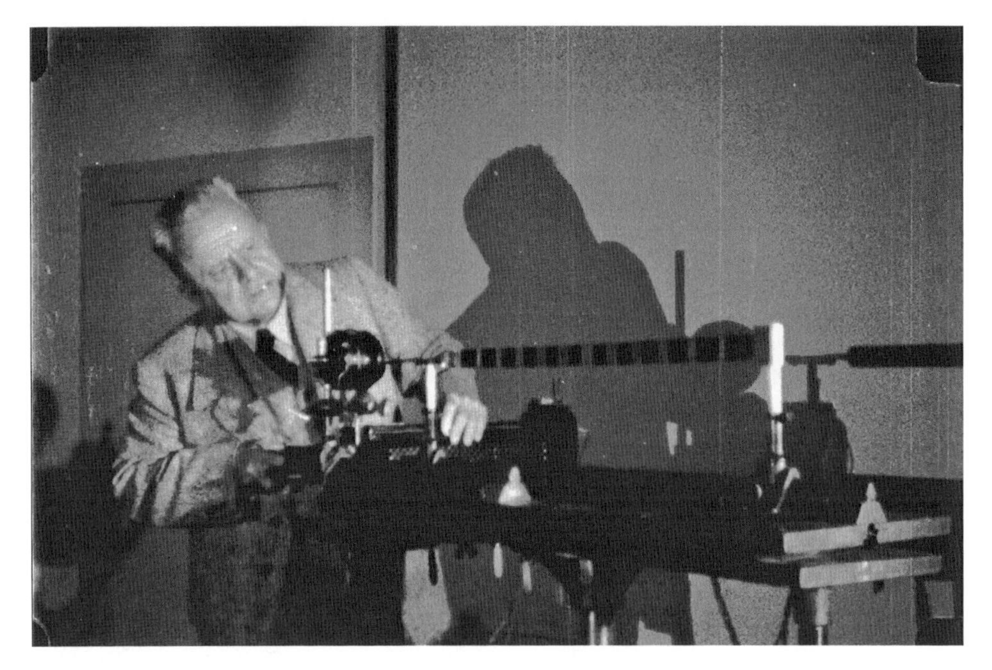

Fig. 7.1. Robert Wichard Pohl during his lecture in 1952, film still Fritz Lüty, *[Bunsenstraße 9]*, 16 mm film (Göttingen, 1952). Family archive Pohl, Göttingen.

article focuses on a case study of extensive augmented media use in the form of shadow projection within twentieth-century university physics teaching.

The protagonist of this case study is Professor Robert Wichard Pohl (1884–1976) at Göttingen University, who would conduct lectures in a bright spotlight placed in front of the audience to throw enlarged shadows of him and the experimental apparatus against the front wall (see Fig. 7.1).

This unconventional type of technology and performance bore a strong resemblance to shadow theatre and was employed by Pohl in all his lectures on experimental physics from 1919 to the 1950s. As a pioneer in this form of pedagogy, Pohl further invented new demonstration experiments or modified well-established ones to make them suitable for shadow projection. Black and white silhouettes of the actual lecture presentations were then used as illustrations in his seminal three-part textbook that was released from 1927 onward in more than 20 editions up to now.[13] Pohl's astounding 'performance-lecture',[14] as his style of teaching would be called today, was profoundly influential to physics teachers and lecturers around the globe. His two-semester basic lecture on experimental physics was attended every year by hundreds of students of physics, medicine, biology, chemistry, pharmacy, and

13 *Mechanik, Akustik und Wärmelehre*, ed. by Klaus Lüders, and Robert Otto Pohl, 21., gründlich überarbeitete Auflage (Berlin: Springer Spektrum, 2017); *Elektrizitätslehre und Optik*, ed. by Klaus Lüders, and Robert Otto Pohl, 24., gründlich überarbeitete Auflage (Berlin: Springer Spektrum, 2018).

14 Ladnar, Daniel, 'The Lecture Performance: Contexts of Lecturing and Performing' (unpublished Ph.D., Aberystwyth University, 2014) <http://hdl.handle.net/2160/2fe1aceb-2458-4f32-90e0-9b3caed4982d> [accessed 9 March 2021].

agriculture, as well as future physics teachers. Members of this last group played a crucial role as advocates for the broad implementation of Pohl's didactic method in secondary school education in Germany and elsewhere, supported by his eminently successful textbooks.[15]

Next to students, physics teachers were the second target audience of Pohl's books.[16] Indeed, one of the first reviews of *Einführung in die Elektrizitätslehre*, published in 1927, was framed as a discussion regarding whether the school curriculum should be adjusted in line with Pohl's methods, particularly his restructuring of the content on electromagnetic fields.[17] That same German book was reviewed in the British journal *Nature*, calling it 'a delightful book, one of the most pleasant features of which is the large number of beautiful illustrations, diagrammatic and photographic, which adorn almost every page'.[18] The English translation of the second edition published in London was also highly praised and welcomed:[19] 'Here is given, with a completeness and mastery unapproached in any English work of the kind known to the reviewer, a lucid and systematic account of the physical principles of electricity and magnetism'.[20]

At that time, some of Pohl's experimental setups had already been distributed as commercial teaching aids by the Göttingen company Spindler & Hoyer (see Fig. 7.2). By 1925, the director of that company had come to Pohl with the idea to produce his teaching aids for schools and universities. Although Pohl was not convinced that this might be a successful business idea,[21] by the beginning of the 1930s, apparatuses 'in accordance with Pohl (*nach Pohl*)' were delivered throughout Europe as well as to China, India, Indonesia, Russia, and the USA, as order books in the company's archive show.[22]

At Göttingen University, objects made by the institute's mechanics for Pohl's lectures in the 1920s remain in use today as part of the teaching collection of the Physics Department, including prototypes like that of the streamline apparatus shown in Fig. 7.2.

As suggested by this short introduction, the case of Pohl and his lectures not only allows the researcher to specify the role of projection media in physics education but also to explicate the relevance of teaching efforts for the historical actors, their institutes, and disciplines and highlight the interconnectedness of school and university education. In the next section, I introduce Pohl's work and the basic conditions of the development of his teaching concept throughout his career at Göttingen University from 1919 to 1952 and

15 Achilles, Manfred, 'Vergessenes und Unvergessenes aus dem didaktischen Werk von R. W. Pohl (1884–1976), mit Experimenten', in *Deutsche Physikalische Gesellschaft. Fachausschuss Didaktik der Physik. Vorträge der Frühjahrstagung 1980*, ed. by A. Scharmann, A. Hofstaetter, and W. Kuhn (Gießen, 1980), pp. 512–17; Achilles, Manfred, 'Vergessenes und Unvergessenes aus dem didaktischen Werk von R. W. Pohl (1884–1976) II, mit Experimenten', in *Deutsche Physikalische Gesellschaft. Fachausschuss Didaktik der Physik. Vorträge der Frühjahrstagung 1981*, ed. by A. Scharmann, A. Hofstaetter, and W. Kuhn (Gießen, 1981), pp. 512–17.

16 Pohl, Robert Wichard, *Einführung in die Mechanik und Akustik* (Berlin [u.a.]: Springer, 1930), p. preface.

17 Hillers, W., 'Bemerkung zur Didaktik des Physikunterrichtes auf der höheren Schule und der Hochschule aus Anlaß der Darstellung im Lehrbuche von R. W. Pohl Einführung in Die Physik', *Zeitschrift für mathematischen und naturwissenschaftlichen Unterricht*, 59 (1928), 229–33.

18 G. H. L., '(1) Applied Magnetism (2) Einführung in die Elektrizitätslehre', *Nature*, 121.3042 (1928), 240 <https://doi.org/10.1038/121240a0>.

19 Pohl, Robert Wichard, and Winifred M. Deans, *Physical Principles of Electricity and Magnetism* (London and Glasgow: Blackie & Son Ltd, 1930) <//catalog.hathitrust.org/Record/102101038>.

20 o. A., 'Physics for Students', *Nature*, 128.3221 (1931), 134–35 <https://doi.org/10.1038/128134a0>.

21 Meinhardt, Günther, *75 Jahre Spindler & Hoyer* (Göttingen: Spindler & Hoyer, 1973), p. 122 (p. 12).

22 Dep. 104, Nr. 115 Bestellbuch für Apparate, City Archive of Göttingen.

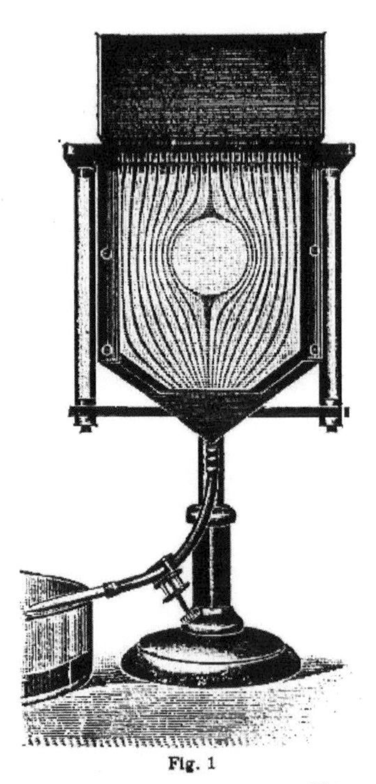

Zwei Apparate zur Hydro= und Aerodynamik

nach Prof. R. Pohl.

Die Hydrodynamik ist im Elementarunterricht lange Zeit hindurch neben der Hydrostatik zu kurz gekommen. Erst neuerdings scheint sich ein gewisser Wandel anzubahnen. Dabei gibt die Hydrodynamik Gelegenheit, eine Anzahl sehr eindrucksvoller Versuche vorzuführen und auf ihre Bedeutung in der technischen Nutzanwendung hinzuweisen (Flugzeug, Drachen, Kreuzen beim Segeln, Schiffsschraube usw.). Der Natur der Sache nach wird man sich im allgemeinen auf qualitative Erklärungen beschränken müssen. Diese erfordern im wesentlichen, daß der Anfänger mit den Begriffen „Stromlinien" und „Wirbelbewegung" vertraut gemacht ist. Diesem Zweck dienen die beiden im Nachfolgenden beschriebenen Apparate.

Stromlinien=Apparat*)

nach Prof. R. Pohl.

Fig. 7.2. Pohl's 'streamline apparatus (Stromlinienapparat)' in a catalogue of company Spindler & Hoyer. Spindler & Hoyer, Liste 52: Zwei Apparate zur Hydro- und Aerodynamik (Göttingen, n. d.), p. 4 <http://vlp.mpiwg-berlin.mpg.de/references?id = lit18183> [accessed 5 August 2020].

beyond. Afterwards, I outline which projection techniques were used in physics education back then and in what way Pohl's approach differed from that of his contemporaries. In the follow-up section, the other part of Pohl's teaching apparatus is examined — the tools and setups that were projected on the lecture hall's front wall. Subsequently, I portray the extensive reception of Pohl's methods at schools and universities and, finally, characterise his teaching style as a kind of 'innovative anachronism', using well-established concepts but in a creative, highly adoptable, and effective way.

Pohl and his Teaching Concept

Robert Wichard Pohl was born in 1884 in Hamburg as the son of a ship engineer. His mother was the daughter of the founder of a middle school, and his uncle was one of his first teachers there.[23] Pohl, who studied physics at Heidelberg and Berlin, got his Ph.D. in

23 Mollwo, Erich, 'ROBERT WICHARD POHL 70 Jahre', *Physikalische Blätter*, 10.8 (1954), 375 (p. 375) <https://doi.org/10.1002/phbl.19540100805>.

1906, and in 1912, he qualified as a professor with a monograph on the physics of X-rays, a kind of radiation that is famously used for producing 'shadow images' in medical and technical contexts since Pohl's school days.[24] Two years before, he had written another monograph on the physics of imaging technology with *Die Elektrische Fernübertragung von Bildern*; his expertise in this area may have thus been a starting point for his interest in projection devices.[25]

At the time of his appointment as an extraordinary professor at Göttingen University in 1916 — at the age of just 32 — it was unforeseeable that Pohl would become such an influential lecturer working on his teaching concept his whole life, as he saw himself primarily as a researcher. As he wrote to his mother while negotiating the conditions of his new position at Göttingen University in 1915:

> If I could only follow my preferences, I wouldn't mind a position with lesser income as long as I could focus on special courses for a small circle of major students and wouldn't have to give the annual beginner's lecture.[26]

As it turned out, he himself was surprised about his growing interest in exactly that kind of teaching.[27]

Although appointed in 1916, he could not accept the position until 1919 because of the First World War. Commencing work in 1919 and promoted to full professor and head of the *I. Physikalische Institut* in 1920, Pohl came upon an extraordinary situation at the Göttingen Physics Department. Four of the five professorships at the five physical institutes were re-staffed by 1921, with Pohl, Max Reich (1874–1941), and the later Nobel laureates Max Born (1882–1974) and James Franck (1882–1964), who was also a college friend of Pohl in Berlin, assuming those positions (see Fig. 7.3).[28]

Under Born and Franck, the Göttingen Physics Department developed into a centre of quantum mechanics, attracting physicists from all over the world interested in theoretical physics — at least until the takeover of the National Socialist Party in 1933, which led to a massive brain drain with Born and Franck leaving Göttingen soon after. The innovations in theory at his neighbouring institutes did not bother Pohl at all. Being an experimentalist to the core, he created a headstrong research programme focused on measuring the optical and electrical properties of solid materials, especially artificial crystals. Hence, during the heydays of Göttingen quantum mechanics, the 'Pohl school' was an isolated, self-referential research group, with Pohl himself rejecting any theoretical interpretation of his experimental results. However, in 1938, when Pohl presented almost twenty years of research at his institute at a conference in Bristol, the works of him and his students instantly

---

24 Pohl, Robert Wichard, *Die Physik der Röntgenstrahlen* (Braunschweig: Vieweg, 1912).

25 Pohl, Robert Wichard, *Die Elektrische Fernübertragung von Bildern* (Braunschweig: Vieweg, 1910).

26 'Koennte ich nur meinen Neigungen nachgehen, waere mir eine, wenn auch etwas geringer bezahlte Stellung ebenso lieb, in der ich keinen alljaehrlich wiederkehrenden Anfaengerkurs zu lesen haette, sondern Spezialkollegs in kleinem Kreise, fuer solche, die Physik als Hauptfach betreiben wollen. 'Letter of Robert Wichard Pohl to his mother Martha, Berlin, 23.12.1915. Family archive Pohl, Göttingen.

27 Letter of Robert Wichard Pohl to his sister Margot, Göttingen, 06.03.1919. Family archive Pohl, Göttingen.

28 Rammer, Gerhard, 'Die Nazifizierung und Entnazifizierung der Physik an der Universität Göttingen' (Georg-August-Universität Göttingen, 2004), pp. 28–29 <https://ediss.uni-goettingen.de/handle/11858/00-1735-0000-0006-B49F-4> [accessed 28 April 2020].

Fig. 7.3. Portrait of the four new professors in 1923. From left to right M. Reich, M. Born, J. Franck, and R. W. Pohl. Family archive Pohl, Göttingen.

were received as the experimental support for widely discussed theoretical concepts, with Pohl later being called 'the real father of solid-state physics' by Nobel laureate Sir Nevill Francis Mott (1905–96).[29]

Back in the 1920s and early 1930s, when physicists focused almost exclusively on sophisticated theory, nobody outside Göttingen was interested in Pohl's research.[30] While virtually irrelevant as a researcher until around 1940, Pohl would instead be perceived as a celebrated university teacher, innovator of demonstration experiments (see Fig. 7.4), and author of unprecedented textbooks within only a few years. Historian of physics Roland Wittje, who did fundamental research on what he calls the 'System Pohl', describes it as an easy-to-use, reliable, vivid, and modular concept of experimental demonstrations whose elements were sold as teaching aids that, together with Pohl's textbook, helped to foster

29 Braun, E., 'The Contribution of the Gottingen School to Solid State Physics: 1920–40', *Proceedings of the Royal Society of London. A. Mathematical and Physical Sciences*, 371.1744 (1980), 104–11 <https://doi.org/10.1098/rspa.1980.0063>; Sir Mott, Neville, 'Bristol Physics in the 1930s', 2020, p. 2 <http://www.bristol.ac.uk/physics/media/histories/11-mott.pdf> [accessed 18 November 2020].

30 Teichmann, Jürgen, *Zur Geschichte der Festkörperphysik: Farbzentrenforschung bis 1940, Boethius* (Stuttgart: Steiner-Verl. Wiesbaden, 1988), p. 162 (p. 130).

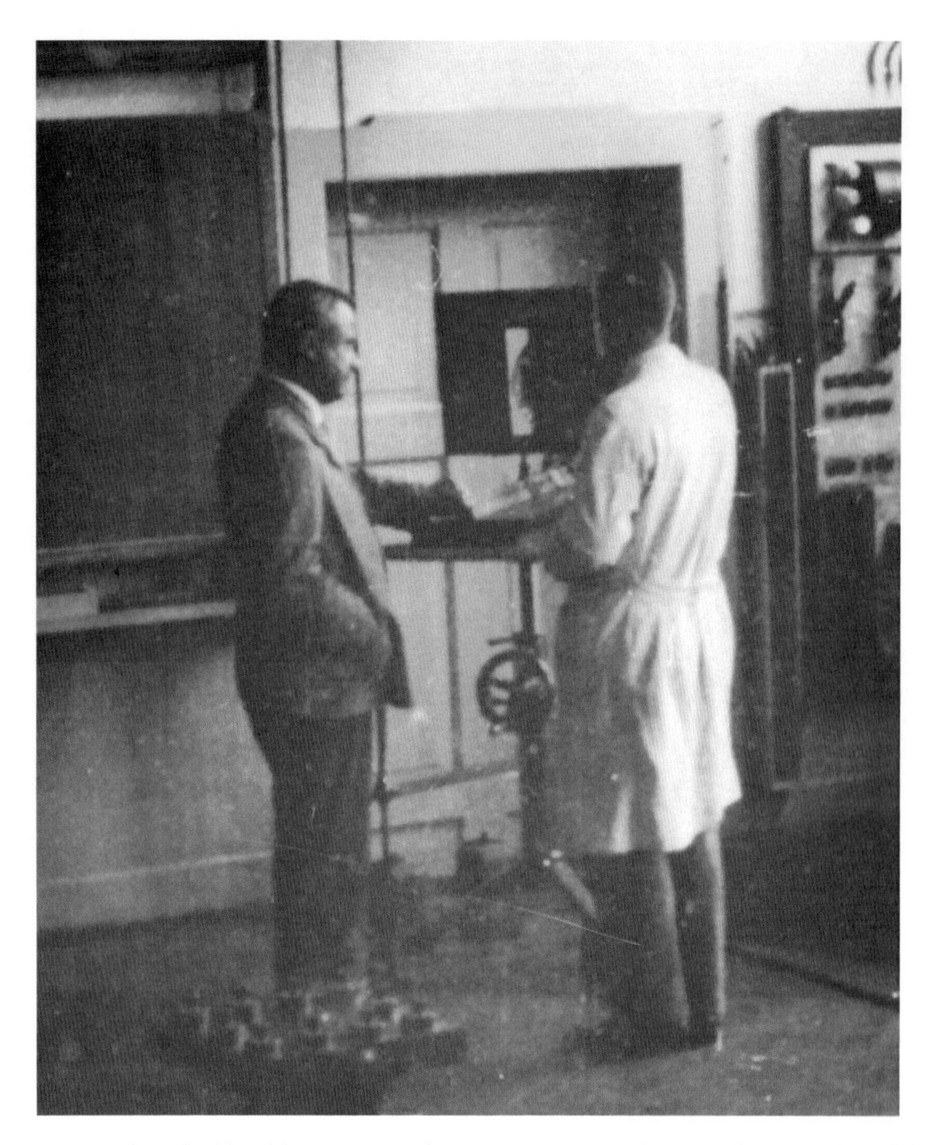

Fig. 7.4. Snapshot of Pohl and his assistant in the pre-1926 Göttingen lecture hall adjusting one of the demonstration experiments (source unknown; the picture is mentioned as an anonymous gift in a letter of Pohl to his wife in 1931). Letter of Robert Wichard Pohl to his wife Tussa, February 17 1931. Family archive Pohl, Göttingen.

the broad reception of the 'System'.[31] For Wittje, 'Robert Wichard Pohl, without doubt, has been the most influential figure in the transformation and renewal of experimental physics teaching in interwar and postwar Germany.'[32]

---

31 Wittje, Roland, "Simplex Sigillum Veri'. Robert Pohl and Demonstration Experiments in Physics after the Great War', in *Learning by Doing. Experiments and Instruments in the History of Science Teaching*, ed. by Peter Heering, and Roland Wittje (Stuttgart: Steiner, 2011), pp. 317–48.

32 Wittje, p. 341.

One important factor of this *longe durée* of Pohl's influence might have been that he neither supported nor opposed the Nazis in his words or actions.[33] Not even the emigration of his old friend Franck as a protest against anti-Jewish laws in 1933 was commented on by Pohl in any way. Certainly, because of this 'neutral' position, Pohl's institute remained unaltered, while there was an almost complete staff turnover at the neighboring institutes in 1933 — those of Born and Franck.[34] As Pohl was exonerated without any hassle after the war, he was able to continue his research and teaching activities without pause or limitations.

After his retirement in 1952, Pohl, who stayed in Göttingen, repeatedly reissued his textbooks until his death in 1976, when the latest edition of his *Einführung in die Optik* was published.[35] When his son Robert Otto Pohl (born 1929) — former professor of experimental physics like his father — published the eighteenth edition of *Einführung in die Mechanik, Akustik und Wärmelehre,* he was able to use Robert Wichard Pohl's extensive revision notes.[36]

Considering the great effort Pohl expended in revising his works and the dozens of admiring book reviews, the textbook doubtlessly plays a vital role in his teaching concept, but the written text alone comprises only part of the story. Throughout Pohl's books, several hundred black-and-white images are used — much more than the usual university textbooks of that time. These images were originally made using photographs of real experiments in Pohl's lecture hall; they were processed on high contrast positive paper that created an almost silhouette-like appearance that could easily be transferred to a printing block.[37] As Pohl and his lecture technician sometimes were part of experiments, they are also shown as silhouettes, for example, as when demonstrating Newton's third law (Fig. 7.5).

These silhouettes serve different purposes. First of all, they display experimental setups and the connected physical phenomena — another departure from textbooks of that time, at least to the extent exhibited by Pohl. Second, as representations of the authentic lecture situation at Göttingen, they function as a memory aid for Pohl's students and are the most suitable visual depiction of the shadow projection technique used by Pohl during his lectures. Last but not least, in mediating between the book's text and a common physical demonstration apparatus, they act as a visual index of useful experiments for physics educators.

How to Project (Shadows) in a Lecture Hall

It is not by chance that many of the contributions in this anthology on 'teaching with light projection' focus upon the magic lantern, which was established as the standard technology

33 For the Göttingen Physics Department during the Third Reich see Rosenow, Ulf, 'Die Göttinger Physik unter dem Nationalsozialismus', in *Die Universität Göttingen unter dem Nationalsozialismus*, ed. by Heinrich Becker, Hans-Joachim Dahms, and Cornelia Wegeler, Zweite, erweiterte Auflage (München [u.a.]: Saur, 1998), pp. 552–88; Rammer.

34 Rammer, pp. 31–34.

35 Pohl, Robert Wichard, *Optik und Atomphysik*, 13., neubearb. Aufl. (Berlin [u.a.]: Springer, 1976).

36 Pohl, Robert Wichard, *Mechanik, Akustik und Wärmelehre*, 18., überarb. Aufl. (Berlin [u.a.]: Springer, 1983), p. preface.

37 Achilles, Manfred, 'Göttinger Geschichten für das Erste Physikalische Institut', 2012, p. 20 <http://wwwuser.gwdg.de/~macadmin/G%C3%B6ttinger_Geschichten__Aufl.2012-2.pdf> [accessed 6 May 2020].

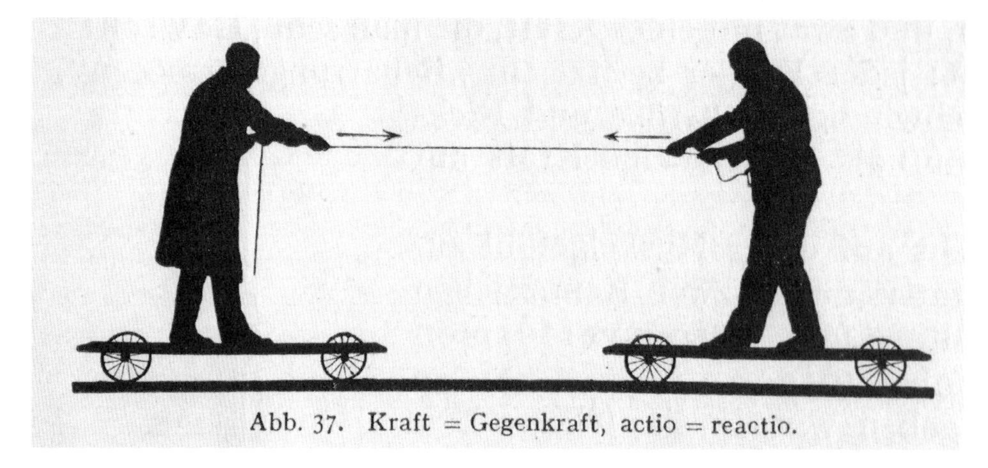

Fig. 7.5. Sperber (left) and Pohl (right) during an action-equals-reaction demonstration in Pohl's textbook. Robert Wichard Pohl, *Einführung in die Mechanik, Akustik und Wärmelehre*, 3. und 4., umgearb. und erg. Aufl. (Berlin [u.a.]: Springer, 1941), p. 20.

for still images in educational settings and environments in the late nineteenth century. In the form of the so-called 'epidiascope', an easy-to-handle, universal projector, the magic lantern was a conventional feature of lecture halls around 1900 and in later classrooms as well. The epidiascope could handle both translucent and opaque objects and was mostly used for lantern slides and illustrations on paper or in books, respectively.

However, scientific presentations include a much broader range of objects to be projected. In his analysis of the usage of magic lanterns throughout the history of scientific presentation, Willem Hackmann distinguished three different applications that were common practice around 1900. According to Hackmann, magic lanterns were utilised for presenting images on glass slides, for microscopic slides with preserved or freshly prepared specimens, and for downscaled demonstration experiments.[38]

Without a doubt, the third application was the most ambitious from a technical point of view, but as Lewis Wright pointed out in 'Optical Projection. A Treatise on the Use of Lantern in Exhibition and Scientific Demonstration' in 1906,

> [a]ny good ordinary lantern can, by a little modification, be made effective in almost any class of demonstration. It may have a good microscope fitted to it [...] interchangeable with the ordinary front. With a slide-stage open at the top, it will project chemical and other tank experiments. With accessory apparatus such as is presently mentioned, it will perform optical experiments. And by withdrawing the front, and arranging small apparatus in the rays from the condenser, it will project that apparatus.[39]

As there is not much space between the condenser and the front lens in a 'conventional' magic lantern, rackwork slides with movable parts like gear wheels or rotating glass discs

38 Hackmann, pp. 122–23.

39 Wright, Lewis, *Optical Projection: A Treatise on the Use of the Lantern in Exhibition and Scientific Demonstration*, ed. by Russel S. Wright, 4th edition (London: Longmans, Green, 1906), pp. 1–2438 (p. 153).

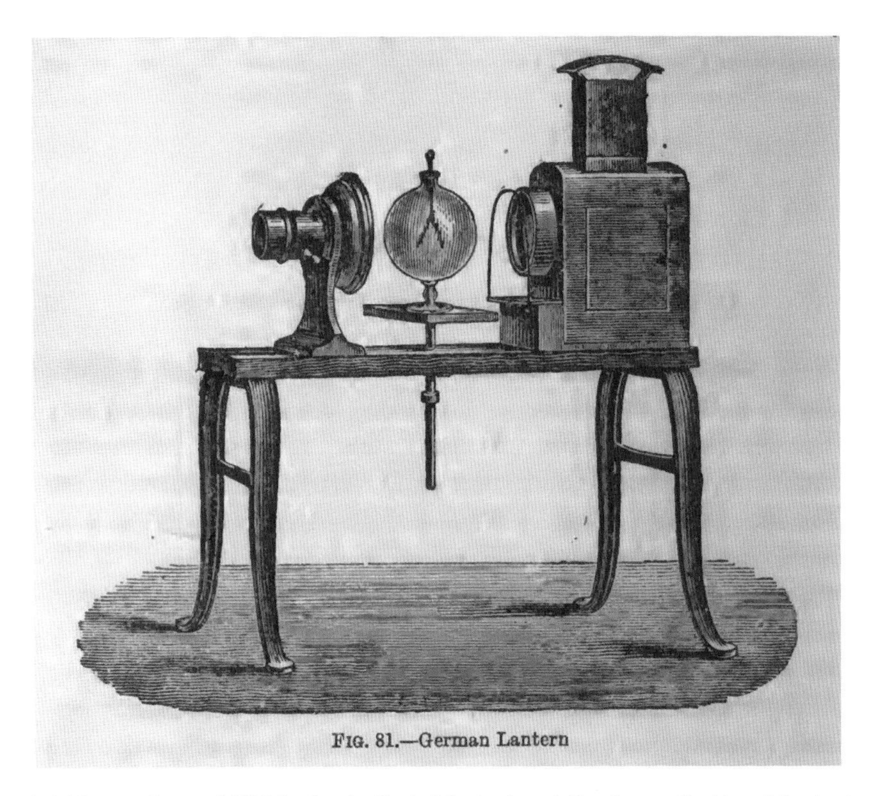

Fig. 81.—German Lantern

Fig. 7.6. A 'German Lantern'. Wright, Lewis, Optical Projection: A Treatise on the Use of the Lantern in Exhibition and Scientific Demonstration, ed. by Russel S. Wright, 4th edition (London: Longmans, Green, 1906), p. 154.

were developed to demonstrate complex phenomena like the steam engine or the revolution of planets.[40] To be more flexible and able to utilise larger, more complex setups, the magic lantern was broken down into its two functional parts with a gap in-between. One part was behind the object to be projected and consisted of the light source itself and a condenser to form an even, uniformly distributed light beam from a source like a limelight, spirit lamp, or arc light (see below). The other one was between the object and the projection screen and often consisted of only a simple but movable lens to achieve a sharp image on screens of varying distances.

Wright called this type of device 'The German Lantern' (see Fig. 7.6) as, according to him, 'it was first systematized in the schools of Germany'.[41] Wright contrasted this with the more compact, table-less, and mobile version he termed 'The American Lantern', which suggests that the designation might be due more to it being permanently mounted on a surface than the construction itself.[42] This points to the fact that such a flexible type of magic lantern was not only very common but existed in countless variations fitted to

40 Hackmann, pp. 128–30.
41 Wright, p. 153.
42 Wright, p. 153.

Fig. 442.

Fig. 7.7 A projecting device for demonstrations in lectures on chemistry. Stock, Alfred, 'Ein Projektionsapparat für die Chemievorlesung', Zeitschrift für Elektrochemie und Angewandte Physikalische Chemie, 17.23 (1911), p. 997 (with kind permission of Wiley-VCH and the Deutsche Bunsen-Gesellschaft für physikalische Chemie e.V.).

the needs of the respective application scenario. An example thereof is a do-it-yourself projecting device produced in 1911 that was based on the well-known epidiascope and was particularly suited for lectures on chemistry (see Fig. 7.7).[43]

---

43 Stock, Alfred, 'Ein Projektionsapparat für die Chemievorlesung', *Zeitschrift für Elektrochemie und Angewandte Physikalische Chemie*, 17.23 (1911), 995–1002 <https://doi.org/10.1002/bbpc.19110172306>.

All these types of lanterns may have worked fine for teaching lessons in most disciplines and classrooms, but lectures at universities, particularly in physics, put high requirements on projection infrastructure. First of all, some phenomena were extremely delicate to show due to a complex experimental setup or them being visible only on a sub-microscopic level. Whilst biological objects often are enlarged up to 1000 times, in physics, some phenomena must be enlarged by a factor of 60,000 for projection purposes.[44] Secondly, in lecture halls, there are often several hundred viewers, and thus large projection screens are needed to render phenomena visible to students even in the last row of seats. Against this backdrop, the article on the architecture of physics institutes in the comprehensive *Handbuch der Architektur* discusses the usual types of projecting equipment in-depth, as well as their integration into the lecture hall, and how to effectively block daylight, stating that electrical light sources are the first choice and have widely replaced older technologies like limelight or gaslight.[45] When the newly-built Physics Institute at Göttingen University was opened in 1905, the head of the Institute, Eduard Rieke (1845–1915), proudly explicated the extensive electrification of the whole building in great detail and especially highlighted the large experimental lecture hall.[46] As Rieke pointed out, lightbulbs were used to light the workspaces throughout the building, except for the large lecture hall, which was illuminated using arc lights.[47]

Arc lights were the most bright and reliable electrical light source at the turn of the century and were utilised in lecture halls as well as cinemas and theatres as a replacement for the dangerous and high-maintenance limelight. By using a direct current — as the alternating current would lead to a loud and annoying hum — a bright arc of light could be generated between two carbon rods that were mounted at a right angle, close together but not touching each other. With a current of 20 to 50 amperes, powerful models at the turn of the century (Fig. 7.7) got very hot and needed large housings; however, there were also considerably smaller arc lights with a current of 5 amperes available when Pohl started lecturing at Göttingen University in 1919. Some years later, he had one of these small devices permanently installed in his lecture hall for slide projection only (see Fig. 7.8). Although occupying a single seat in row nine of the auditorium, it was bright enough to project the usual 9 x 12 centimetre glass slides onto an area of 12 square metres or 4 x 3 metres, respectively.[48]

As can be seen in the figure, the formerly black-boxed magic lantern was fully dismantled with its elements mounted onto a profile made of steel. This 'Normalprofil' was used for optical benches in scientific laboratories until the late twentieth century and is still

44 Junk, Carl, 'Physikalische Institute', in *Handbuch der Architektur. Vierter Teil: Entwerfen, Anlage und Einrichtung der Gebäude. 6. Halbband: Gebäude für Erziehung, Wissenschaft und Kunst. 2. Heft, a: Hochschulen, zugehörige und verwandte wissenschaftliche Institute*, ed. by Eduard Schmitt, Zweite Auflage (Stuttgart: Alfred Kröner Verlag, 1905), pp. 164–236 (p. 193).

45 Junk, pp. 186–92.

46 Riecke, E., 'Das neue physikalische Institut der Universität Göttingen', *Physikalische Zeitschrift*, 6 (1905), 881–92.

47 Riecke, p. 884.

48 Pohl, Robert Wichard, 'Der physikalische Hörsaal der Universtität Göttingen', *Physikalische Zeitschrift*, 34 (1933), 408–10 (p. 410).

Fig. 7.8. Projecting device at Pohl's lecture hall around 1930. Family archive Pohl, Göttingen.

available by suppliers of teaching aids.[49] The physical elements of the magic lantern were now fully visible: an arc light on the left, a large condenser lens in front of it, a mounting frame for slides, and a projection lens on the right. It was, therefore, possible to literally see how projection devices work, rendering the magic lantern a teaching aid in and of itself. Moreover, the open construction was completely built out of standard laboratory equipment, which allowed for fast adaption if another slide size or projecting technique were needed. This 'amalgamation (*Verschmelzung*)' of an optical bench and a projection

49 *Dreikantschienen, Aufbaumaterial, Lichtquellen: Katalog SH 234*, ed. by Spindler u. Hoyer GmbH u. Co (Göttingen, 1980), p. 113; Conatex Lernsysteme, 'Optische Bank Mit Zubehör — Eco | Conatex Lehrmittel', 2020 <https://www.conatex.com/catalog/physik_lehrmittel/optik/optische_banke_zubehor/product-optische_bank_mit_zubehor_eco/sku-M3064> [accessed 30 November 2020].

Fig. 7.9. Physics lecture hall at Göttingen as opened in 1905. Göttinger Vereinigung zur Förderung der angewandten Physik und Mathematik, Die Physikalischen Institute der Universität Göttingen: Festschrift im Anschlusse an die Einweihung der Neubauten am 9. Dezember 1905 (Leipzig: Teubner, 1906), p. 61.

apparatus to a 'universal apparatus (*Universalapparat*)' was already described in 1905 by Volkmann as a standard technology of that time.[50]

Apparently, the small, mobile, and flexible arc lights came in so handy that Pohl wrote to the trustee of the University that he often used five of them at a time for demonstrations; fittingly, this letter was written in 1927,[51] the year the first of his three textbooks — the one on electricity — was published.[52] But to use so many high-quality light sources necessitated an appropriate projection area. When Pohl came to Göttingen in 1919, the institute's building had met the highest standards at its opening 15 years prior. Seating 150 students, the large lecture hall provided a substantial technical infrastructure with a sink, a fume hood, a large switchboard, and four wall-mounted instruments. But projection-wise, there was only a small permanent screen in front of the gallery and another, much smaller one as a roll-up curtain that covered the chalkboard when in use (see Fig. 7.9).

With a teaching style extensively incorporating projection, Pohl soon pushed the lecture hall beyond its limits. Student numbers had generally risen in the first decades

50 Volkmann, Wilhelm, *Der Aufbau physikalischer Apparate aus selbständigen Apparatenteilen* (Berlin: Springer, 1905), p. 48.
51 Göttingen University Archive, Kur. 7467, p. 383.
52 Pohl, Robert Wichard, *Einführung in die Elektrizitätslehre* (Berlin: Springer, 1927).

Fig. 7.10. The lecture hall after rebuilding in 1926. Family archive Pohl, Göttingen.

of the twentieth due to changes in the German educational system around 1900.[53] At Göttingen University, they were just reaching a new peak in 1919 with up to 600 physics students, five times more than before World War I.[54] Fortunately, Pohl had successfully applied for funding through the Rockefeller Foundation to enlarge the lecture hall, which happened between 1925 and 1926. He took this opportunity to fundamentally rebuild the lecturer's working area according to his needs. Thus, the turn-of-the-century-style lecture hall housing 150 people with its crowded area for demonstrations was replaced by a much larger one for 400 spectators with a white front wall and a large, empty stage with an oak parquet floor, 12 metres wide and 5 metres deep (see Fig. 7.10).[55]

Gone was not only most of the wall-mounted equipment but also the large, stationary table that carried the demonstration equipment and which, prior to Pohl, served as the de facto focal point during experimental lectures. Its function and position were thus taken over by a new type of equipment that Pohl developed for his projection-based teaching style, as characterised in the following section.

---

53 Pyenson, Lewis, and Douglas Skopp, 'Educating Physicists in Germany circa 1900', *Social Studies of Science*, 7.3 (1977), 329–66 <https://doi.org/10.1177/030631277700700304>.

54 Letter of Robert Wichard Pohl to his sister Margot, on 9 May 1919. Family archive Pohl, Göttingen.

55 Pohl, Robert Wichard, 'Der physikalische Hörsaal der Universtität Göttingen', p. 408.

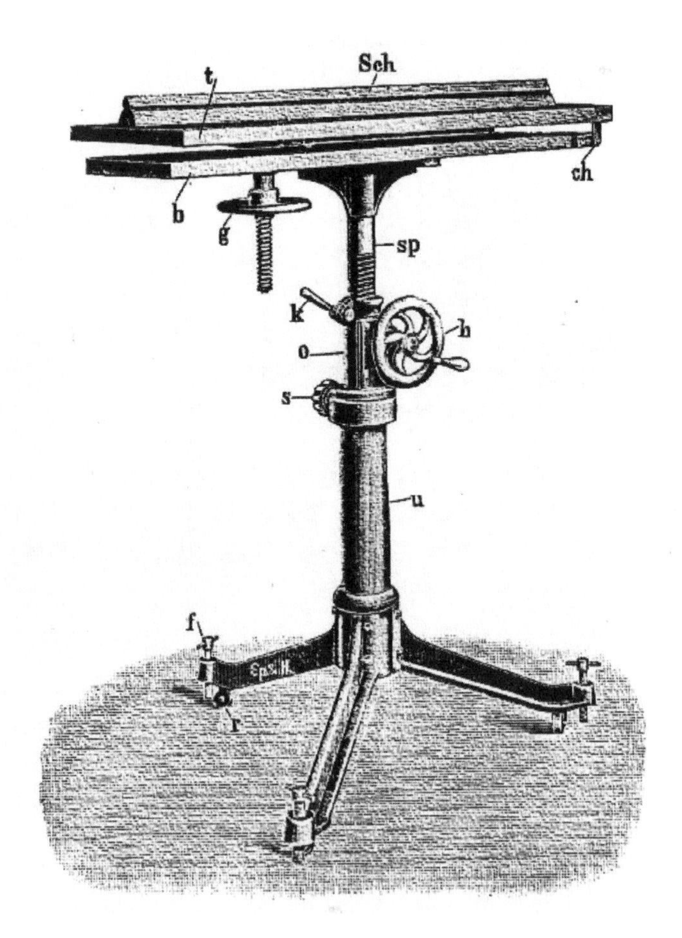

Fig. 7.11. Pohl's Experimental Table as sold by Spindler & Hoyer. Spindler & Hoyer, *Liste 50*, p. 1.

Projections on the Front Wall: Phenomena, Demonstration Apparatus, and Lecturers

The earliest depiction of Pohl during his lectures to date is a photograph that he himself found on his desk after a lecture in 1931, showing him and his mechanic Wilhelm Sperber (born 1892) preparing a demonstration (see Fig. 7.4). The picture must have been taken between 1919 and 1926, as the front wall is the one before the conversion of the lecture hall, although Pohl had already got rid of the large wooden tables that blocked the stage. Both men were standing in front of a three-legged rack with a large crank in its middle and a plate mounted atop carrying some equipment. The rack could be found in a Spindler & Hoyer catalogue from 1926 on and is called an 'experimental table in accordance with (Experimentiertisch nach) Professor R. Pohl'.[56] As one can see in the catalogue's illustration (see Fig. 7.11), the table features a Normalprofil on its top plate as well as transport castors.

56 Spindler & Hoyer, Liste 50: Der drehbare Experimentiertisch nach Prof. R. Pohl (Göttingen, n. d.), p. 4 <http://vlp.mpiwg-berlin.mpg.de/references?id = lit18183> [accessed 5 August 2020].

Furthermore, it can also be adjusted in both height (using crank 'h') and angle (using crank 'g') and is rotatable (using wheel 's'), as the table's full name suggests.

Although mobile tables for experimental demonstrations in lecture halls have been utilised earlier and might even have influenced Pohl, this particular construction seems to have been novel at the time of publication.[57] As the catalogue's description and the product name state, its most important feature is its ability to rotate. While a demonstration on a fixed table could not be seen equally well from every perspective in the auditorium, the lecturer could now rotate the setup in every direction while describing its parts before bringing it into a proper position for execution. Moreover, being rotatable, inclinable, and mobile, this type of table facilitates demonstrations that are most suitable for projection in every direction and onto every convenient surface.[58] And last but not least, with the integration of the Normalprofil from optical benches, a wide range of experimental devices on normalised sliders that were omnipresent in physics laboratories like lenses, diaphragms, prisms, or light sources could easily be utilised, as was the case with Pohl's already described slide projection apparatus. By 1910, this type of optical bench was already used for demonstration experiments as a publication of one of the most important German producers of optical equipment, Carl Zeiss Jena, shows.[59]

However, Pohl not only mounted the Normalprofil on his newly designed table, which was particularly useful in lectures, but also developed his own demonstration devices for it that partially worked like the downscaled 'rackwork' experiments used with a magic lantern that Hackmann described. One such device was the previously mentioned 'streamline apparatus' (*Stromlinienapparat*) that was used to demonstrate flow behaviour with the help of streaming water, ink, and small objects shaped like wings and ship trunks (see Fig. 7.2). While water from a tank above the apparatus would stream between two tightly mounted glass plates, ink from another much smaller tank would enter the stream through little holes; these would form coloured lines, while small objects could be placed in the stream to change the flow conditions.

While this apparatus might have been an innovative device in itself, the whole demonstration setup was not that far away from a slide projector in Pohl's manner (see Fig. 7.8), only this time installed on his experimental table. As can be seen in a photograph taken by the lecture-mechanic for his preparation (circa the late 1960s; see Fig. 7.12), the thin streamline apparatus was placed at the position of the slide frame (and still is during today's lectures on that topic). As a result, when the arc light was switched on, and the optical axis was oriented towards the front wall, an enlarged image of the streaming water-ink-blend was projected onto it. The only technical difference between a slide projector and this setup was an inverting prism at the front of the setup as lenses turn the (projected) image upside down. In a slide projector, one has to turn the slide upside down to overcome this effect, but a prism is necessary as the water would otherwise stream to the ceiling when projected.

Even though this setup consisted of only a few, mainly normalised parts, it had to be explained in detail to students from the wide range of disciplines that usually attended

57 Gerlach, Walther, 'Robert Wichard Pohl, 10.8.1884–5.6.1976', in *Jahrbuch der Bayerischen Akademie der Wissenschaften*, 1978, pp. 214–19 (p. 216).
58 Spindler & Hoyer, Liste 50, p. 3.
59 Zeiss Jena, Carl, *Gebrauchsanweisung für die Projektion optischer Versuche: Mikro 398* (Jena, 1910).

Fig. 7.12. Picture of the complete apparatus taken by an earlier mechanic for his lecture preparation. Teaching collection of the I. Physikalisches Institut, Göttingen University, unknown photographer.

Pohl's lectures. Accordingly, Pohl rotated the setup ninety degrees and used another arc light — this time without any optical elements — to generate a shadow image of the whole demonstration apparatus. As all elements on the Normalprofil are aligned on a single axis, the parts cannot overlap and thus are clearly visible on the wall in an icon-like fashion. Additionally, standard parts like the arc light with its characteristic housing or the condenser next to it are easily recognisable during other lecture sessions.

Consequently, Pohl did not restrict himself to lecture demonstrations in the field of optics when using his experimental table and projection technology to demonstrate the setup and the generated phenomena. Even experiments from the field of electricity were aligned on the optical bench to be shown in shadow projection using electrical parts that were permanently mounted on pins to fit the sliders of the Normalprofil.[60] Here as well, new types of devices were developed to obtain easy to grasp shadow images, as in the case of an electrostatic generator using translucent isolators that were shaded in print (see Fig. 7.13).

While bringing small but more or less complex devices into the optical axis of a projecting apparatus was a common practice during physics lectures at the turn of the century, projecting the whole demonstration apparatus as a shadow image before Pohl

60 Spindler & Hoyer G.m.b.H, *Liste 60/32: Apparate zur Elektrizitätslehre nach Prof. Dr R. W. Pohl* (Göttingen, 1932) <http://vlp.mpiwg-berlin.mpg.de/references?id = lit18152>.

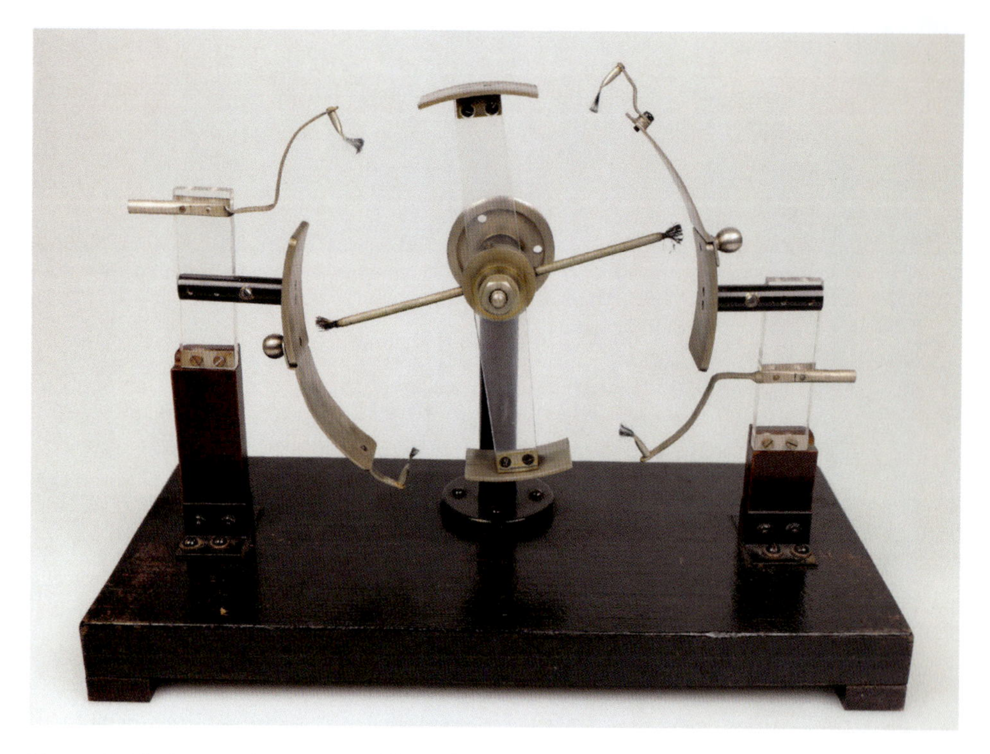

Fig. 7.13. Pohl's electrostatic generator with translucent isolators particularly suited for shadow projection. Teaching collection of the I. Physikalisches Institut, Göttingen University, photographer: Marie-Luise Ahlig & Lara Siegers.

was rarely done. Indeed, twenty years earlier, Wright, in his already quoted treatise on optical projection, tried to convince his contemporaries by addressing their reservations and lack of experience:

> I fear that this simple means of producing screen projection will be regarded with some scepticism, if not derision, by those ignorant of it. I can only state that the sharpness and precision of outline thus attainable will astonish those who try it for the first time; and that in all cases where a silhouette representation of an operation or piece of apparatus is sufficient, as it is in a vast number of experiments, it is perfectly satisfactory, and unequalled in its convenience for dealing with apparatus generally, of the usual size, as it comes to hand in most physical laboratories.[61]

In Germany, one additional factor might have been that the audiences were much smaller, and thus there was no need to make demonstration setups clearly visible to students more than ten metres away from the lecturer's desk. But, as there also was only a little space for this type of projection in front of an audience due to the omnipresent large, wooden experimental desk and the wall-mounted equipment, it required the commitment to

61 Wright, p. 218.

Fig. 7.14. Professor Ansgar Reiners and lecture-mechanic Michael Hillmann performing the demonstration from Fig. 5 during the introductory lecture on experimental physics in winter semester 2020/2021. Still from the streamed and recorded lecture, 18 November 2020.

getting rid of familiar teaching equipment and completely restructuring the way that demonstration experiments were constructed and performed.

Particularly attractive, at least to Pohl, was the opportunity to be part of the demonstrations instead of just commenting on a self-contained experiment from a distance. He developed dozens of mechanical demonstrations using his own body and that of his lecture-technician, as already shown in Fig. 7.5, with some of them still shown during introductory lectures at the Göttingen Physics Department today (see Fig. 7.14).

Reception

In 1951, the German-American geneticist Ernst Caspari (1909–88) shared his memories of the lecture by Pohl he had attended twenty years earlier in the *Journal of the American Association of Physics Teachers.* He specifically emphasised its performative aspect:

> A particularly impressive feature of Pohl's demonstrations is the extensive use of the human body. In a large number of experiments Pohl uses himself and his assistant as 'masses', and it is particularly these experiments which are long remembered by the students.[62]

Although the eminently positive reaction to Pohl's teaching style was shared by most of his students from Germany and abroad, some of his colleagues were less convinced and criticised the lectures as 'theatre' or 'circus'.[63] A more solid objection was that Pohl's

---

62 Caspari, Ernst, 'The Lectures of Professor Robert Pohl in Göttingen', *American Journal of Physics*, 19.1 (1951), 61–63 (p. 62) <https://doi.org/10.1119/1.1932703>.

63 *Crystals, Electrons, Transistors*, ed. by Michael Eckert and Helmut Schubert (AIP-Press, 1989), p. 99 <https://www.springer.com/gp/book/9780883186220> [accessed 3 April 2020].

pronounced focus on perfectly working demonstration experiments meant that the 'actual physics (*eigentliche Physik*)' got lost during the lectures.[64] Some experimental physicists of the day even extended this negative criticism to the research programme at his institute, stating that 'his results might have been wonderful theater plays but not serious research'.[65]

Pohl's performance strategy, however, seems to have been perfectly suited to his teaching goals. He sought to fill non-physics students with enthusiasm for physical knowledge and thinking as much as he worked to inspire the next generation of physics teachers who visited his lectures and used his textbooks. When Pohl won the Oersted Medal in 1959 — an American prize for notable contributions to physics education — the chairman of the Committee on Awards quoted some of the American students who had studied physics under Pohl in Göttingen:

> Let me assure you that during all my years, I have never seen more interesting lectures performed. Dr Pohl is a true artist. [...]
>
> Many of the methods which he introduced and which I first saw in Goettingen have since become widely used, though not as widely as they deserve. The use of shadow projection, for example, as a means of obliterating the unimportant and showing the essentials of a demonstration is one of the most magnificent ways to show what makes physics tick. All of us who have had the benefit of attending some of Pohl's lectures have absorbed some of the techniques of this master teacher.[66]

To practice and perform Pohl's teaching techniques was particularly easy as his experimental tables and many of his devices, including the streamline apparatus, were sold as teaching aids to schools and universities around the globe from 1926 until at least the 1970s. Even if one did not utilise shadow projection to show the apparatus, the plain setups with their few, simple parts were perfectly usable as didactic tools, especially in smaller spaces like classrooms where an enlarged image projected onto the wall was not necessary at all. In this context, Pohl's textbook, with its black-and-white silhouettes as a kind of setup instruction, could be treated as a demonstration manual — and was so — at least at the university level in Germany during the 1960s.[67]

In 1970, nearly fifty years after Pohl started teaching at Göttingen and sixteen years after his retirement, Pohl was asked to publish an article on the technique and usage of shadow projection in a comprehensive two-volume handbook on demonstration experiments focusing on the American educational system.[68] At that time, more modern media — closed-circuit television, film, and overhead projectors — were already well-established in physics demonstrations at universities, colleges, and schools, and for this reason, they were

---

64 Pick, Heinz, Oral history interview with Heinz Pick, 1981 October 2, 1981, p. 24 [transcript], American Institute of Physics. Niels Bohr Library & Archives. One Physics Ellipse, College Park, MD 20740, USA <http://repository.aip.org/islandora/object/nbla%3A270962> [accessed 22 April 2020].

65 'Seine Ergebnisse seien wunderbare Theatervorführungen, aber eigentlich keine ernsthafte Forschung.' Pick, p. 23 [transcript].

66 Overbeck, C. J., 'Robert Wichard Pohl: Oersted Medalist for 1959', *American Journal of Physics*, 28 (1960), 528–29 (p. 529) <https://doi.org/10.1119/1.1935872>.

67 Ruppel, W., 'Elektrizitätslehre', *Physikalische Blätter*, 20 (1964), 497–98 (p. 498).

68 Pohl, Robert Wichard, 'Shadow Projection', in *Physics Demonstration Experiments*, 1: *Mechanics and Wave Motion*, ed. by Harry F. Meiners (New York: Ronald Press, 1970), pp. 27–41.

Fig. 7.15. The new physics lecture hall of the Technische Universität Berlin in 1930. P.-L. Flouquet, 'Le Mobilier Moderne a Lécole', BATIR, No. 16, 1934, 598–600 (p. 600). The University's name is neither mentioned in the caption nor the article itself. However, a comparison of that image with a photograph from another source (F 8106, TU Architekturmuseum) indicates the location.

covered in articles in that same handbook.[69] It is hard to tell whether shadow projection was a widespread teaching aid, particularly in the US at that time, as newly-built lecture halls were adjusted to the needs of closed-circuit television and presumably never had the large, uniform white front walls needed for projections of several experimental setups.[70] Shadow projection apparently was unusual in physics lectures in Great Britain as well, as illustrated by Sir Lawrence Bragg (1890–1971), Emeritus Professor of the Royal Institution in 1970:

> [Shadow projection] is of course a well-known device, but in my experience lecturers are often surprised at its clarity when we suggest its use in an experiment. The clear-cut black and white is artistically effective, and one gets a surprisingly good three-dimensional illusion.[71]

---

69 Hoyt, Rosalie, 'Closed-Circuit TV in Physics Demonstrations', in *Physics Demonstration Experiments*, II: *Heat, Electricity and Magnetism, Optics, Atomic and Nuclear Physics*, ed. by Harry F. Meiners (New York: Ronald Press, 1970), pp. 655–69; Eppenstein, Walter, 'The Overhead Projector in the Physics Lecture', in *Physics Demonstration Experiments*, II: *Heat, Electricity and Magnetism, Optics, Atomic and Nuclear Physics*, ed. by Harry F. Meiners (New York: Ronald Press, 1970), pp. 719–30; Leitner, Alfred, 'Film as a Lecture Aid', in *Physics Demonstration Experiments*, II: *Heat, Electricity and Magnetism, Optics, Atomic and Nuclear Physics*, ed. by Harry F. Meiners (New York: Ronald Press, 1970), pp. 670–97.

70 e.g. White, Harvey E., 'New Physical Sciences Lecture Hall', *American Journal of Physics*, 33.12 (1965), 1050–55 <https://doi.org/10.1119/1.197148> on closed-circuit television; see *Modern Lecture Theatres*, ed. by Cyril John Duncan, Oriel Academic Publications (Oriel P., 1966) for an overview on contemporary western lecture hall architecture.

71 Bragg, Lawrence, 'Lecture Demonstrations in England', in *Physics Demonstration Experiments*, I: *Mechanics and Wave Motion*, ed. by Harry F. Meiners (New York: Ronald Press, 1970), pp. 13–26 (p. 18).

At least in Germany, however, the situation was markedly different. As early as 1931, a physics lecture hall 'arranged following the principles established by R. W. Pohl (*nach dem von R. W. Pohl eingeführten Prinzip eingerichtet*)' seating 1,000 students, with an unobstructed stage thirty-two metres wide and equipped with retractable chalkboards to enlarge the projection area was opened at the Technische Universität Berlin (see Fig. 7.15).[72] Pohl even wrote an article about the architectural principles he had established in Göttingen two years later, claiming that, 'The lecture hall had been a model for multiple reconstructions and new buildings'[73] elsewhere.

While Pohl's teaching aids are no longer available and lecture halls today may follow different design principles, his textbooks are still published.[74] As previously mentioned, in 1983, Robert Wichard Pohl's son, Robert Otto Pohl, published a revised edition of his father's mechanics based on an unfinished revision copy. Later on, he and colleague Klaus Lüders created editions of the complete works in German and English, with the latest published in 2017 and 2018.[75] In 2002 and 2003, with support of the Institut für Wissenschaftlichen Film, which was primarily aimed at producing educational movies for institutions of higher education,[76] Robert Otto Pohl and Klaus Lüders appeared in more than sixty short films presenting some of Pohl's demonstration experiments from the fields of mechanics, acoustics, electricity, and optics, likely because Pohl's demonstrations were rare in modern introductory lectures to experimental physics at least outside Göttingen. The movies were added to the reissued textbooks on DVD and today are provided as weblinks therein, and most of them are independently available online.[77] As mainly historical equipment and even the original lecture hall was used, and Robert Otto Pohl himself lectured together with his father, the technical side of the performance seems to be very authentic.[78] On the production end, however, the edits and close-ups disturb the impression of taking part in a lecture, and there is no commenting voice as these were added during postproduction in German and English. As moving images, though, these films are important didactical additions to textbooks based on a specific lecture in experimental physics that is rich in demonstrations.

Pohl's Teachings Style As an Innovative Anachronism

From today's perspective, it may seem obvious to capture demonstration experiments on film. How well this could have worked out even in the days of Pohl, however, can be

72 Westphal, Wilhelm H., 'Das physikalische Institut der TU Berlin', *Physikalische Blätter*, 11.12 (1955), 554–58 (p. 556) <https://doi.org/10.1002/phbl.19550111206>.

73 'Auch hat der Hörsaal schon mehrfach bei Um- und Neubauten als Vorbild gedient.' Robert Wichard Pohl, 'Der physikalische Hörsaal der Universtität Göttingen', p. 408.

74 Since 2003 one can find a slightly downscaled copy of Pohl's original lecture hall at the newly built Physics Department at Göttingen University.

75 Lüders and Pohl, *Mechanik, Akustik und Wärmelehre*; Lüders and Pohl, *Elektrizitätslehre und Optik*.

76 *Wissenschaftlicher Film in Deutschland*, ed. by Institut für den wissenschaftlichen Film (Göttingen: Institut für den Wissenschaftlichen Film, 1981), p. 56.

77 TIB AV-Portal, 'The Physics Experiments of Robert Wichard Pohl (1884–1976)', 2020 <https://av.tib.eu/series/305/the+physics+experiments+of+robert+wichard+pohl+1884+1976> [accessed 4 December 2020].

78 Wissler, B. F., 'Lecture Demonstrations', *Physics Today*, 12.11 (1959), 32–34 (p. 32) <https://doi.org/10.1063/1.3060566>.

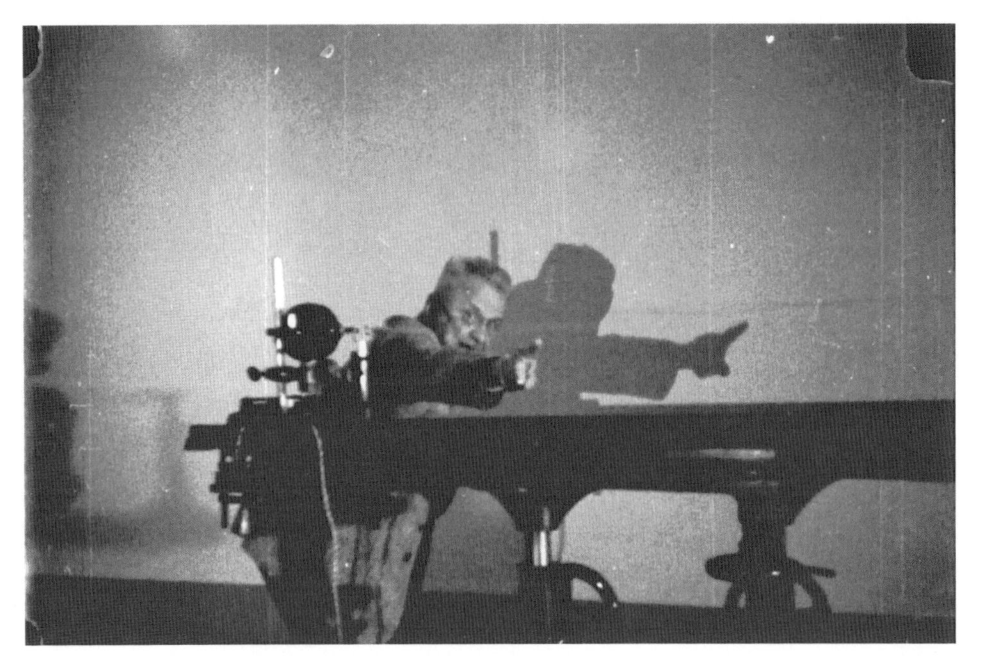

Fig. 7.16. Pohl presenting stationary waves using an animated rubber band. Filmstill, Lüty 1952, copy from the Family archive Pohl, Göttingen.

seen in a silent film that was taken in 1952 in one of Pohl's final lectures. The filmmaker, doctoral student Fritz Lüty (1928–2017), would later become a professor for physics at the University of Utah.[79] Lüty recorded snippets less than three minutes in total of a lecture Pohl gave on mechanical oscillations. To my knowledge, this is the only footage that exists of Pohl in performative action, presenting a setup using an animated rubber band to demonstrate stationary waves, among other things (see Fig. 7.16).

This document on Pohl's practice of lecture-demonstration, although short and fragmentary, gives a convincing impression of the usefulness of shadow projection in lectures and the attention that is achieved using spotlights, with the lecturer walking around on-stage and drawing the spectators' gaze to the apparatus and its operation. With proper camera work and editing, performance-lectures like this would have made excellent material for films, a medium already in use for educational purposes in the 1910s and 1920s.[80]

Unsurprisingly, as early as 1931, American physicist Robert W. Wood (1868–1955) visited Göttingen and encouraged Pohl to produce sound films of his lectures, especially for the American market as teachers there would learn 'through imitating the lecturer (*durch direktes Kopieren des Dozenten*)'.[81] However, there is no evidence so far that Pohl was interested in the medium of film for teaching, even when it came to mediating his particular

79 URL: http://www.edge-cdn.net/video_1153644?playerskin=37016 [accessed 4 December 2020].

80 For Germany e.g. Lampe, Felix, 'Der Lehrfilm. Sein Wesen und seine Verwendung', *Jahrbuch des Zentralinsitiuts für Erziehung und Unterricht, Vierteljahrschrift für Sozial- und Wirtschaftsgeschichte. Beihefte*, 3 (1921), 122–67, see also Hostettler in this volume for Switzerland.

81 Letter of Robert Wichard Pohl to his wife Tussa, July 26 1931. Family archive Pohl, Göttingen.

teaching style. In 1959, Pohl was a speaker at a conference on lecture-demonstrations in Middletown, Connecticut; despite the popularity of film projectors as teaching aids at schools and universities at the time, Pohl staunchly opposed the idea 'to let educational films take the place of real experiments (*Unterrichtsfilme an die Stelle wirklicher Experimente treten zu lassen*)'.[82] For him, lecturing experimental physics was about material setups producing actual phenomena supported by media — and not about replacing them with pre-recorded facsimiles. While in shadow projection, the apparatus is always in the audience's field of vision with its shadow behind; according to Pohl's didactical frame of reference, using live images from a camera would have led to a distracting attention shift from the demonstration setup to the monitor. This might explain why the historical glass slides within the teaching collection of the Göttingen Physical Department mostly show tables and charts rather than phenomena, apparatuses, or sketches of experiments, although thousands of respective photographs were taken for the textbooks and can still be found in the historical deposit. These 'info graphics' purely showing abstract data could be projected onto the lecture hall's front wall without competing with the physical setups and their pedagogical meaning.

But as already mentioned, a magic lantern-type of projection was used not only for slides, with one such device permanently mounted in the ninth row of seats in Pohl's lecture hall. As a modular setup on an optical bench suitable for projecting small demonstration apparatuses, the magic lantern was omnipresent in Pohl's lectures, albeit hard to identify as such. In combination with shadow projection using simple arc lights well-known from the theatre, the technical core of Pohl's teaching style was widely established around 1900 in the media-intensive sectors of entertainment and education. Thus, seen from a media history perspective, it was not so much the projection techniques being used that were innovative, but rather the creativity, scope, and, ultimately, the persistence of their application throughout all educational fields of experimental physics.

In Pohl's educational setting, which was grounded on physical apparatuses and their manipulation, shadow projection was the most commonly used medium, and it might have been a perfect choice. In his monograph on 'The semiotics of light and shadows: Modern visual arts and Weimar cinema', Piotr Sadowski states that a shadow is simultaneously an indexical sign physically caused by an object and an iconic sign resembling the forms of its referent.[83] As he writes, 'It is thus the combined emotive power of indexicality and iconicity that accounts for a truly 'magical', compelling effect of immediacy and suggestiveness produced by fleeting cast shadows or the fixed images of photography'.[84] In the 'live' situation of Pohl's lecture, the powerful, fleeting index-icon was thus able to mediate between the present apparatus representing physical reality and its abstract, mathematical, formulaic interpretation. To facilitate this transition, Pohl suggested drawing a diagram of the apparatus on the chalkboard while using shadow projection so that the 'students

82 Letter of Robert Wichard Pohl to his wife Tussa, June 23 1931. Family archive Pohl, Göttingen.

83 Sadowski, Piotr, *The Semiotics of Light and Shadows: Modern Visual Arts and Weimar Cinema*, Bloomsbury Advances in Semiotics (London: Bloomsbury Academic, 2018), pp. 13, 23.

84 Sadowski, p. 26.

can see clearly the transition from the diagram to the shadow projection and the brightly illuminated setup'.[85]

Not only did Pohl use projection techniques, he also built up a comprehensive physical environment attuned to the media he used. With his large sphere of influence due to the specific institutional situation at the Göttingen Physics Department with Pohl being one of four young, innovative professors and the only one with a continuous activity in research and teaching during Word War II and beyond, he enriched the educational reality of teaching physics. Pohl also extended the curriculum's spatial and material capabilities, rebuilding the lecture hall and designing new devices based on familiar laboratory equipment like the optical bench and its elements. This combination of a consistent use of standard tools and a strong orientation towards shadow projection led to simple setups that made his teaching principles highly adaptable to different settings, whether at another university, a Gymnasium, or a foreign media environment focused on overhead projection and film.

Thus, back in 1920s Germany, when the cinema attracted one million visitors each day,[86] and magic lanterns had arrived at every school, the most influential concept for teaching experimental physics was enacted in a darkened lecture hall, albeit not based on immersive media but on shadow projection instead. These shadows not only augmented an already complex teaching environment consisting at least of physical demonstration equipment, chalkboards, lantern slides, textbooks, and lecturers but might also be understood as a kind of 'umbrella medium' connecting the said elements on stage. This view of teaching aids and educational media as an 'ensemble' brought together to achieve specific pedagogical goals sheds new light on the history of teaching. While up to now, teaching aids like textbooks or films are often analysed separately within the research field of education history, it is apparent from the case of Pohl that they are inextricably linked for courses in physics education, as well as in other disciplines.[87] From a history of science perspective, meanwhile, Pohl's commitment to teaching and its far-reaching influence on physics education throughout the twentieth century at schools and universities alike is a strong argument that the history of lecture halls, teaching equipment, and performance practices is equally as important as the history of the laboratory, its instruments, and its actions themselves.

---

85 Pohl, Robert Wichard, 'Shadow Projection', p. 29.

86 Kaes, Anton, 'Film in der Weimarer Republik', in *Geschichte des deutschen Films*, ed. by Wolfgang Jacobsen, Anton Kaes, and Hans Helmut Prinzler (Stuttgart: J. B. Metzler, 2004), pp. 38–98 (p. 39) <https://doi.org/10.1007/978-3-476-02919-5_2>.

87 For natural history education see Klinger, Kerrin, and Michael Markert, 'In den sauren Apfel beißen. Naturkundliche Lehrmittelensembles in Historischer Perspektive', *Medienimpulse*, 4, 2016 <http://www.medienimpulse.at/articles/view/1004>; for biology education see Degler, W. and others, 'Staging Nature in Twentieth-Century Teacher Education and Classrooms', *Paedagogica Historica*, 56.1–2 (2020), 121–49 <https://doi.org/10.1080/00309230.2019.1675731>.

JOACHIM SCHÄTZ

DIY versus Ditmar 1006: The Economics, Institutional Politics, and Media Ecology of Classroom Projectors Made in 1950s Austria

In February 1950, the Austrian Ministry of Education's section for educational media had big news for Austria's teachers. The state organization, called the *Bundesstaatliche Hauptstelle für Lichtbild und Bildungsfilm* (the 'State Office for Slides and Educational Film' or SHB-Film), presented new, domestically produced models of slide and film projectors. These had been designed according to the specifications of SHB-Film, along with its financial support, to accommodate classroom use. As was touted on the cover page of the organization's monthly journal *SHB-Film-Post*, 'These devices only take into account the needs of the classroom and do not attempt a disruptive compromise with the special requests of home cinema, photo amateurs and other interested parties.'[1]

The aim of this paper is to take this claim seriously and investigate what 'the needs of the classroom' concerning slide and film projection were understood to be at this time and in this place. How were these needs defined in comparison to other modes of media use? While keeping slide projection in mind, I focus mainly on the new models for film projectors produced by Austrian companies in collaboration with SHB-Film. Chief among these is the 16 mm silent film projector that got top billing in the cited article: the Ditmar 1006. The projector marked the beginning of a decade-long, close collaboration on classroom media equipment with the company Austria Email. In 1960, both Austria Email and its only Austrian competitor, Eumig, stopped production of these 16 mm classroom projectors. This case study thus tracks a limited, but crucial, period of time for the renewed institutionalization of educational film in a number of European countries. German annexation and occupation had upended national efforts at educational film distribution

* This research was funded in whole by the Austrian Science Fund (FWF) P 32343-G (*Educational film practice in Austria*). For the purpose of open access, the author has applied a CC BY public copyright licence to any Author Accepted Manuscript version arising from this submission.

1 '*Diese Geräte nehmen ausschließlich auf die Bedürfnisse des Klassenzimmers Rücksicht und versuchen kein störendes Kompromiß mit den Sonderwünschen des Heimkinowesens, der Photoamateure und anderer Interessentenkreise.*' Anonymous, 'Ditmar 1006. Der neue österreichische Schulfilmprojektor. Pantax — Wica II. Die neuen österreichischen Kleinbildwerfer', *SHB-Film-Post*, 21 (1950), 1–5 (p. 1).

Joachim Schätz • University of Vienna, Joachim.schaetz@univie.ac.at

Learning with Light and Shadows: Educational Lantern and Film Projection, 1860-1990, ed. by Nelleke Teughels and Kaat Wils, TECHNE-MPH, 8 (Turnhout, 2022), pp. 197-216
© BREPOLS PUBLISHERS 10.1484/M.TECHNE-MPH-EB.5.131500

and presentation in a number of countries, even as it also established structures that some institutions would build upon after World War II.

However, taking the abovementioned claim seriously does not mean taking it at face value. There are two major caveats to the apparent transparency of the claim to cater to 'the needs of the classroom' that inform the argument and methodology of this article. Firstly, what these needs were perceived to be was contested at the time. Like the cited promotional text, many of the historical documents this article draws on (e.g., articles in *SHB-Film-Post*, guidebooks, slides on projector use, institutional correspondence) privilege the position of executive members of the national institution SHB-Film. Opposing positions are aired, however, and contended with in the pages of *SHB-Film-Post* as well. Persuasion rather than prescription was a necessity, given that both SHB-Film and the (independently organized) educational media offices of Austria's nine provinces were set up as services with only an advisory and supportive function for schools and teachers. These different institutions, along with their relative possibilities of influence, reflect the administrative structure of Austria's educational system. While the federal ministry is responsible for funding and supervising education, Austria's primary and secondary schools were at the time and are still administered by the country's nine *Bundesländer* (provinces). These primary and secondary schools — a four-year *Volksschule* (elementary school) and a four-year *Hauptschule* (middle school) — were just as prominent as tertiary education in SHB-Film's focus on classroom projection.[2]

While decrees concerning educational media were occasionally made at the level of individual provinces, SHB-Film as a national institution mostly exerted its influence by offering advice, services, and an infrastructure of distribution, but also equipment which was tailor-made to encourage certain ways of using media and discourage or prevent others. I thus propose to examine the projectors and their accompanying promotional and instructional media as part of efforts to steer teachers' educational film use within a range of options. These endeavours were typical for the push-and-pull between the makeshift and DIY usage of film by individual educators, and efforts towards its centralization and professionalization — a tension that was characteristic of educational film culture not just in Austria. My observations on state-supported, Austrian-produced, classroom film projectors aim to contribute to the European history of educational film practice in the 1950s. They are in line with and flesh out some aspects that have been brought up regarding the Netherlands or Switzerland, in interpreting the 1950s as a phase of consolidation and centralization of educational film practice.[3] In this period, national institutions were willing to forsake some of the possibilities of film use that had been promoted during the late 1910s and 1920s, the first period of institutional organization of educational film practice in Europe.[4] Ditmar 1006 was to some degree a counter to the DIY spirit enacted by individual schools and teachers' collectives.

---

2 Engelbrecht, Helmut, *Geschichte des österreichischen Bildungswesens. Erziehung und Unterricht auf dem Boden Österreichs. Band 5* (Wien: Österreichischer Bundesverlag, 1988), pp. 366–71, 410–31.

3 Gertiser, Anita, 'Schul- und Lehrfilme', in *Schaufenster Schweiz. Dokumentarische Gebrauchsfilme*, ed. by Yvonne Zimmermann (Zurich: Limmat, 2011), pp. 384–471 (pp. 447, 453); Eef Masson, *Watch and Learn. Rhetorical Devices in Classroom Films after 1940* (Amsterdam: Amsterdam University Press, 2012), pp. 70–74.

4 On this earlier wave of institutionalization, see: *The Institutionalization of Educational Cinema: North America and Europe in the 1910s and 1920s*, ed. by Marina Dahlquist and Joel Frykholm (Bloomington: Indiana University Press, 2019).

Secondly, an analysis of projector models (such as Ditmar 1006) as direct expressions of institutional policies (such as those of SHB-Film) threatens to abstract from the messiness, constraints, and compromises inherent in the construction, maintenance, and use of a projection device. This carries the risk of flattening the materiality of a machine into clear-cut meaning. The insight that projection equipment is not merely a neutral technological fact has been an essential contribution of 'apparatus theory' to film and media studies, condensed in the title of a 1980 Stephen Heath article, 'The Cinematic Apparatus: Technology as Historical and Cultural Form'. I will not follow classical apparatus theory here, though, or its recent historio-pragmatic reformulations. As immensely valuable as the latter have proven to be to the study of useful cinema (encompassing films used in contexts other than entertainment and art, including sponsored, industrial, scientific as well as educational films),[5] they retain from the earlier theories and their context of Althusserian ideological critique the implied viewer's subject position as their main point of interest.[6]

Instead, I want to take some modest cues from recent research into media practice, which centres not so much on the implied viewer as on the positioning of the teacher as the projectionist, which is arguably as prominent in SHB-Film's considerations. Also, as far as an analysis of technical equipment is concerned, concepts of practice derived from Isabelle Stengers and Andreas Reckwitz help to complicate any one-way reading of equipment as an 'expression' of ideas. Drawing on Stengers and Reckwitz, media scholars such as Petra Löffler, Karin Harrasser, and John Postill have stressed the non-rationalist and underdetermined dynamics of media practices, exploring the implications of this concept with different media technologies ranging from the piano to Flickr groups. They argue for a focus on the mutual definition and often blurry overlap of institutional policies with material limitations and affordances, budget considerations, divisions of labour, and allocations of information and agency.[7] In practices, these heterogeneous factors come together and mutually define each other in what Stengers has called 'reciprocal capture', which may result in stable relations for some periods as well as unexpected detours or failures (such as when a piece of equipment disappears from use).[8]

In this view, the material conditions of projection are neither just the expression of an institutional policy nor brute fact that an institution needs to plan around. Rather, they interact with other factors, such as choices and conflicts about how agency is distributed between a number of agents, from individual teachers to schools to provincial media offices, which were usually allied with but independent from SHB-Film due to the province-based administration of most schools. For instance, the latter were responsible for the instruction

5 Kessler, Frank, 'Notes on Dispositif', Unpublished paper, Utrecht Media Research Seminar, <http://frankkessler.nl/wp-content/uploads/2010/05/Dispositif-Notes.pdf>; Masson, *Watch and Learn*, pp. 99–125.

6 Schneider, Alexandra, 'Carole und Brenda. Das Paradox des Stativs oder eine Archäologie des Amateurfilms', in *Abenteuer Alltag. Zur Archäologie des Amateurfilms*, ed. by Siegfried Mattl, Carina Lesky, Vrääth Öhner, and Ingo Zechner (Vienna: Synema, 2015), pp. 58–70 (pp. 60–62).

7 Löffler, Petra, 'Ökologien medialer Praktiken', in *Materialität der Kooperation*, ed. by Sebastian Gießmann, Tobias Röhl, and Ronja Trischler (Wiesbaden: Springer, 2019), pp. 359–83; Harrasser, Karin, 'Vom kontingenten Kosmos zur Kosmopolitik', in *Kosmos & Kontingenz. Eine Gegengeschichte*, ed. by Philipp Weber, Tim Sparenberg, and Reto Rössler (Paderborn: Fink, 2016), pp. 276–87; Reckwitz, Andreas, 'Grundelemente einer Theorie sozialer Praktiken. Eine sozial- theoretische Perspektive', *Zeitschrift für Soziologie*, 4 (2003), 282–301; Stengers, Isabelle, *Cosmopolitics I* (Minneapolis: University of Minnesota Press, 2010).

8 Stengers, *Cosmopolitics I*, pp. 28–41.

of teachers in handling projection equipment. In 1948, this spawned a debate in the pages of *SHB-Film-Post* about how much technical expertise should be taught to teachers in these courses. Ferdinand Geiger, the head of the *Bezirksbildstelle* (the educational media office of the district) of Salzburg, argued that teachers should have a working knowledge of the machinery so as to be able to fix minor technical problems.[9] Against this plea for thorough technical schooling, the head of another province's *Landesbildstelle* (or provincial educational media office) cautioned that the confident tinkerers among the faculty often did the most damage to projection equipment with their 'repairs'. In the mind of this administrator, the repairs were supposed to be left to office staff or contract technicians.[10] Geiger touched upon another point that was less controversial in *SHB-Film-Post*: There were not only too few film projectors owned by schools or at least available to rent from the provincial educational media offices, as much equipment had been lost or sold off in the immediate post-war period.[11] Their use was complicated by the variety of types among the remaining projectors.[12] Standardization of projector types — both by acquiring new ones and by collecting identical types in the same province's educational media office — was seen as a prerequisite to careful, skilled use of machinery.[13]

Tending to a Niche Market: Economics and Ecologies

These discussions stressed the need for robust and not-too-varied projection equipment. SHB-Film's first concerted push for new equipment followed a few months later in the autumn of 1948: While the Ditmar 1006 was being designed, one hundred newly fabricated 16 mm projectors of the pre-war type Eumig PII were offered to schools at special rates. This was promoted as a first measure to answer the urgent need for equipment, and it may have helped in getting Eumig on board for the development of new projector types.[14] This first initiative already bore important hallmarks of later *Geräteaktionen* (equipment sales)[15] to come during the 1950s. For one, both film and slide projector equipment were offered — one hundred items each in the first initiative. While they were offered independently of each other, with no package deals, it was stressed that they were designed for joint use. Secondly, SHB-Film offered schools financial incentives to make the equipment affordable. The first initiative offered discounts, options for instalment payments, and a cash contribution by SHB-Film to incentivize schools and parents' associations to collect the necessary funds.[16] Similar offers and aid would continue to be part of the different

9 Geiger, Ferdinand, 'Über Technik und Methodik des Unterrichtsfilmes', *SHB-Film-Post*, 8 (1948), 4–7.

10 Köchl, Karl, 'Zur Frage der technischen Schulungen', *SHB-Film-Post*, 10 (1948), 1–3.

11 Schwarz, Adalbert, 'Die Lehrmittel Lichtbild und Film an den oberösterreichischen Pflichtschulen', *SHB-Film-Post*, 64 (1957), 1–4 (p. 1); Potsch, Oskar, 'Erfahrungen eines Bildstellenleiters', *SHB-Film-Post*, 13 (1948), 5–6 (p. 5); Haustein, Johann P., 'Vorläufer und Anfänge der Bundesstaatlichen Hauptstelle für Lichtbild und Bildungsfilm', *Sehen und Hören*, 48 (1970), 155–58 (p. 158).

12 Geiger, 'Über Technik und Methodik des Unterrichtsfilmes', pp. 5–6.

13 Anonymous, 'Vereinheitlichung der Gerätetypen', *SHB-Film-Post*, 8 (1948), 11.

14 Anonymous, 'Neue Geräte', *SHB-Film-Post*, 13 (1948), 7–8.

15 *Geräteaktion* (equipment sale) was the name given to campaigns in which new or revised equipment was presented by SHB-Film and offered to schools for a certain price until a declared end date some months later.

16 Anonymous, 'Neue Geräte'.

waves of equipment sales during the 1950s. In this period, both the pages of SHB-Film and individual school chronicles abounded with reports of local fundraising initiatives and negotiations with parent associations resulting in the purchase of a school projector, including free servicing by the provincial educational media office.[17] By 1952, SHB-Film already boasted a higher average of school film projectors per pupil than Britain, France, Italy, or Sweden, although its numbers (self-calculated from UNESCO data) should be taken with a grain of salt.[18]

More importantly than these individual measures, the overall design and production of film projectors exclusively for school use was presented not just as a pedagogical consideration, but as a price-saving strategy by SHB-Film. Starting with Ditmar 1006 in 1950 and Eumig P 25 S in 1953, these projectors were sold exclusively via SHB-Film, and exclusively to schools, thus not necessitating an advertising budget on the part of their producers. (Accordingly, there is little paid advertising to be found for these types of projectors.). Instead, SHB-Film presented itself as the only necessary broker between the producing companies and interested schools. They stressed that aside from any extra deals they could offer schools, they had already kept prices low by financing development with advances to the companies and guaranteed pre-orders (as far as I could reconstruct, this did not mean that SHB-Film bought all the projectors and sold them on — only that they had committed to do so if too few projectors were sold).[19] On SHB-Film's part, Technical Division Head Paul Ullmann maintained the collaboration with Austria Email and Eumig.[20] In a 1956 article in *SHB-Film-Post*, he paints both companies, as well as those concerned with slide projectors, as none-too-willing partners. Still, the two were the only domestic companies that had proven to have the necessary expertise with film projection equipment, so SHB-Film pressed on. Given the existence of domestic companies capable of delivering the equipment wanted, this state institution deemed it, in Ullmann's words, 'unacceptable and irresponsible to import school equipment to Austria' from abroad.[21] According to Ullmann, the Austrian companies had to be convinced to adapt their projectors to classroom needs in a piecemeal and informal process that he contrasts with the more formalized requirements for equipment made in West Germany. SHB-Film's West German equivalent, the *Institut für Film und Bild in Wissenschaft und Unterricht* (the Institute for Film and Image in Science and Education, or FWU), published a periodically updated document on its technical requirements for school equipment. If a company wanted approval for school use by FWU, it sent in a piece of equipment to be tested

17 Anonymous, 'Ein burgenländisches Beispiel', *SHB-Film-Post*, 16 (1949), 8; Anonymous, 'Ein Eumig P 25 S aus Kartoffelspenden!', *SHB-Film-Post*, 45 (1953), 4; Anonymous, 'Salzburg', *SHB-Film-Post*, 46 (1954), 10; Hübl, Adolf, 'Es geht vorwärts', *SHB-Film-Post*, 68 (1957), 1–2. For a report in a school chronicle from the 1950s, see: Anonymous, *Volksschule Brückl*, <http://wwwu.uni-klu.ac.at/elechner/schulmuseum/schulchroniken/vbr%FCckl1945.PDF>, p. 8.

18 Anonymous, 'Lesenswerte Zahlen', *SHB-Film-Post*, 33 (1952), 5.

19 H., 'Unsere Geräteaktionen 1957/58', *SHB-Film-Post*, 73 (1958), 3–4.

20 On Paul Ullmann's collaboration with the equipment companies, see also his personnel file at the ministry of education in Österreichisches Staatsarchiv: OeStA/AdR BMUKK SM 151: 'Betrifft: Kustos I. Kl. Dr Dipl. Ing. Paul Ullmann Beförderungsantrag', 4 August 1960.

21 *'Da fähige Firmen im Inland bestehen, wäre es ein Unding und auch kaum zu verantworten gewesen, Schulgeräte nach Österreich zu importieren.'* Ullmann, Paul, 'Schulgeräte II. Und so geschah und geschieht es bei uns', *SHB-Film-Post*, 63 (1956), 3–4 (p. 3).

according to these guidelines in FWU's media laboratory. SHB-Film did not operate such a laboratory, and it did not publish official technical requirements either. According to Ullman, individual 'wishes' for improvement were communicated to cooperating companies via letters or phone calls. He argues that in contrast to West Germany, there was far less business competition in the smaller pond of Austrian projector manufacturers, and this competition had to be tended cautiously. When the first, newly constructed models had met some demand, other companies wanted to compete with new equipment, forcing the original companies to revise their models.[22] Ullmann argues that SHB-Film's task was to slowly but steadily direct this competition towards technical improvements: 'Demands had to be measured out carefully according to the state of technological possibilities. Pushing demands too far would have made the companies resign.'[23] While Ullmann describes all this cautious and low-priced improvement as being managed by SHB-Film in the name of the schools, there is no mention of school educators' ongoing 'ground-level' demands being fed into the process. In contrast to the description of attentive cooperation with private companies, schools figure as an afterthought in this and other top-down descriptions in the pages of *SHB-Film-Post*.

Ullmann's claims about the hesitancy of the manufacturers is given some credibility by the fact that by 1958, both Eumig and Austria Email had decided to gradually stop producing 16 mm projectors altogether.[24] Instead of a clear-cut model of supply and demand, the production of Austrian-made classroom film projectors hews closer to a model of collaboration that Petra Löffler extracts from Stengers for what she calls an ecological approach to researching media practices. An ecological approach towards media practices in this context means stressing the relations between institutions, material properties, economic conditions, and social practices as constitutive of practices, without giving one of these elements overarching explanatory power.[25] I argue that this is an appropriate framework for understanding, for instance, the production and commissioning process described by Ullmann in which economic incentives are so much bound up with matters of institutional strategy and considerations of technological feasibility.

This approach, which stresses relations, also seems appropriate for many of the texts in *SHB-Film-Post* wherein teachers often blur didactics and logistics or shift between talking as classroom practitioners, institutional functionaries, or media pedagogues. Investigating media practices, Löffler argues, needs to take into account the heterogeneous interests, incentives, and obligations of the parties involved. Thus, she stresses the open-ended nature of collaborations, with the end or breakdown of collaboration being a constant option even without dramatic breaks.[26] In the case of Austrian-produced 16 mm projection equipment, the collaboration ended, as far as I can tell, for no other reason than that the two producing companies (which had been hesitant to get involved with SHB-Film in the first place)

22 Ullmann, Paul, 'Schulgeräte I. So macht man es in der Deutschen Bundesrepublik', *SHB-Film-Post*, 62 (1956), 1–2; Ullmann, 'Schulgeräte II', pp. 3–4.

23 '*Die Forderungen, entsprechend dem jeweiligen Stand der technischen Möglichkeiten, mußten richtig dosiert werden. Ein Überspannen der Ansprüche hätte Resignation bei den Firmen verursacht.*' Ullmann, 'Schulgeräte II', p. 3.

24 H., 'Unsere Geräteaktionen 1957/58', p. 4.

25 Löffler, Petra, and Florian Sprenger, 'Medienökologien. Einleitung in den Schwerpunkt', *Zeitschrift für Soziologie*, 4 (2003), 10–18 (pp. 10–14).

26 Löffler, 'Ökologien medialer Praktiken', pp. 360–63.

re-evaluated their product portfolio in light of the European Common Market established in 1957.[27] For Austria Email, which mostly produced lamps, metalware, and enamelware, film projectors had been an outlier to begin with. The company was closely affiliated with the Austrian state after World War II (a majority of Austria Email's shares were owned by the nationalized *Creditanstalt-Bankverein*).[28] But this affiliation had recently become less powerful, with *Creditanstalt-Bankverein*'s state share being reduced starting in 1956.[29]

SHB-Film's willingness to commission new equipment from domestic companies rather than importing existing equipment from foreign companies had more general functions, of course, than supporting companies affiliated to the state. It also made the issue of classroom films and slides, considered somewhat niche during the ongoing economic recovery of the post-war years, into a factor of domestic economic support.

Commissioning equipment from Austrian companies (slide projector equipment was produced by Optimar in Salzburg, among others)[30] helped SHB-Film to differentiate itself from West Germany's school media policies. In many respects, West German FWU, which possessed vastly greater resources to address German-language schools, was an important point of reference for SHB-Film. For instance, film subjects were chosen by SHB-Film to supplement, rather than duplicate, what FWU had on offer.[31] Moreover, both institutions still used a lot of films from their Nazi German predecessor organization, the *Reichsstelle für den Unterrichtsfilm* ('Reich Office for Educational Film' or RfdU, which was renamed *Reichsanstalt für Film und Bild in Wissenschaft und Unterricht*, the 'Reich Institute for Film and Image in Science and Education' or RWU, in 1940).[32] FWU was also an important model for SHB-Film concerning the technical requirements and priorities in its own school equipment. But while cordial relations were stressed, SHB-Film also occasionally made a point to set apart its priorities in school equipment, demonstrating that it was not simply an Austrian offshoot of FWU. In 1956, for instance, Ullmann took the first page of *SHB-Film-Post* to argue against FWU's preference for slide projection in daylight and for its own system of slide projection in dimmed light.[33]

The good deal for schools proposed by SHB-Film (low prices for tailor-made equipment) was supposed to cut the other way as well: SHB-Film officials constantly stressed the state organization's responsibility to support and foster national companies. By 1957, the school board of the Austrian province of Salzburg even decreed that schools were allowed to buy foreign-produced school projectors only in well-founded exceptions, after consulting the province's educational media office.[34] Opportunities for export, on the other hand,

---

27 H., 'Unsere Geräteaktionen 1957/58', p. 4.

28 Czeike, Felix, 'Austria Email AG', *Wien Geschichte Wiki*, 27 September 2017, <https://www.geschichtewiki.wien.gv.at/Austria_Email_AG>.

29 Anonymous, 'Creditanstalt-Bankverein AG, CA', *AEIOU — Austria-Forum, das Wissensnetz*, 25 March 2016, <https://austria-forum.org/af/AEIOU/Creditanstalt-Bankverein_AG,_CA>.

30 Further companies involved were Gerstendörfer, Pani, and Kahles. Haustein, 'Vorläufer und Anfänge', p. 158.

31 For this complementary use of FWU films, see for instance the FWU entries in *SHB. Bild — Film — Ton. Verleihprogramm 1971*, ed. by Franz Hubalek, Alois List, and Johann Schrodt (Vienna: SHB-Film, 1971).

32 In the 1963 catalogue for SHB-Film's film distribution, several hundred titles come from RWU/RfdU, including their original archival signature from that context: *SHB-Film. Filmverzeichnis 1963*, ed. by Hannes Schmid (Vienna: SHB-Film, 1963), pp. 39–58, 87–95, 103–04, 106.

33 Ullmann, 'Schulgeräte I', p. 1.

34 Anonymous, 'Ein beachtlicher Erlaß', *SHB-Film-Post*, 65 (1957), p. 11.

were constantly promoted. When the head of SHB-Film, Adolf Hübl, was tasked by UNESCO with helping build a school media organization in Turkey in the early 1950s, this was reported as an opportunity to spread its portfolio of Austrian-made classroom projectors abroad among other proud announcements of Ditmar projectors being sold abroad, including to Australia and India.[35]

Even taking into account the promotional character of SHB-Film's reports in its own journal, the sales campaigns since 1949 seem to have had impressive results in getting film projectors into Austrian schools until the late 1950s. In late 1946, only 1189 small-gauge projectors were in use in Austrian schools; by December 1954, there were 3601. Towards the end of Eumig and Ditmar school projector production, it was declared that Austria's 5425 schools were about to be equipped with a total of 5124 small-gauge film projectors — bringing Austria near the often declared goal of having one film projector in every school.[36]

Cultivating Classroom Film Projection: Technology and Practice

As the quote at the beginning of this text established, when the Ditmar 1006 projector was presented to educators by *SHB-Film-Post* in February of 1950, it was not merely an economic proposition. Rather, the first batch of equipment was promoted as focusing on 'the needs of the classroom' exclusively. To some degree, the distinction between classroom film projection (meaning projection not in a designated room or hall, but in any normal classroom) and other modes of film presentation was already baked into the most basic technical decisions following international protocols. Both Ditmar 1006 and all subsequent film projectors until 1960 had a 16 mm gauge. This had been chosen as the small-gauge format for educational films by the European association *Internationale Lehrfilmkammer* (International Chamber of Educational Film) in 1932.[37] By 1950, this format marked the projection as semi-professional rather than amateur (for which 8 mm was more widespread), while the lack of sound equipment on most of the Ditmar and Eumig classroom projectors distinguished them from many other semi-professional 16 mm uses. Again, this decision followed a broad international consensus found especially in German-speaking countries and held unwaveringly since the 1930s, that truly effective classroom film had to be silent.

Within this framework, then, what decisions concerning 'the needs of the classroom' were expressed via the 16 mm silent film projector Ditmar 1006? Two main decisions were heavily promoted by SHB-Film. Firstly, an unusually bright 100 volt and 500 watt lamp was used, according to the promotion 'erstmalig im Schulfilmgebrauch' (for the first time

---

35 Hübl, Adolf, 'Drei Monate als UNESCO-Delegierter in der Türkei', *SHB-Film-Post*, 28 (1951), 1–2; Hübl, Adolf, 'Ögretici Filmler Merkezi', *SHB-Film-Post*, 32 (1952), 1–2; Anonymous, 'Unser Schulfilmprojektor Ditmar 1006', *SHB-Film-Post*, 39 (1953), 12. While Turkish demand for Austrian film projection equipment seems to have been low, two different models of gas-fuelled slide projectors were designed in Austria specifically for Turkish schools not connected to electricity. One was made by Austria Email and the other one, clunkily named Türkilux, by Optimar. Anonymous, 'Türkilux', *SHB-Film-Post*, 36 (1952), 7; Anonymous, 'Kleinbildwerfer mit Petrolgaslicht', *SHB-Film-Post*, 44 (1953), 14.

36 Anonymous, 'Unsere Eumig-Aktion', *SHB-Film-Post*, 16 (1949), 7; Anonymous, 'Stand der Schulfilmgeräte', *SHB-Film-Post*, 55 (1955), 5; H., 'Unsere Geräteaktionen 1957/58', p. 3.

37 Filip, Josef, 'Aus der Pionierzeit des Unterrichtsfilmes', *SHB-Film-Post*, 19 (1949), 4–6 (pp. 5–6).

Fig. 8.1. The display of the voltmeter built into the Ditmar 1006 projector, with the attached series resistor on the left. – Josef Sikora, *Der 16mm-Schmalfilmprojektor 'Ditmar 1006'*. (Vienna: Bundesstaatliche Hauptstelle für Lichtbild und Bildungsfilm, 1950), slide 10 of 15. Collection of Pädagogische Hochschule Steiermark.

in school projection). This was to accommodate a clearly lit image one and a half metres wide during projection from the back of an eight metres long classroom.[38] As the lamp was one of the more expensive parts of the projector, each projector was delivered attached to a resistor (as was common at the time in order to make sure not too much voltage was used and to accommodate the voltage fluctuations that occurred frequently in the post-war era).[39] Each projector also had a voltmeter built into it (see Fig. 8.1).[40] A strong

38 Anonymous, *Der 16mm-Schulfilmprojektor 'Ditmar 1006'* (Vienna: Bundesstaatliche Hauptstelle für Lichtbild und Bildungsfilm, 1951, four-page informational brochure), p. 2.

39 Sikora, Josef, *Schmalfilmprojektion. Bildbeschreibung und Erläuterungen zur Lichtbildreihe U 1002* (Vienna: Bundesstaatliche Hauptstelle für Lichtbild und Bildungsfilm, 1948), pp. 15–17.

40 Sikora, Josef, *Der 16mm-Schmalfilmprojektor 'Ditmar 1006'. Bildbeschreibung und Erläuterungen zur Lichtbildreihe U 1117* (Vienna: Bundesstaatliche Hauptstelle für Lichtbild und Bildungsfilm, 1950), pp. 7–9.

projector lamp was considered imperative to the productive use of film during classes, because the image could be seen in dim (rather than thoroughly dark) classroom light. This was seen as a cost-saver as blackout curtains were another cost factor for schools. It was also considered a pedagogical advantage because dim light allowed for notetaking during projection and distinguished the work atmosphere of classroom projection more clearly from the commercial cinema experience.[41]

As with many prescriptions about educational film practice, this pedagogical consideration shaped the technological set-up of projection at some cost. But it also seems to have become more pointed as a principle only when it became practically feasible. When bright projector lamps were not available and a dark classroom was necessary for visibility, the darkness was described on the pages of the same publication as a productive precondition for student concentration on the moving image.[42] While not arbitrary in their relation, technological solutions and pedagogical recommendations mutually shape each other in 'reciprocal capture'.[43]

This may seem to reflect poorly on the makeshift nature of concepts and arguments advanced in journals and guideline books aimed at teachers as communities of practice. However, many texts are instructive precisely for their fine-tuned sense of education as a set of situated practices. Educational media practice is presented in these texts as determined as much by the practicalities of how to quickly blackout different types of windows[44] or where to mount which type of screen on the classroom wall[45] as by ideas about student attention and contemplation. Similar to recent ethnographic research into educational practice, these practical texts on the pages of *SHB-Film-Post* stress the dynamics and ramifications of school spaces as well as the participation of manifold objects (beyond the obvious teaching aids) in processes and protocols of classroom teaching and learning.[46]

The second innovation SHB-Film claimed for its set of projectors was the coordination between slide and film projectors. Slide projectors were fitted with lenses with an unusually large focal length (16 to 18 cm). This was done so that they would result in images that were the same size as the film projector's while being operated by the teacher from the same position at the back of the classroom (see Fig. 8.2).[47] Many conventional projectors would produce larger slide images from the same position. This was perceived by the chief pedagogues at SHB-Film and several provincial educational media offices as a hindrance

41 Anonymous, '10 Jahre SHB-Film', *SHB-Film-Post*, 55 (1955), 2–4 (p. 4); Anonymous, 'Schul-Kleinbildwerfer', *SHB-Film-Post*, 63 (1956), 7; Ullmann, Paul, 'Projektion in der verdämmerten Schulklasse', *SHB-Film-Post*, 71 (1958), 6–9.

42 Köchl, Karl, 'Vom Wesen und der pädagogischen Bedeutung des Unterrichtsfilms', *SHB-Film-Post*, 8 (1948), 1–4 (p. 4); Albrecht, Josef, 'Zur Film- und Bildmethodik', *SHB-Film-Post*, 20 (1950), 8–9 (p. 8).

43 Stengers, *Cosmopolitics I*, p. 36.

44 Anonymous, 'Die Frage der Verdunklung', *SHB-Film-Post*, 17 (1949), 4–5; H., 'Die leidige Verdunkelung', *SHB-Film-Post*, 73 (1958), 5.

45 Sikora, Josef, 'Vom mangelhaften Schirmbild', *SHB-Film-Post*, 23 (1950), 6–7; Albrecht, Josef, and Josef Sikora, *Lichtbild und Schmalfilm in Schule und Volksbildung* (Vienna: Verlag für Jugend und Volk, 1950), pp. 73–74, 83–86.

46 Cf. Röhl, Tobias, 'Unterrichten. Praxistheoretische Dezentrierungen eines alltäglichen Geschehens', in *Praxistheorie. Ein soziologisches Forschungsprogramm*, ed. by Hilmar Schäfer (Bielefeld: transcript, 2016), pp. 326–37.

47 Anonymous, 'Neue Geräte', pp. 7–8; Ullmann, Paul, 'Der Zusammenhang zwischen der Bildwurfweite und der Schirmbildbreite bei den österreichischen Schulprojektoren', *SHB-Film-Post*, 41/42 (1953), 19–20; Anonymous, 'Ditmar 1006', p. 5.

Fig. 8.2. The Ditmar 1006 projector together with a slide projector on a projection table, as recommended by SHB-Film's pedagogues. – Josef Sikora, *Der 16mm-Schmalfilmprojektor 'Ditmar 1006'*. (Vienna: Bundesstaatliche Hauptstelle für Lichtbild und Bildungsfilm, 1950), slide 15 of 15. Collection of Pädagogische Hochschule Steiermark.

to the smooth combination of films and slides, which was a goal of their educational use. The core of this argument was less didactic than topological. In short, either student learning from slides would suffer from projected images too big to take in as an entity, or teachers would be forced to change their position and cross the distance between the film and slide projector.[48]

This seemingly minor point was consistently expounded on in the pages of *SHB-Film-Post*. The discussions stressed that pedagogically useful classroom projection needed to consider

48 Sikora, Josef, 'Über Brennweite und Schirmbildgröße bei Schmalfilm- und Kleinbildprojektoren', *SHB-Film-Post*, 12 (1948), 12–13; Hummer, Franz, 'Kombinierte Verwendung von Steh- und Laufbild (Dia und Film). Stundenbild: Tunesien', *SHB-Film-Post*, 20 (1950), 10–11 (p. 10).

both the positioning of the students and the teacher, who was supposed to be operating the machines. For the students to have an unobstructed view of the projected image, the projection equipment was supposed to be at the back of the classroom. Yet with many normal slide projectors, according to Ullmann, this position far back in the room would have made for an image too large in size to take in well from the first rows.[49] From the teachers' perspective, the position behind the students is described as one of overview and control of both the projected image and the class. This positioning next to *both* the film and slide projector was also seen as affording the teacher free choice to switch between slides and film projection without changing their position (which would have been necessary, had the slide projector been positioned closer to the screen). The differentiation from amateur film projection came into play when Ullmann argued that in contrast to the amateur projection of home equipment for their own enjoyment, the projectionist-teacher did not need to see the screen image well in class as they were already prepared for the content.[50] These arguments were doled out repeatedly during the 1950s, occasionally to warn teachers off new-fangled, expensive amateur equipment that was not attuned to the abovementioned standards of classroom projection, such as slide projectors offering automatized alternation of images via remote control. While Ullmann stressed in this article that schools were independent in their choice of equipment, the admonishing tone of his text worked hard towards keeping them in line with SHB-Film's recommendations.[51]

As far as these heavily promoted features are concerned, the Austrian-produced film and slide projectors can be understood as a manifestation of SHB-Film's tenets of preferable classroom film practice. In other aspects, it is hardly possible to decide whether decisions about the Ditmar 1006 were made pointedly or merely to cut corners. For instance, the Ditmar 1006 was sent out with a rather short bobbin arm that could only accommodate a film length of 120 meters, or eleven minutes of film at a speed of twenty-four frames per second. This was in keeping with longstanding distinctions between the *Unterrichtsfilm* as a true teaching aid for classroom use, understood to be no longer than a few minutes, versus longer and more 'generally educational' types of *Lehrfilm*, as well as with the frequent length of films both produced by SHB-Film and from the large catalogue of former RfdU/RWU films.[52] This distinction had been developed in the 1920s and was one of the common tenets among the members of the *Internationale Lehrfilmkammer* (International Chamber of Educational Film), an early international association of educators interested in film centred in Basel with a strong contingent in Austria. Many of this generation were still active in film education in 1950s Austria, including SHB-Film's director Adolf Hübl, and

49 Ullmann, Paul, 'Über die Schirmbildgröße bei der Klassenprojektion', *SHB-Film-Post*, 66 (1957), 6–10.

50 Ullmann, 'Über die Schirmbildgröße', p. 7.

51 Sikora, Josef, 'Brauchen unsere Schulen ein "Maschinengewehr" für den Unterricht mit Stehbildern?', *SHB-Film-Post*, 64 (1957), 4–7. For another warning about equipment on offer, see: Anonymous, '"Tageslicht-Projektionswand" für Durchlicht- und Auflichtprojektion', *SHB-Film-Post*, 62 (1956), 6.

52 On the programmatic differentiation between *Lehr-* and *Unterrichtsfilm*, see: Wiener Arbeitsgruppe der V. Kommission (Lehrfilmprüfung), 'Neue Einteilung der Filme, Begriffsbestimmungen, Anforderungen an Unterrichtsfilme, Richtlinien für Filmprüfung', in *Fortschritte Österreichs im Lichtbild- und Lehrfilmwesen. Bericht für die III. Internationale Lehrfilmkonferenz in Wien (26. Bis 31. Mai 1931) auf Grund von Einzelberichten und amtlichen Erhebungen*, ed. by Gustav Adolf Witt (Vienna and Leipzig: Österreichischer Bundesverlag für Unterricht, Wissenschaft und Kunst, 1931), pp. 158–61.

the concept of *Unterrichtsfilm* still informed the choice of subjects deemed appropriate for classroom film. These were subjects that stressed discrete visible processes in motion, such as biology, crafts, and industries or even (via animation) geometry, rather than matters like history or ethics, which were dependent on an argument made by editing and/or a commentary. Films on the latter two matters would only become less marginal in Austrian classroom projection after 1960.[53] So, was the short length of the bobbin arm meant to curtail the projection of films other than short *Unterrichtsfilme*? Maybe, but for a small extra fee, bobbin arms of double the length were also issued, and became the standard when a revised model called Ditmar 1006N was released in 1954.[54]

A more substantial confluence of pedagogical and financial considerations marked the issue of film sound. The preference for silent (or, as Eef Masson suggested for classroom film, mute)[55] film projection that SHB-Film shared with many national educational institutions in Europe throughout the 1950s was strongly motivated by the concept of the teacher's voice, personality, and knowledge of their class guiding student perceptions.[56] But given the big price tag for sound projection equipment, it also seemed strategically prudent. In 1953, Austria Email and SHB-Film presented their only 16 mm sound film projector for classroom use, not so imaginatively called Ditmar 1106. This optical sound projector's special price for schools was five times the amount asked for the silent Ditmar 1006N projector in 1955 (16,957 in contrast to 3350 Austrian Schillings).[57] Rather than totally bypass sound films, it was suggested that they be projected silently in class.[58] This was something Ditmar 1006 was equipped to do, according to SHB-Film rather uniquely among the other silent film projectors used by Austrian schools at the time. Unlike most silent projectors, the Ditmar 1006's film transport was sprocketed only on one side; this allowed it to run 16 mm sound prints through the projector without destroying the soundtrack on the side.[59] Like many properties described for Ditmar 1006, this was also kept on for the new Eumig model presented in 1953: The Eumig P 25 S was mostly distinguished by the use of lighter materials that predisposed it for mobile use, especially in terms of being sent out to different schools by a district educational media office.[60]

Around 1960, SHB-Film's chief pedagogue Adolf Hübl hesitantly followed international trends in conceding that sound film could have a place in classroom education.[61] Mostly,

---

53 For this shift, see the two *SHB-Film* catalogues from 1963 and 1971, respectively: *SHB. Bild — Film — Ton. Verleihprogramm 1971*, ed. by Hubalek and others; *SHB-Film. Filmverzeichnis 1963*, ed. by Schmid.

54 Anonymous, 'Unsere Geräteaktionen', *SHB-Film-Post*, 55 (1955), 4–5.

55 Masson, *Watch and Learn*, p. 261 n. 29.

56 On the preference for silent film, see for instance: Anonymous, 'Auf einer Filmtagung in Bern', *SHB-Film-Post*, 15 (1949), 7; Köchl, Karl, '"Bilder bilden!" Gedanken zur Pädagogik der Film- und Lichtbildarbeit in der Schule', *SHB-Film-Post*, 29 (1951), 1–3 (p. 1). For Switzerland, see: Gertiser, 'Schul- und Lehrfilme', pp. 468–69.

57 Anonymous, 'Unsere Geräte', *SHB-Film-Post*, 57 (1955), 1–12 (pp. 3–5). The former price equalled ten times the annual salary of the average Austrian industrial worker in 1956. Anonymous, 'In 50 Jahren gewaltige Preissteigerungen', *kaernten.orf.at*, 10 October 2006, <https://ktnv1.orf.at/stories/142722>.

58 Despite the preference for silent films, a number of sound films were distributed by SHB-Film from the early 1950s on, including films gifted to SHB-Film from Marshall plan promotion campaigns and other image films.

59 Anonymous, 'Ausgabe von Tonfilmen an die Landesbildstellen.', *SHB-Film-Post*, 33 (1952), 4.

60 Anonymous, 'Ein neuer Schulfilmprojektor: Eumig P 25 S', *SHB-Film-Post*, 43 (1953), 1; Anonymous, 'Die österreichischen Schulfilmprojektoren', *SHB-Film-Post*, 44 (1953), 6–11 (p. 11).

61 Hübl, Adolf, 'Tonfilm als Unterrichtsfilm — wo?', *SHB-Film-Post*, 86 (1960), 1–5; Hübl, Adolf, 'Die pädagogische Bedeutung des Tonfilms an der Pflichtschule', *SHB-Film-Post*, 95 (1962), 5–10; Hübl, Adolf, 'Die pädagogische Bedeutung des Tonfilms an der Pflichtschul (Fortsetzung)', *SHB-Film-Post*, 96 (1962), 6–8.

though, sound film was seen as the domain of educational efforts towards a more general audience. Accordingly, the sound film projector Ditmar 1106 was pitched by SHB-Film not exclusively to schools, but also to not-for-profit sites of popular education.[62] With bobbin arms long enough to carry a feature-length 1200 metres of sound film, this projector type was better equipped for the overlap between educational film and a broader culture of film education and appreciation that institutions of popular education like Uranias and adult education centres specialized in.[63]

In contrast to this broader appeal, the silent projectors were carefully tuned to flexible use in school classrooms in the domain of the teacher. Relevant aspects ranged from the absence of sound and the length of the bobbin arm to the strength of the light source and even the lens of the accompanying slide projectors. This did not mean, however, that these film projectors were aimed at giving the teacher a maximum of flexibility in how to use film.

Shifting the Limits of Projection: Reel and Performance

What is most striking about Ditmar 1006 and its accompanying texts is that the agency of the individual teacher (which was given as a main reason for the lack of film sound and the combination of film and slide projection) was not an absolute in the design of the projector. One instructive trade-off in the projector's construction involved a rather prosaic element, the rotary disc shutter. The shutter was designed to be narrower than usual, with two shutter wings instead of the customary three, so the image would lose less light during projection. As a consequence, this slender shutter meant that the moving image would start flickering if projected with a speed of less than twenty-four frames per second. This was not deemed a problem, though, because, as one informational text in *SHB-Film-Post* put it, 'all Austrian educational films' (meaning here, all films centrally commissioned or produced by SHB-Film) 'are being shot at twenty-four frames per second and meant to be projected at that speed'.[64]

This may read like a quaint statement, but it stands in stark contrast to the ideas of exploratory film use that had been envisioned for classroom projection during the 1920s. In Austria, the *Schulkinobund* (Federation of School Cinemas), the national association for film in school, had explored such ideas in its journal *Das Bild im Dienste der Schule und Volksbildung* (The image in the service of schools and popular education, or *Das Bild*, which ran from 1924 to 1930). During this period, classroom projection was an idea more than an actual practice in Austria. Due to fire protection regulations that were in part even stricter than in Germany, film screenings for classes mostly took place in designated *Schulkinos* (school cinemas) with separate projection booths, which were usually filled with several classes at once.[65]

Even in that period, classroom projection was openly described as the best environment for film use in schools. Only in class, the argument went, would it be possible to really fit

62 Anonymous, 'Letzte Nachricht', *SHB-Film-Post*, 44 (1953), 2.

63 Anonymous, 'Unsere Nachricht', *SHB-Film-Post*, 63 (1956), 4–9 (p. 6).

64 '*Da alle österreichischen Unterrichtsfilme mit 24 Bildern je Sekunde gedreht und vorgeführt werden, ist trotz der Zweiflügelblende eine flimmerfreie Projektion möglich.*' Anonymous, 'Die österreichischen Schulfilmprojektoren', p. 7.

65 Albrecht, Josef, 'Zur Didaktik und Psychologie der Verwendung des Lichtbildes im Unterricht', *Das Bild im Dienste der Schule und Volksbildung*, 1 (1929), 2–6 (pp. 2–3).

films into an individual class's curriculum and the flow of a lesson with active tasks for the students. Apart from early experimentation, this practice was eventually facilitated by the use of small-gauge film around 1930. Indeed, this understanding of classroom film projection as the contextualized presentation and active discussion and consideration of a short film (which Eef Masson has fittingly called 'embedded use' of educational film)[66] was also the consensus best-practice model among the educators writing in *SHB-Film-Post*. For instance, this short, programmatic text was repeatedly printed in a 1955 special issue of *SHB-Film-Post* doubling as a catalogue of titles distributed by SHB-Film: 'To utilize films correctly means educating the students to observe them independently and to think logically! Incorporate the film into class, place it at the center of a lesson!'[67] While occasional, unstructured screenings to a couple of classes were damningly compared to colourful travelling cinema programmes, *SHB-Film-Post* published detailed examples of how to construct a lesson around a new film under the rubric *Methodik und Technik* (Methodology and technology).[68]

Still, the limits of what this classroom use of film entailed had narrowed in the period between the interwar journal and the post-war one. In the earlier period of expectation and experimentation, manipulation of the projector by the teacher-projectionist was stressed, rather than robust and easy usability. In the pages of *Das Bild im Dienste der Schule und Volksbildung*, there was agreement that classroom film projection should include flexible use of the film strip, including rewinding and replaying sequences, projecting them at slower speeds or (especially) halting the image for closer inspection. A lot of ink was spilled over innovative technologies cooling the projector light while projecting an individual image, so as not to harm the print, often with the assumption that such flexible handling was a necessary precondition for the educational use of film.[69] In 1929, Federation of School Cinemas Chairman Josef Filip deemed devices for freezing the moving image mid-projection feasible and almost indispensable even for the more collective, disruptive screening practice in 'school cinemas'.[70]

Like many demands made by Austria's proponents of classroom projection, this was in line with opinions held by other European associations (and often dependent on engineering know-how in other countries). Moreover, the demands for freezing an image and rewinding mid-projection were not left off as soon as classroom projection became feasible on a broad

---

66 Masson, *Watch and Learn*, p. 112.

67 '*Filme richtig auswerten, das heißt die Schüler zur selbständigen Beobachtung, zu logischem Denken erziehen! Den Film in den Unterricht einbauen, ihn in den Mittelpunkt einer Unterrichtsstunde stellen!*' Anonymous, 'B. Verwendbarkeit der Filme nach Schulstufen und Gegenständen', *SHB-Film-Post*, 52/53/54 (1955), 67–103 (pp. 68, 79, 87).

68 For instance, see Albrecht, Josef, 'Bericht über Lehrproben', *SHB-Film-Post*, 9 (1948), 9–11; Schwarz, Ferdinand, 'Wie ich den Unterrichtsfilm verwende', *SHB-Film-Post*, 22 (1950), 4–5.

69 Wiedemann, Karl, 'Stillstand- und Rücklaufeinrichtungen an Kinoprojektoren und deren Bedeutung für Schulen und Volksbildungsstätten', *Das Bild im Dienste der Schule und Volksbildung*, 1 (1927), 15–18; Filip, Josef, 'Die 8. Deutsche Bildwoche in Dortmund', *Das Bild im Dienste der Schule und Volksbildung*, 12 (1927), 209–13 (p. 213); Sellnick, G., 'Kinematographischer Unterricht im Eisenbahnwesen', *Das Bild im Dienste der Schule und Volksbildung*, 5 (1927), 86–89 (p. 87); Jordan junior, Ernst, 'Feuerschutz im Bildwerfer', *Das Bild im Dienste der Schule und Volksbildung*, 4 (1928), 75–77; Zwiener, Bruno, 'Filmband und Zeitlupe im neuzeitlichen Zeichenunterricht', *Das Bild im Dienste der Schule und Volksbildung*, 1 (1928), 1–5 (pp. 3–4).

70 Cf. Filip, Josef, 'Einrichtung und Organisation von Schulkinos', *Das Bild im Dienste der Schule und Volksbildung*, 3 (1929), 56–60 (p. 57).

basis in the mid-1930s. For instance, a 1936 leaflet by the Nazi German educational film organization RfdU prominently noted which of the recommended classroom projectors featured devices for halting and rewinding mid-projection.[71] The RfdU's guidebook on the technical instruction of teachers included instructions for these features as well.[72]

This enthusiasm for the flexible use of film via projection had waned considerably by 1950. In that year, the new official teachers' manual for instruction courses in Austria stated that projecting a halted image, while no longer a fire hazard due to safety film and cooling technology, was still likely to make the print brittle and damage the image. Likewise, the backwards projection of film was said to frequently damage the print, while usually achieving nothing more than the fleeting amusement of pupils.[73] In a later text, SHB-Film's chief technician and Ditmar projector mastermind Paul Ullmann also stated his suspicion that teachers would not stop the image for educational effects, but for the sake of amusement — thus being in line both with the use of film for entertainment and with an amateur culture of home projection that SHB-Film was working hard to distance itself from.[74]

Accordingly, the Ditmar 1006 and the Eumig P 25 S, as well as subsequent revised models, were constructed so the projection light would switch off automatically, with no way of circumventing it, as soon as the film was halted or run backwards. In Ullmann's consideration, this was very much not a mistake or compromise, but an intended feature. A promotional text bluntly called these projection options 'superfluous and undesirable in school use' and expressly stated that the necessary parts were 'left off with full intent'.[75] Indeed, Ullmann went so far as to state that repeating an individual section of a film was not within the purview of the teacher operating the projector — only the editor working on the film print.[76] Even rewinding to review a certain passage was thus more than the teacher-projectionist was allowed to do.

Synthesizing these points, an analysis of the Ditmar 1006 suggests SHB-Film's vision of a clear division of labour in educational film practice in 1950s Austria. While the teacher's voice still reigns supreme over the mostly silent film projection, the film as produced or edited by the national institution is a closed-off pedagogical unit, to be accepted as such. This development, since the 1920s, concerning teacher agency is in line with some observations offered by Eef Masson for the Netherlands' SHB-Film counterpart NOF for the same timeframe.[77] From this vantage point, the stress on having slide images the same size also seems a compensation for the foregone promise of the halted film image. Accompanying slides, which were sometimes produced simultaneously with new film, offered a centrally pre-selected set of still images to choose from, managing the teacher choices. Obviously not everybody was happy with that option, as Ullmann in 1954 saw the

---

71 Anonymous, 'Merkblatt' (reprint), in *Neue Dokumente zur Geschichte der Schulfilmbewegung in Deutschland II*, ed. by Malte Ewert (Hamburg: Dr Kovač, 2003), p. 199.

72 Lorentz, H., *Die technische Ausbildung der Lehrer in der Handhabung von Schmalfilm udn Schmalfilmgeräten für Unterrichtszwecke* (Stuttgart and Berlin: W. Kohlhammer, 1937), p. 43.

73 Albrecht and Sikora, *Lichtbild und Schmalfilm*, S. 138.

74 Ullmann, Paul, 'Warum kein Bildstillstand in der Schulfilmprojektion', *SHB-Film-Post*, 47 (1954), 4–5.

75 *'[I]m Schulbetrieb überflüssi[g] und unerwünsch[t]'; 'mit voller Absicht vermieden'.* Anonymous, 'Die österreichischen Schulfilmprojektoren', pp. 6, 11.

76 Ullmann, 'Warum kein Bildstillstand in der Schulfilmprojektion', p. 5.

77 Masson, *Watch and Learn*, pp. 95–96.

need to justify his position against freezing the moving image at length in *SHB-Film-Post* to answer 'repeated requests'.[78]

On the one hand, this shift from the 1920s to the 1950s reflects broad changes in institutional policy. Without disregarding small-scale and independent initiatives throughout this time period, this shift roughly corresponds to an ongoing institutionalization of educational film and the consolidation of centralized national distribution to allow for classroom projection on a wider scale in Europe. More specific to the case of Austria, the shift also reflects that after 1945 the Austrian state had a greater stake in educational film activities and debates than before the Nazi annexation in 1938 — with the Nazi German predecessor organization RfdU/RWU not only providing a library of films to build upon, but also a model for organization. The broad shift of educational film activities from independent associations to the state in Austria is already represented in the contrast between the two cited journals: While *Das Bild* (1924–1930) documented the work and exchange within the Federation of School Cinemas, an independent registered association which was liquidated in 1939,[79] *SHB-Film-Post* was issued by SHB-Film, the ministry of education's section of educational media.

But while centralization within national organization explains some of the features of the Austrian-made projectors, I argue that this more centralized division of labour, which takes some tasks and choices away from the teacher, is not just an expression of the will to power in an institution or the way it is shaped by its pedagogical concepts. It should be understood in its reciprocal capture with other factors, such as the material properties of film prints and projection equipment along with the logistics of film and projector management and maintenance. The issue of freezing the film image mid-projection and its changing prioritization in educational film discourse demonstrates what happens when expected technological solutions arrive, yet do not fully deliver on their promise — stranding users on a middle ground between old and new directives.

Apparently, this state persisted for a while. Even in 1967, during the Congress of the *International Council for the Advancement of Audio-Visual Media in Education* (ICEF) in Vienna, it was still lamented that heat protection filters, while functional, necessarily impaired the projected image due to loss of contrast and brightness.[80] In this sense, to cite sociologist Andreas Folkers, the role of materiality for practice should be considered not simply as glue holding certain practices together (as in reading the Ditmar 1006 as an expression of institutional policy), but also as a lubricant forcing the emergence of new dynamics.[81] In the case of the Austrian projectors, this can be seen in the shift away from halting the image to a new obsession with providing slide and film projection with identical image sizes — a different way to arrive at a similar solution, which produces new

78 Ullmann, 'Warum kein Bildstillstand in der Schulfilmprojektion', p. 4.

79 On the liquidation, see Schulkinobund's file in Österreichisches Staatsarchiv: OeStA/AdR BKA I BP-Direktion VB Signatur XIV 1070 Österreichischer Schulkinobund 1929–1939: 'Aktenübersicht', p. 2.

80 Hübl, Josef, 'Das Tonband im Unterricht', in *Die Integration der audio-visuellen Medien. Bericht des 17. Kongresses der ICEF. Wien, 25.–30. September 1967*, ed. by Alois List (Vienna: Bundesstaatliche Hauptstelle für Lichtbild und Bildungsfilm, 1969), p. 99.

81 Folkers, Andreas, 'Was ist neu am neuen Materialismus? — Von der Praxis zum Ereignis', in *Critical Matter. Diskussionen eines neuen Materialismus*, ed. by Tobias Goll, Daniel Keil, and Thomas Telios (Münster: Edition assemblage, 2013), pp. 16–32 (p. 22).

technological challenges, educational protocols, and material objects (e.g., the slender wooden tables that were designed to carry a slide and a film projector).

One main reason brought forth against pausing and backwards projection was the damage done to the film strip and projection equipment.[82] These were problems especially for the provincial educational media offices, which were tasked with maintaining both their print libraries and the equipment owned or rented by schools. As mentioned earlier on the matter of technical instruction, these institutions saw the distribution of agency between teachers and educational media offices not least as a matter of protecting equipment and films. This concern for careful handling even comes through in the slide series about Ditmar 1006 produced by SHB-Film to accompany instructional courses for teachers. The slides were designed by Josef Sikora, the technical manager of the joint educational media office for the provinces Vienna, Niederösterreich, and Burgenland. They aimed to provide additional information not covered by the manual or by hands-on practicing on the actual projector.

Still, the fifteen-part series is bracketed by two slides, the second and the penultimate one. These show not just machine elements in photography or schematic drawing, but hands manipulating the equipment as if during a course exercise (no such images had appeared in the comparable 1948 slide series of forty-two images that showed an overview of small-gauge projectors in school use).[83] In the latter image (see Fig. 8.3), this involves the rewinding of a film reel, which is done manually. The depicted movements (of cranking the projector while running the film strip through the fingers of the other hand) are also described in some detail in the brochure accompanying the slide series, which stresses the tactile care needed to rewind without damaging the print.[84] What is striking to me about this image is not the archival truism that film preservation and film presentation work at cross purposes. Rather, this slide image of rewinding as a manual process stresses that material media objects, the people who handle them, and their institutional infrastructures are entangled with each other not least via practices that are both knowledge-based and embodied.[85]

Conclusion: Managing the Boundaries of Classroom Projection

Investigating the employment of film in business consulting and industrial efficiency management in the 1910s, Florian Hoof has argued for the merits of the concept of the 'boundary object' for understanding the uses of film in different contexts. Coined by

82 Albrecht and Sikora, *Lichtbild und Schmalfilm*, S. 138; Ullmann, 'Warum kein Bildstillstand in der Schulfilmprojektion', p. 5.

83 Sikora, Josef, *Schmalfilmprojektion. Bilder zur Einführung in den Gebrauch von Schmalfilm-Projektionsgeräten* (Vienna: Bundesstaatliche Hauptstelle für Lichtbild und Bildungsfilm, 1948), 42 slide images. (Collection of Pädagogische Hochschule Steiermark).

84 Sikora, Josef, *Der 16mm-Schmalfilmprojektor 'Ditmar 1006'*. (Vienna: Bundesstaatliche Hauptstelle für Lichtbild und Bildungsfilm, 1950), slides 2 and 14 of 15. (Collection of Pädagogische Hochschule Steiermark). Sikora, Josef, *Der 16mm-Schmalfilmprojektor 'Ditmar 1006'. Bildbeschreibung und Erläuterungen zur Lichtbildreihe U 1117* (Vienna: Bundesstaatliche Hauptstelle für Lichtbild und Bildungsfilm, 1950), p. 12.

85 Postill, John, 'Introduction: Theorising Media and Practice', in *Theorising Media and Practice*, ed. by Birgit Bräuchler and John Postill (New York and Oxford: Berghahn, 2010), pp. 1–32 (pp. 9–11); Schäfer, Hilmar, 'Einleitung. Grundlagen, Rezeption und Forschungsperspektiven der Praxistheorie', in *Praxistheorie. Ein soziologisches Forschungsprogramm*, ed. by Hilmar Schäfer (Bielefeld: transcript, 2016), pp. 9–25 (pp. 10–13).

Fig. 8.3. How to carefully rewind a reel by hand. – Josef Sikora, *Der 16mm-Schmalfilmprojektor 'Ditmar 1006'*. (Vienna: Bundesstaatliche Hauptstelle für Lichtbild und Bildungsfilm, 1950), slide 14 of 15. Collection of Pädagogische Hochschule Steiermark.

Susan L. Star and James R. Griesemer in the field of Science and Technology Studies, this concept stresses how certain concepts and objects can provide common points of reference and contact points for participation to different interest groups. Hoof argues that film worked as such a boundary object in the context of business consulting as it meant something — if not different things — to scientists (producing and storing data) as well as industrialists (promotion and record) and factory workers (entertainment and glamour). It was precisely this diffuse common reference, Hoof argues, that provided loose linkage between these groups.[86]

86 Hoof, Florian, 'The Boundary Objects Concept: Theorizing Film and Media', in *Media Matter. The Materiality of Media, Matter as Medium*, ed. by Bernd Herzogenrath (New York: Bloomsbury Academic, 2015), pp. 180–200; Hoof, Florian, 'Ist jetzt alles "Netzwerk"? Mediale "Schwellen- und Grenzobjekte"', in *Jenseits des Labors.*

For the case study of classroom projection investigated in this article, the opposite seems to have been true. Rather than providing common references, the promotional and instructional texts surrounding the Ditmar 1006, for instance, mostly stressed distinctions from other modes of film projection and viewing. Classroom projection was defined by being different from amateur film and slide projection at home (which would have the projector closer to the screen), or from travelling cinema projection with its uncontextualized sequence of attractions. The temporarily dimmed classroom is described as distinct from the dark cinema auditorium of commercial film exhibition, which demands silent absorption. And 'true' silent, short *Unterrichtsfilm* for the classroom is even distinguished from the more broadly educational, entertaining fare that would be shown at sites of popular and adult education.

Yet while the borders of classroom projection are fortified to define it as a practice belonging to school more than to the imagination of cinema, there is also a countervailing movement in writings in *SHB-Film-Post* and elsewhere. If it fits in with the promotion of classroom projection, affective fascination for the moving image in a dark room is willingly incorporated into descriptions and depictions.[87] As I have noted, sometimes a blurring of the boundaries was not just rhetorically but strategically necessary: The comparatively expensive sound film projector Ditmar 1106 was as much pitched to sites of popular education as to schools. But educators' policing of the boundaries of classroom projection would shift even more drastically in the decade to come. With new advances made by Super-8 projection systems from the mid-1960s on, the film amateur as an ally would make a rousing comeback in educational film debates: *SHB-Film-Post*'s successor publication, *Sehen und Hören* (published from 1962) is filled with numerous reports on amateur technology and practice as models for new ways of using film in the classroom.[88] The growing acceptance of Super-8 as a complementary format for classroom projection also strengthened the notion of teacher-produced films, which had been considerably more costly in the framework of the 16-mm format.[89]

In this light, the concept of film as a *boundary object* may also be useful for tracking the underdetermined (but not arbitrary) shifts of alliances, overlaps, and distinctions in media history.[90] In this case, the shifting alliances also involved a conspicuous return. After ending its collaboration with SHB-Film, Eumig specialized in Super-8 technology in the mid-1960s. This once reluctant partner turned into the institution's sponsor for the abovementioned ICEF conference, which included an opportunity to promote its product line to international guests on a factory tour.

Transformationen von Wissen zwischen Entstehungs- und Anwendungskontext, ed. by Florian Hoof, Eva-Maria Jung, and Ulrich Salaschek (Bielefeld: transcript, 2011), pp. 45–62 (pp. 53–58).

87 For example: Zinnecker, Otto, 'Aus der Praxis', *SHB-Film-Post*, 16 (1949), 5–6 (p. 5); Puschej, Max, 'Unterrichtsfilm, — ein Klassenerlebnis', *SHB-Film-Post*, 52/53/54 (1955), 1.

88 For instance: Stoiber, Albert, 'Vom Anfänger zum Amateur', *Sehen und Hören*, 50 (1970), 66–67; Anonymous, 'Filmen leicht gemacht!', *Sehen und Hören*, 31 (1967), 10.

89 Weiner, Hermann, 'Das "Ureigene" am Unterrichtsfilm', *Sehen und Hören*, 58 (1972), 105–06.

90 On the benefits of the 'boundary object' to media history, see: Schüttpelz, Erhard, 'Elemente einer Akteur-Medien-Theorie', in *Akteur-Medien-Theorie*, ed. by Tristan Thielmann and Erhard Schüttpelz (Bielefeld: transcript, 2013), pp. 9–67 (pp. 38–41).

The Political Made Visual

SABRINA MENEGHINI

Lantern Slides in Geography Lessons: Imperial Visual Education for Children in the British Colonial-Era

In 1907 the Colonial Office Visual Instruction Committee (COVIC) hired the British artist Alfred Hugh Fisher (1867–1945) to create a photographic documentation of the peoples and lands within the British Empire. The same year Fisher set off on an ambitious three-year journey which took him from India, Burma, Aden and Somaliland to Cyprus, from Canada to Weihaiwei, Hong Kong, Borneo, the Malay Peninsula and Singapore, from Gibraltar and Malta to Australasia. COVIC used many of the images he captured for the production of lantern slide sets and textbooks in children's geography lessons.

At the turn of the twentieth century, various patriotic movements and organizations in Britain, by recognizing the importance of education to instil notions in children, started to direct their activities towards the propagation of imperial ideas.[1] COVIC was established to instruct schoolchildren in colonial education. The Committee emerged from an historical and political climate in which imperialists perceived a dangerous lack of knowledge of the Empire's history and geography among the British population. COVIC developed a scheme to contribute to the dissemination of colonial knowledge through the use of lantern slides in the classrooms.

I am concerned with the question of how the Committee intended to use the visual material produced and collected by Fisher. How was imperial visual education going to be delivered? While I do not intend to explore the potential effect of the images on their audience, I look at the effect COVIC wanted to generate and how the visual material

* Materials used in this chapter are part of my PhD research in the Photographic History Research Centre at De Montfort University Leicester, titled *Classroom Photographic Journeys: Alfred Hugh Fisher and the British Empire's Development of Colonial-era Visual Education*, which has been fully funded by De Montfort University. I would like to thank Nelleke Teughels and Kaat Wils, as well as my supervisor Gil Pasternak for their insight and helpful feedback.

1 To state a few: the Royal Colonial Institute, the Victoria League, and the League of the Empire. For further reading on these organizations, see: Greenlee, James Grant, *Education and Imperial Unity, 1901–1926*, Routledge Library Editions: Education 1800–1926, 6 (London: Taylor and Francis, 2016).

Sabrina Meneghini • De Montfort University, sabrina.meneghini@my365.dmu.ac.uk

Learning with Light and Shadows: Educational Lantern and Film Projection, 1860-1990, ed. by Nelleke Teughels and Kaat Wils, TECHNE-MPH, 8 (Turnhout, 2022), pp. 219-244

10.1484/M.TECHNE-MPH-EB.5.131501

was geared towards triggering it. The classroom experience created by COVIC with the assistance of the lantern device is at the heart of this chapter. Indeed, as Gabrielle Moser has punctuated, COVIC's plan differed from other projects as 'its focus [was] on the classroom as a key site where the empire's health could be measured and a feeling of imperial unity strengthened'.[2] To reconstruct the classroom experience and what it might have entailed, I analyse written passages excerpted from the six textbooks COVIC produced, along with the corresponding images from the Fisher Photograph Collection.

Many scholars have contributed to the study of colonial-era British visual culture including COVIC's project. In historical studies and cultural geography, for example, John M. MacKenzie studied imperial propaganda, especially its impact on popular British culture, and James Ryan examined photography's role in the domestication of foreign landscapes in the nineteenth and early twentieth century.[3] In his analysis of Halford John Mackinder, and his role in the theory and practice of imperialism, Gerry Kearns discussed Mackinder's writings in geography, politics and education as well the textbooks which he prepared for COVIC.[4] In 2019, Gabrielle Moser published a comprehensive examination of COVIC's scheme, suggesting that the lectures attempted to promote a feeling of imperial citizenship amongst the audience in the metropole and in the colonial periphery.[5] Moser contends that COVIC's lectures were intended to encourage a self-reflexive spectatorship, enabling viewers to identify themselves as similar to the people depicted to understand that they were all imperial citizens. Alternatively, I suggest that the lectures on the Empire prepared for the British audience were aimed at fostering a feeling of collectiveness within the classroom, not with the subjects depicted, to strengthen national and imperial identity, and to emphasise the difference between colonizer and colonized.

I expand the work of such scholars by reflecting on COVIC's intended use of the lantern device as a strategic educational tool. In following the narrative of the lectures, I bring new understandings of the role the lantern device played in COVIC's educational scheme for conveying a particular visual ideology within the classroom. Drawing on material from the COVIC archive and the Fisher Photograph Collection, held at the Royal Commonwealth Society Library (RCS) in Cambridge University Library, I investigate the pedagogical power of lantern slides in attempts to indoctrinate children. In particular, I focus on the lantern device as an integral part of the teaching process to elaborate understanding on how visual media was used in British colonial education.

The Fisher Photograph Collection consists of 30 albums containing 4091 photographs; 22 files with 959 pictures collected by COVIC; 150 paintings and 64 notes by Fisher; his diary letters; correspondence between COVIC and members of the British Association; 37 sheets of watercolours; and additional written material related to Fisher. As this chapter

2 Moser, Gabrielle, *Projecting Citizenship. Photography and Belonging in the British Empire* (University Park, Pennsylvania: The Pennsylvania State University Press, 2019), p. 47.

3 MacKenzie, John M., *Propaganda and Empire: The Manipulation of British Public Opinion 1880–1960* (Manchester: Manchester University Press, 1986); Ryan, James R., *Picturing Empire: Photography and the Visualization of the British Empire* (Chicago: University of Chicago Press, 1997).

4 Kearns, Gerry, *Geopolitics and Empire: The Legacy of Halford Mackinder* (Oxford: Oxford University Press, 2009).

5 Moser, Gabrielle, *Projecting Citizenship. Photography and Belonging in the British Empire* (University Park, Pennsylvania: The Pennsylvania State University Press, 2019).

is concerned with lantern slides, it is relevant to mention that the collection includes only two out of 2580 slides that accompanied the six textbooks produced by COVIC. To date, no additional surviving slides have been found.[6] I therefore analyse the textbooks along with the images in the collection that were reproduced as lantern slides.[7] I start with a consideration of COVIC's imperial vision and its visual ideology to explain its pedagogical approach and perception of the capabilities of the lantern device. I then explore the teaching activities prepared by COVIC to influence the knowledge of children and teachers, providing insights into the visual and written material that COVIC produced for that purpose. By addressing the interaction of visual and verbal information forms I show how the imperial visual education provided through the textbooks facilitated national and imperial identity formation in British children.

COVIC's Pedagogical Approach

In 1902 Joseph Chamberlain, Secretary of State for the Colonies, established the Visual Instruction Committee. After the evaluation of a memorandum on the use of visual education written by Professor Michael E. Sadler, he appointed a group of gentlemen connected with imperial and educational organizations with the purpose of educating the young generation of the British Empire through photography.[8] Sadler was the Director of Special Inquiries and Reports at the Board of Education, an office established in 1895 to improve Britain's educational system.[9] With this in mind, he gathered reports on methods of schooling outside Britain. He was particularly impressed by the use of lantern slides for educational purposes in the schools of the State of New York conceived by the American Museum of Natural History. Its educational plan offered a well-organized system of lectures, illustrating the scenery, social life, and the economic situation of the United States as well as views from different parts of the world. Sadler highlighted the notable program of lending slides to schools throughout the State and proposed to use the same technique for the creation of lectures dedicated to colonial education.[10] As he observed

6 In 2018 these two lantern slides, both illustrating Canadian scenes, were found in the RCS Archive, mixed in a box of slides with geographical subjects suggesting the different uses that had been made of COVIC's visual material over the years. Since 2018 the RCS and I have been searching for the COVIC lantern slides hoping to find and trace the history of these visual records.

7 This analysis is possible by comparing the slides' list provided in the textbooks with the captions to the photographs and paintings in the collection.

8 The original Members of the Committee as listed by Sir Charles Lucas were: the 12th Earl of Meath (Reginald Brabazon, founder of Empire Day); Sir Cecil Clementi Smith, as having had special experience of Crown Colonies and as being connected with the Royal Colonial and the Imperial Institute; Mr Mackinder, as representing the Victoria League; Dr R. D. Roberts, as Secretary to the Gilchrist Trustees; Mr Sadler, as representing the Board of Education; Sir John Struthers, as representing the Scotch Education Department; Sir C. Lucas, as representing the Colonial Office. Papers of the Colonial Office Visual Instruction Committee, GBR/0115/RCS/RCMS 10/3 Misc. No. 265 (1911) p. 1. Cambridge University Library.

9 Greenlee, James Grant, *Education and Imperial Unity, 1901–1926*, Routledge Library Editions: Education 1800–1926, 6 (London: Taylor and Francis, 2016), p. 63; For more on Michael Sadler's work in education, see: Phillips, David. 'Michael Sadler and Comparative Education', *Oxford Review of Education*, 32.1 (February 2006), 39–45.

10 Colonial Office (CO) Papers 885/8/8 Misc. No. 150, Memorandum, Michael Sadler, (December 1902), 4.

in the memorandum, the intention of the project was to illustrate life in the British Empire 'as an educational means of strengthening the feeling of Imperial unity and citizenship'.[11] Sadler stressed the potential advantages arising from a broader acquaintance with the Dominions and Colonies but also on the dangers due to the limited knowledge of their geography and appreciation of their value. His idea of creating lantern lectures on the Empire reflected a belief in the power of visual experience to impress notions more clearly upon the audience. Sadler highly recommended a system of visual instruction as seen in New York as 'much can be learnt through the eye as well as through the ear'.[12] The lantern slides had the advantage of reaching a large number of people and most importantly to 'vividly' present the world to them.[13] Sadler's original plan was on a grand scale, for a large audience, aspiring to reach 'the masses of the people in the United Kingdom' to give them acquaintance with life and economic conditions throughout the Empire. The images of the 'Mother Country' to be shown in the Colonies on the other hand, were aimed at being 'interesting' life's scenes.[14] COVIC's formation followed Sadler's suggestion to create a visual overview of the vast Empire, with photographs and texts at their core.

The illustrated lectures were to be used in the United Kingdom, the Colonies and India. The project's cultural aims were twofold: British schools would receive images of the colonies, while colonial school children would see slides from the 'Mother Country'. COVIC developed the scheme from Sadler's proposal with some modifications, for example its intended audience was not the general public, but schools. From 1903 until 1905 the Committee focused on the production of lectures on the United Kingdom for use in the colonies, and the well-known British geographer and future politician, Halford John Mackinder prepared the first set of seven lantern slide lectures on Britain. Mackinder himself was a passionate imperialist and strongly believed in the important role education should play in disseminating patriotism, knowledge and awareness of the Empire. In a meeting on Empire and Education held in 1904, Mackinder stated: 'The building up of empire was to be achieved not only by an army and navy, and through policy, but also by a united, designed, carefully-planned effort in all the schools of the Empire for a generation'.[15] The emphasis of the Committee's scheme was on the visual rather than the text, 'the lectures were delivered and written with a view to the slides, rather than the slides procured to supplement the lectures'.[16] The importance of the visual narrative was stressed once again in the description of the lectures' style which was 'simple and straightforward, the object being — it must be repeated — to attempt to put before the eyes'.[17]

With the completion of the first part of the scheme, by providing children in the colonies with lectures on the United Kingdom, the Committee began preparing the ones aimed

11 CO 885/8/8 Misc. No. 150, p. 1.
12 CO 885/8/8 Misc. No. 150, p. 1.
13 CO 885/8/8 Misc. No. 150, p. 4–5.
14 CO 885/8/8 Misc. No. 150, p. 1.
15 [Anonymous], 'Mr Brodrick On The Empire And Education,' *The Times*, 19 December 1904, p. 14.
16 CO 885/17/8 Misc. No. 188, Circular, Mr Alfred Lyttleton to The Governors and High Commissioners, (April 1905), 2.
17 [Anonymous], 'Education and Empire,' *The Times*, 23 April 1907, p. 11.

at the children in the metropole, which is the focus of this chapter.[18] COVIC entrusted the lectures and the project's supervision to Mackinder and decided to employ one single photographer to visit the overseas territories. In 1907 the Committee hired Alfred Hugh Fisher who travelled for the following three years to photograph and to paint the peoples and landscapes of the British Empire.

COVIC's interest in the pedagogical potential of the lantern slides to create an imperial vision fits into the debate on their use in education at the turn of the twentieth century. Before the magic lantern was accepted as an instrument of instruction and education, its use, as the name suggests, was mainly to entertain people.[19] As Jennifer F. Eisenhauer observes, photography was 'the most important invention to impact the transition of the magic lantern to an instrument of scientific vision'.[20] This transition from *magic lantern* to *optical lanterns*, that is from leisure to instruction, characterized the debate on the use of this technology in education in the last decades of the nineteenth century. COVIC's employment of the lantern technology was in geographical education. Emily Hayes, writing on the development of the lantern practices within the Royal Geographical Society (RGS), refers to 'mixed emotions engendered by the lantern' within members of the Society regarding its use.[21] The technology produced different perceptions in the RGS fellows, negative ones because it was associated with entertainment and religious advocacy, and positive as it could encourage geographical teaching and learning. In the 1880s, as a consequence to the perceived lack of geographical knowledge in British schools, the RGS began to focus its attention on education and gradually started to adopt the lantern medium. As Hayes has discerned, the approval of the lantern slide is linked with the acceptance of photography as a tool to gather and promote geographical knowledge.[22] One of the three case studies Hayes analyses to demonstrate the gradual engagement of the RGS with the lantern is the famous lecture 'On the Scope and Methods of Geography' given by a young Mackinder to the RGS in 1887. The geographer had been a key figure in the acceptance and employment of the lantern projector to promote geographical knowledge. In his presentation, in which images and text were synchronized, Mackinder demonstrated he was ready to fully embrace the technology. The debate on the use of lantern slides in geographical education saw Mackinder as one of the main supporters, who recognized the pedagogical potential of the technology to instruct children. COVIC's decision to employ him, in addition to his being an expert lecturer and, to state it in Bernard Porter's

---

18 COVIC's original plan to adapt the textbooks on the colonies to be used in the overseas territories was never accomplished.

19 For further reading on the history of the Magic Lantern, see: Crangle, Richard, Mervyn Heard, and Ine van Dooren, *Realms of Light. Uses and Perceptions of the Magic Lantern from the 17th to the 21st Century* (London: Magic Lantern Society, 2005); Crangle, Richard, Stephen Herbert, and David Robinson, *The Encyclopedia of the Magic Lantern* (London: Magic Lantern Society, 2001).

20 Eisenhauer, Jennifer F., 'Next Slide Please: The Magical, Scientific, and Corporate Discourses of Visual Projection Technologies', *Studies in Art Education*, 47.3 (Spring 2006), 201.

21 Hayes, Emily, 'Geographical Light: the Magic Lantern, the Reform of the Royal Geographical Society and the Professionalization of Geography *c.* 1885–1894', *Journal of Historical Geography*, 62 (2018), 24. For further reading on the significance of the magic lantern in geographical education, see: Hayes, Emily, 'Geographical Projections: Lantern-slides and the Making of Geographical Knowledge and the Royal Geographical Society *c.* 1885–1924' (PhD diss., University of Exeter, 2016).

22 Hayes, 'Geographical light', p. 29.

terms, 'the leading geographer among the imperial zealot', was inspired by his effective use of the medium to disseminate a shared ideology.[23]

The intrinsic connotation of projecting images lay in the audience's whole experience of seeing, looking at pictures in a dark room while hearing, listening to the speaker's voice and being able to focus entirely on the visual. The lantern device had the power to engage the whole room at once. As Mike Simkin observes, 'whereas children might lose their way in a book through lack of concentration, the impact of the lantern with a single brightly-focused image could capture their attention and hold their concentration'.[24] In this sense, the lantern performance differs from the cinema show as by projecting one single image at a time it enhances the audience's experience and connection with it. As beautifully expressed by Martyn Jolly: 'It [a magic lantern slide] was intended to be shown in a particular sequence, for a particular reason, in a particular darkened room, at a particular place, on a particular time, to a particular group of people sitting shoulder to shoulder, in the dark, collectively immersed in a multisensory experience'.[25] Through the slides, enlarged and projected onto a big screen, the visual material produces a multisensorial experience creating the ideological space essential to prepare children to absorb COVIC's imperial imagery. The Committee saw the lantern projector as the perfect technical device that could instruct students more effectively by creating in the classroom an environment able to generate a sense of belonging and to provoke feelings of emotional identification. It is likely then that COVIC perceived the darkness of the room, in which only one image at a time is made visible, and the voice of the teacher breaking the dead silence, as the only elements needed to create an environment capable of realizing the classroom journey.

COVIC's aim to impart a colonial mindset through images had to be supported by textual instructions to shape teachers' and children's interpretation. Mackinder had the responsibility of writing the texts and of providing the lecture's structure; the narration combined history and geography within an imperial vision. Although he only authored the first textbook *India*, the succeeding ones followed the same pattern.[26] In a later lecture, *The Teaching of Geography from an Imperial Point of View, and the use which could and should be made of Visual Instruction*, he reflected on the use of visual instruction in geography teaching.[27] Mackinder delivered the paper in 1911, after the Indian lectures were published, and in it promoted COVIC's work and his own. He explained the Committee's belief in the power of visualisation as the best system for teaching Empire topics. Gearóid Ó Tuathail refers to Mackinder's works in geographical education as 'works of imperialist pedagogy' and indeed, his role in COVIC's project allowed him to create an imperial

23 Porter, Bernard, *The Absent-Minded Imperialists* (Oxford: Oxford University Press, 2004), p. 184.

24 Simkin, Mike, 'The Magic Lantern and the Child', in *Realms of Light. Uses and Perceptions of the Magic Lantern from the 17th to the 21st Century*, ed. by Richard Crangle, Mervyn Heard, and Ine van Dooren (London: Magic Lantern Society, 2005), p. 28.

25 Jolly, Martyn, 'The Magic Lantern at Work. Witnessing, Persuading, Experiencing and Connecting', in *The Magic Lantern at Work. Witnessing, Persuading, Experiencing and Connecting*, ed. by Martyn Jolly, and Elisa deCourcy (London: Routledge Studies in Cultural History, Taylor and Francis, 2020), p. 8.

26 Mackinder, Halford J., *India. Eight Lectures Prepared for the Visual Instruction Committee of the Colonial Office* (London: George Philip & Son, 1910).

27 Mackinder, Halford J., 'The Teaching of Geography from an Imperial Point of View, and the Use Which Could and Should be Made of Visual Instruction', *The Geographical Teacher*, 6.30 (1911), 79–86.

narrative, supported by geography and history, and visually delivered.[28] As Ryan has observed, 'COVIC realized in institutional form Halford Mackinder's conception of an imperial geography based on the power of sight'.[29] As Mackinder himself stated: 'you can make geography a scientific study and a discipline that will contribute vitally to the education of the nation'.[30]

Colonial Visual Education in COVIC's Textbooks

Fisher's work reflects colonial policies recording the British culture, people and economy at the beginning of the twentieth century within the Empire. Before his departure, Fisher received exhaustive photographic training. Furthermore, Mackinder provided the artist with instructions to follow during his journey, which comprised the places to document and how to document them. It is interesting to note Mackinder's attention to details, given that he had in fact never travelled to many of the places himself. In reminding Fisher that the main aim of the project was to produce educational lectures on the Empire, Mackinder recommended him to represent 'buildings and natural objects as are essential to a broad and vivid impression of the country'.[31] To facilitate the task, he suggested to depict the Indian colony from two points of view: its distinctive characteristic and the novel ones carried over by the British. In his directions, Mackinder always had COVIC imperialist aims and the impact it wanted the images to have on its audience in mind, and instructed the photographer accordingly. When Mackinder referred to places and subjects in the Indian subcontinent, he provided Fisher with a list of categories to project, which would also apply to every other country.

> While you will present typical street scenes in the native quarters of the great cities, scenes in a typical village, pageants in the Native States, and portrait sketches of great native rulers (where you are fortunate in your opportunities), and such great historic monuments as those of Agra and Trichinopoly, you will also, on the other hand, show a military cantonment, the Viceroy's traveling camp, if you happen to see it, scenes in railway stations and railway carriages, great engineering works, either connected with irrigation or communications, and the European quarters of such cities at Calcutta and Bombay.[32]

Mackinder emphasised the importance of subjects which could support and disseminate imperial values and at the same time demonstrate the impact of the Empire. Fisher was thereby instructed to copiously highlight the expansion of transport infrastructure. By asking the photographer to depict the Indian colony from two viewpoints Mackinder perpetuated the construction of the 'Other', crucial for the consolidation of colonial power. This polarised distinction is also decisive in the construction of the Empire in

28 Tuathail, Gearóid Ó, 'Putting Mackinder in His Place: Material Transformations and Myth', *Political Geography* 11.1 (January 1992), 114.

29 Ryan, p. 208.

30 Mackinder, Halford J., 'Geography in Education', *The Geographical Teacher*, 2.3 (October 1903), 101.

31 CO 885/17/8 Misc. No. 188, Correspondence, Mr H. J. Mackinder to Mr Fisher (October 1907), 135.

32 CO 885/17/8 Misc. No. 188, p, 135.

British culture and society. As Edward Said has argued, the West defines itself through its representation of the Orient, and in order to do it, contrasts have to be emphasised.[33]

Fisher worked during a memorable historical period in which the British Empire was at the pinnacle of power and there was worldwide stability before the sweeping violence of World War I. His images thus do not show war scenes nor cruelty but are focused on the human figure. The lecture's aim was to inculcate a positive enthusiasm for the Empire in British society, thus Mackinder, and the two other authors after him, selected photographs that would reinforce the positive colonial stereotypes propagated by political power. Therefore, entirely absent are images showing war, crime, death, poverty or violence. To inspire imperial sympathy, the textbooks and the images had to convey a positive message tailored for each country. The colonial visual ideology was adapted to serve colonial interests.

The structure of COVIC's teaching activities was aimed at shaping children's interpretation and the six textbooks demonstrate the Committee's strategic use of the visual and written material to promote national and imperial consciousness. But how were the textbooks conceptualized, designed, written and published? Three authors participated in the writing process. In 1910 after the publication of the Indian lectures Mackinder, due to his increased involvement in politics, handed over the remaining lectures to one of his colleagues, Arthur John Sargent, who authored *The Sea Road to the East*; *Australasia*; *Canada and Newfoundland*; and *South Africa*.[34] Algernon Edward Aspinall, Secretary to the West India Committee, who had already published several books on those territories, wrote *West Indies and Guiana*.[35] Fisher did not travel to South Africa or the West Indies, so the last two textbooks do not feature his images but works purchased from different sources.

Although Sadler in his memorandum, described the lecture's audience as both young and adult, he emphasised the importance of promoting and adopting the scheme of visual instruction in the schools of the Empire 'as ideas and associations can generally be better fixed in the mind during the impressionable years of childhood than in later life'.[36] In discussing the material work of photography, one of Elisabeth Edwards's questions, 'what "work" is expected of photographs as objects' seems quite appropriate here.[37] Sadler emphasised the importance of the technology employed for *fixing* ideas in the young users, suggesting that it was not just the photographic image, but the photographic experience that was aimed at provoking emotions in COVIC's young audience.

---

33 Said, Edward W., *Orientalism* (London: Penguin, 2003).

34 CO 885/21/7 Misc. No. 249, Correspondence, Mr H. J. Mackinder to Sir C. P. Lucas, (October 1910), 31; Sargent, Arthur John, *The Sea Road to the East. Six Lectures Prepared for the Visual Instruction Committee of the Colonial Office* (London: George Philip & Son, 1912); Sargent, Arthur John, *Australasia. Eight Lectures Prepared for the Visual Instruction Committee of the Colonial Office* (London: George Philip & Son, 1913); Sargent, Arthur John, *Canada and Newfoundland. Seven Lectures Prepared for the Visual Instruction Committee of the Colonial Office* (London: George Philip & Son, 1913); Sargent, Arthur John, *South Africa. Seven Lectures Prepared for the Visual Instruction Committee of the Colonial Office* (London: George Philip & Son, 1913).

35 Aspinall, Algernon E., *West Indies and Guiana. Six Lectures Prepared for the Visual Instruction Committee of the Colonial Office* (London: George Philip & Son, 1914).

36 CO 885/8/8 Misc. No. 150, Memorandum, Michael Sadler, (December 1902), 1.

37 Edwards, Elizabeth, 'Objects of Affect: Photography Beyond the Image', *Annual Review of Anthropology*, 41.1 (October 2012), 222.

Soon after the first edition of the Indian lectures was published in 1910 by Messrs. Waterlow and Son, it was issued in a popular edition by Messrs. George Philip and Son; the succeeding books were also published in popular editions.[38] The cheaper edition included black and white illustrations taken from the slides, a much smaller number compared to the lantern glasses though.[39] Some of the lantern slides sets that accompanied the textbooks could be purchased in colour, a few of the slides were hand coloured by artists and a small number of the paintings were reproduced in colour using the Sanger Shepherd process.[40] Despite the fact that the six textbooks were written by three different authors, with three different voices and approaches, the prose style is very similar. Each textbook is divided into lectures describing respectively regions, cities or countries of a specific area. Their use was not aimed directly at children but at teachers who would read the text out loud whilst the lanternist would operate the slides. The pedagogical potential of the lantern device to educate children was reinforced by the instructional method followed by the teacher. The visual power is even more effective: the text which accompanies the pictures is said aloud, children could then just focus on looking, absorbing what the teacher explained, conditioning their way of looking and influencing them by directing their gaze. It is this specific mode of lecturing that COVIC required in order to instruct teachers and pupils.

The textbooks are arranged as a descriptive journey throughout the Empire, as if children were travelling with the narrator, with their teacher. From Britain, the 'Mother Country', the classroom would depart to visit the colonies, possessions and Dominions of the Empire; the travel would be by ship, boat or train. The narration in the textbooks does not follow Fisher's original routes; the photographer's journey was far from being linear as it had to conform to Mackinder's requests and to COVIC's economic contingency.[41] The textbooks instead are geographically organized to follow feasible boat routes. The Indian lectures and *The Sea Road to the East* follow the routes from England to those colonies with maps illustrating the journey. The narration and images of the following four books commence by approaching the lands by providing coastal views and maps of the places from which they would begin the journey. The maps would inform children in which part of the Empire they were landing at and the pictures would illustrate what they would see at their arrival. Avril Maddrell, reflecting on the practice in geographical textbooks of following a sea road to the colonies to encourage emigration, argues that they 'served as legitimating and familiarizing as well as an information function'.[42] Indeed, this method allowed the authors to show children the possibility of emigrating within the Empire, just as they

38 CO 885/22/3 Misc. No. 276, Minutes of Meeting of the Visual Instruction Committee, (November 1912), 29. My analysis is focused on this latter and cheaper edition.

39 The printed photographs and maps are less than fifty in each book whereas the slides were on average 350 per book, approximately three per page.

40 CO 885/21/7 Misc. No. 249, Correspondence, Messrs. Newton to Sir C. P. Lucas, (August 1911), 76. In the first decades of the twentieth century the Sanger Shepherd was one of the methods to obtain colour photographs: three separate filters, red, blue and yellow, were exposed at the same time and the result was combined to form the colour image.

41 CO 885/17/8 Misc. No. 188, Correspondence, Mr H. J. Mackinder to Mr C. P, Lucas, (September 1907), 123, 128. CO 885/17/8 Misc. No. 188, Correspondence, Mr C. P, Lucas, to Mr R. L. Forbes, (September 1907), 129.

42 Maddrell, Avril M. C., 'Empire, Emigration and School Geography: Changing Discourses of Imperial Citizenship, 1880–1925', *Journal of Historical Geography* 22.4 (1996), 381.

were doing through the text, but in COVIC's lectures the voyage is visually constructed and performed. It should be recalled that one of the reasons for Sadler to advocate for the employment of the lantern projector was to illustrate the economic possibilities the Empire could offer to citizens (white young British) to promote emigration. Thus, slides were aimed at illustrating what children would encounter at their arrival for instance in Tuticorin (Southern India), or in Newfoundland or again in Gibraltar. The majority of the lectures open with maps or charts which enabled children to locate the places they would visit. This rigid structure imposed an organization and divided peoples and lands into fixed categories. Pupils would then know to situate Hindus in Madras, white fishermen in Newfoundland, and Toda people in Southern India.

Each lecture then provides images with subjects representative of types. The representation of other races was crucial in supporting imperial ideology. The construction of a racial imagery furthered the classification of the colonial subjects into types and into a rigid hierarchy of differences. Images that reinforced gender and race distinctions were crucial in maintaining the domination and control over the colonized. Peter Yeandle, in his work on the impact of popular imperialism in history textbooks, has reminded us of the use of racial discourse in textbooks to highlight the contrast between colonized and the Europeans.[43] The created image of racial differences conveyed the imperialist message of superiority of the race, the British race. COVIC's educational aim to strengthen the concept of different 'types' followed the tendency of the textbooks during the Victorian and Edwardian era to implant the idea of racial hierarchy in children's minds. Visual aid was the perfect tool to demonstrate and reinforce the theory of classification. Education was key in supporting and shaping the narrative of distinct categories for the imperial subjects; as Castle has shown, the racial imagery was a necessity 'to build "character" at home and maintain the Empire abroad'.[44]

In the lecture illustrating New Zealand for example, Sargent's judgmental long description of the Māori accompanies and supports images to affirm racial and religious distinctions. Sargent uses Fisher's photograph of two Māori women in a traditional Māori greeting, to perpetuate a stereotypical narrative that frames the native population as inferior by focusing the attention on the supposed lack of hygiene. 'On the roadway we meet Maoris, tattered and dirty looking. Here we see two of them meeting and saluting one another by touching noses'.[45] *Maori women, greetings* (*Australasia*, Lecture VII, Image 17) is the title which accompanies the slide. Fisher's caption of the image though, *The Hangi* [Hongi] — *greeting between Maggie Papekura and a Maori woman*, tells us more about the scene and the people depicted (Fig. 9.1).[46] Firstly, the artist annotates the Māori definition for this exchange of greeting and, secondly, he writes the name of one of the two ladies, denoting her importance. Maggie Papekura was a well-known Māori guide

43 Yeandle, Peter, *Citizenship, Nation, Empire. the Politics of History Teaching in England, 1870–1930* (Manchester: Manchester Univ. Press., 2015), p. 107.

44 Castle, Kathryn, *Britannia's Children. Reading Colonialism through Children's Books and Magazines* (Manchester: Manchester University Press, 1996), p. 179.

45 Sargent, *Australasia*, p. 107.

46 The Hangi — greeting between Maggie Papekura and a Maori woman, 1910, GBR/0115/RCS/Fisher 27/7016. Cambridge University Library.

Fig. 9.1 Alfred Hugh Fisher, The Hangi – greeting between Maggie Papekura and a Maori woman, 1910, New Zealand, photographic print. GBR/0115/RCS/Fisher 27/7016. Reproduced by kind permission of the Syndics of Cambridge University Library.

who also assisted the future George V and Queen Mary in their visit to New Zealand in 1901. Sargent's deliberate choice not to report Papekura's name and instead to describe the photograph as showing two poor Māori women demonstrates his disregard for the native population and his desire to influence the meaning of the image. By not providing the subject's name, Sargent followed the approved tendency of representing a collective character over individuality, contributing to the construction of a precise colonial imagery which would depict the subjects as part of a group, of a 'type'. Photographing the subjugated people by organizing them into 'types' furthered the process of classification and categorization. The writer dedicates several pages to the Māori and fourteen images to support his accounts of their customs, ways of living and physical appearances. This is a quite long description of a native population compared to other peoples described in the textbooks. Stephen Heathorn has shown how the description of the Māori in school books in the late-Victorian Empire was to some extent positive, though they were still considered in a lower position than Europeans. In contrast with other people, the Māori were admired for their physical superiority, their skills at sea and at war, their bravery, and their advance in social concepts.[47] Sargent's lectures confirm that the Māori are worthy of mentioning as 'He [the Māori] was, when we first met him, on a far higher level than the aborigines of Australia'.[48] After asking his audience 'Who, then, are the Maoris?', Sargent guides the readers through the images by telling them what to see and how to imagine the Māori in present days but also in the past.[49] *Maori Chief, with Staff* (Fig. 9.2) (*Australasia*, Lecture VII, Image 29) follows a portrait of the same subject who, again, is for Sargent without a name despite Fisher having recorded it, *Chief Tikitere* (Fig. 9.3).[50] The close-up portrait calls our attention to the Chief's face with his white long beard and the carved wooden staff he is holding. The bright sun makes him squint showing his difficulty in looking at the camera, but attracting our gaze towards his eyes even more. The attention of the viewer would naturally be taken by his mature face and eyes, Sargent instead skilfully calls attention to his clothes, which are only partially visible and certainly not easy to identify. 'Here again we see him [the Chief] with his *taiaha*, or staff of office: he looks civilised enough now; in fact, dressed as he is, he might be mistaken for an Englishmen, but doubtless he was very different in the old days'.[51] Sargent tells the reader that the Chief is dressed like an Englishman to claim that the Māori has been civilized, not entirely, but certainly more than in the past. The text guides the reader through the image which is aimed at visualizing how much Britain had accomplished in its civilising mission in New Zealand in less than half a century. As Kathryn Castle has observed, in children's texts 'the colonial subject was essentially dehumanized and recast as rationale for the maintenance of the European dominance'.[52]

COVIC's representations of different 'groups' reinforces the concept of otherness found in geography textbooks of the same period; 'the explicit division of humankind into highly generalized 'race types', accompanied by cultural, social, and moral value judgements'.[53] In the lectures the binary separation between colonizer and colonized is fostered through an Eurocentric discourse of knowledge and civilization. The narration is written from the colonisers' point of view, a Eurocentric historical position which, as Richard Price reminds

47 Heathorn, Stephen J., *For Home, Country, and Race: Constructing Gender, Class, and Englishness in the Elementary School, 1880–1914* (Toronto: University of Toronto Press, 2000), pp. 135–37.

48 Sargent, *Australasia*, p. 109.

49 Sargent, *Australasia*, p. 109.

50 Chief Tikitere, and Maoris, 1910, GBR/0115/RCS/Fisher 26/6997 and no. 6996. Cambridge University Library.

51 Sargent, *Australasia*, p. 110.

52 Castle, p. 180.

53 Heathorn, p. 121.

Fig. 9.2 Alfred Hugh Fisher, Chief Tikitere, and Maoris, 1910, New Zealand, photographic print. GBR/0115/RCS/Fisher 26/6997. Reproduced by kind permission of the Syndics of Cambridge University Library.

Fig. 9.3 Alfred Hugh Fisher, Chief Tikitere, and Maoris [Chief Tikitere], 1910, New Zealand, photographic print. GBR/0115/RCS/Fisher 26/6996. Reproduced by kind permission of the Syndics of Cambridge University Library.

us 'continues the intellectual colonization of such places'.[54] Indeed, this was a central point in Mackinder's teaching method as he clearly stated: 'let our teaching be from a British standpoint, so that finally we see the world as a theatre for British activity'.[55] It was crucial to provide knowledge of the colonies and their peoples to the future British rulers, to teach

54 Price, Richard, 'One Big Thing: Britain and its Empire', *Journal of British Studies* 45.3 (July 2006), 614.
55 Mackinder, *The Teaching of Geography from an Imperial Point of View*, p. 83.

history in relation to the contemporary international concerns. In this regard, Mackinder highlighted the need 'that the great human contrasts which are the outcome of universal history should be generally known, and that — to take only a single category, by way of illustration — the distinction of Christian, Mohammedan, Hindu, and Buddhist should be generally realised in some degree of historical perspective'.[56]

The intention of the authors to validate Britain's imperial role is at the core of each textbook. The narration affirmed and confirmed ethnocentric assumptions and attitudes, to support the conception of Great Britain's superiority over the indigenous population. Writing about the Australian territory, Sargent's argument is more than a justification for the British appropriation of the land. It is a proclamation of racial supremacy which does not leave any space for a different interpretation. He informs pupils that:

> An area about equal to that of the United States could not be left on the sole occupation of a few thousand savages. Now, instead of the savage with his primitive tribal system, we have a white race, purely British in origin, with industry and agriculture of the most advanced type, and an elaborate political constitution of federated States. It is the utilisation, by the white man, of the resources of this vast area which we must study.[57]

The imposition of ideas about race is fostered by the consolidation of European cultural assumptions of superiority which excludes any possible different readings. According to Sargent, before the British arrival there was nothing worth studying. What is important to learn is the imperial present and the economic progress of the *white* Dominion. COVIC's textbooks regard the present, how the Empire appeared at the turn of the twentieth century. References to the past are in order to explain how the British conquered and civilized their colonies and possessions; a side of history which supported imperial values and showed the advancement made. These aspects of the culture of imperialism, the exploitation of resources, concepts of racial and civilization hierarchy, and ideas of a collective Europe, recur regularly in the lectures. For instance, the following passage written by Sargent about Cyprus clearly reflects how the production of imperial knowledge relied on the notion of an uncivilized native past and a prosperous, laborious present built by Europeans.

> The Government is undertaking the work [of replanting trees in the inland plain of Cyprus], but the people and the goats are most destructive, so forest guards have to be employed, such as the two picturesque figures who are posing here to our artist for their portraits. Time and money, especially money, are needed to repair centuries of neglect, and the natives will do nothing without European control.[58]

The 'two picturesque figures' Sargent is describing can be seen in the slide *Forest Guards* (*The Sea Road to the East*, Lecture II, Image 48) (Fig. 9.4) replicating Fisher's photograph *Hilarion, the Belvedere*.[59] As the title to this photograph notes, the two Cypriots are at the ruins of Saint Hilarion Castle, on the Kyrenia mountains, and what the two guards are patrolling, thanks to the strategic fortress' position, is the pass road from Kyrenia to

56 Mackinder, *The Teaching of Geography from an Imperial Point of View*, p. 82.
57 Sargent, *Australasia*, p. 14.
58 Sargent, *The Sea Road to the East*, p. 26.
59 Hilarion: the Belvedere, 1908, GBR/0115/RCS/Fisher 8/1179. Cambridge University Library.

Fig. 9.4. Alfred Hugh Fisher, Hilarion: the Belvedere, 1908, Cyprus, photographic print. GBR/0115/RCS/Fisher 8/1179. Reproduced by kind permission of the Syndics of Cambridge University Library.

Nicosia. In addressing the two men as 'picturesque figures', Sargent follows the dominant representational taste within western audiences to depict oriental landscape and people as exotic and picturesque. As Ayshe Erdogdu has observed 'the picturesque was accepted in the Victorian period as the appropriate aesthetic vocabulary for articulating the racial, political, and economic difference of peripheral peoples from the normative English audience'.[60]

In COVIC's textbooks colonial British knowledge about the non-European world was written, prepared and delivered by privileged British white men, and whereas it was crucial for the Committee to use visual material taken from the photographer's field experience 'to provide accurate information and true pictures of the British Empire', the text instead could be, and actually was, composed by writers who mainly had not travelled to the places they were describing.[61] The texts were very detailed and rich in historical and geographical explanation, setting the target audience at upper-middle and upper class

---

60 Erdogdu, Ayshe, 'Picturing Alterity: Representational Strategies in Victorian Type Photographs of Ottoman Men', in *Colonialist Photography. Imag(in)ing Race and Place*, ed. by Eleanor M. Hight, and Gary D. Sampson (London: Routledge, 2002), p. 117.

61 CO 885/21/7 Misc. No. 249, *Geography of Empire. Work of the Visual Instruction Committee.* Extract from the *Morning Post*, 15 February 1911, p. 49.

children, schoolboys who would have an interest in the Empire and in national pride, and who would probably become colonial administrators or military officers in the overseas territories in their adulthood.

Fisher did not take part in any of the selection processes of the visual material he produced. He sent back to England the negatives without seeing the final results of his work. He was also not involved in the editing of the pictures or in the organization or composition of the textbooks which featured his photographs.[62] Indeed, Fisher's complete exclusion illustrates COVIC's intention to have full control on the material he produced and to direct its organization and meaning. The printed photographs had been placed in albums chronologically organized and numbered following Fisher's journeys. The attention devoted to the numeration as well as to accurately recording notes and captions demonstrates the importance of cataloguing them for the subsequent selection process. It also shows COVIC's original intention, in the person of Mackinder. The Committee's aim was to use the images to prepare the lectures and not vice versa, hence the relevance of the album arrangement. From these sources Mackinder, and after him Sargent and Aspinall, selected the images to include in the textbooks and from which to construct the narrative. Even if Fisher's images did not always follow Mackinder's instructions, authors were able through the selection process and the texts to adjust and tailor the message. Words helped placing the images in the appropriate place. In a letter to W. H. Mercer, Crown Agent for the Colonies and one of COVIC's members, Mackinder stressed the importance of the selection of subjects in order to create lectures that would advance COVIC's project in producing works of imperial education:

> The more I work in connection with our scheme, the more I feel what Sir Charles Lucas has indicated to the Committee — that we must rest our credit on the educational character of our work, and that while our slides are thoroughly good, it must be the selection of subjects and their correlation with the text of lectures which lifts our work out of competitions in which at certain points we are bound to be beaten.[63]

Indeed, from the selection process of the photographs we can extrapolate what Mackinder and COVIC wanted to show and what they wanted their audience to understand and see. The lantern slides along with the textbooks testify to COVIC's ideological inspirations and educational aspirations.

In the slide lectures the pictures were continuously supported by the spoken word. Through the narration, Mackinder, Sargent and Aspinall imparted their interpretation, emphasising the authors' important role in developing the narrative and providing meanings to the images. COVIC's whole project was based on the personal testimony and vision

---

62 CO 885/21/7 Misc. No. 249, Correspondence, Mr A. H. Fisher to Sir C. P. Lucas, (August 1910), 17. In July 1909 Fisher toured Great Britain to take photographs for the lectures on the United Kingdom to be used in Canada and South Africa. Mackinder asked the artist to look into the slides on the United Kingdom prepared for use in India to verify weaknesses and strengths of the collection in order to add images. This is the only episode in which Fisher was involved in an active selection process. However, Mackinder instructed him on the subjects and views that had to be recorded in order to show British industrial power especially to the Canadian audience. CO 885/19/8 Misc. No. 218, Correspondence, Mr H. J. Mackinder to Mr A. H. Fisher, (July 1909), 46.

63 CO 885/19/8 Misc. No. 218, Correspondence, Mr H. J. Mackinder to Mr W. H. Mercer, (June 1910), 44.

of one artist sent around the Empire to obtain material through his direct observation. The selection process and the writing of the text though provided a second filtering of the images, a further reinterpretation. COVIC's project was created and developed by British individuals: the Empire is seen through the eyes of British men, the main two figures being Fisher and Mackinder, whose works created a visual and written narrative that perpetuated an Eurocentric mindset. Even images purchased from additional sources were mainly produced and sent by British photographers living in the overseas territories, or by those connected with colonial associations or the Governments of the various colonies.

The Lectures at Work

Many scholars have reflected on the lantern device and the experience it created on the spectators. Martyn Jolly and Elisa de Courcy in their book *The Magic Lantern at Work*, have discerned how the magic lantern was a technology of persuasion that enabled audiences to virtually and collectively experience the image projected on the wall. Ursula Frederick has further emphasised the lecturers' awareness of the affective experience the lantern technology created in the audience. Consequently, it was used as a 'persuasive explanatory device'.[64] It is likely then that COVIC intended to use the technology to further a sensorial experience within the classroom and that the interaction between language and image was aimed at creating and consolidating an imperial consciousness to be embedded in British society. The text had to teach children what and how to observe, where to focus their attention, to visualize what the teacher was narrating. COVIC's textbooks are arranged as a journey throughout the Empire, and expressions like *we land, we arrive, we cross, we see, now let us run*, etc. recur regularly to further pupils' imagination by creating a visual experience of travelling, stimulating interest in the Empire. Thanks to this '"we" that travels', as referred to by Sarah Dellmann, it is likely that students would perceive that the entire classroom was taking the journey, fostering their engagement with the narration.[65] These passages, which were aimed at producing the sensation of moving to explore and envision new lands and people, also included logistical details probably taken from Fisher's letters, as it is evident in the following extract from the Indian lectures:

> As we think over these things we are continuing our journey northward. We must change from train to steamer as we cross the Ganges. The passage of the river occupies about twenty minutes from one low-lying bank to the other. Then, as we traverse the endless rice fields with their clumps of graceful bamboo, the hills become visible across the northern horizon. We run into a belt of jungle, and change to the mountain railway, which carries us up the steep hill front with many a turn and twist.[66]

---

64 Frederick, Ursula K., 'Flights of Fancy. The Production, Reception and Implications of Lawrence Hargrave's Magic Lantern Lecture 'Lope de Vega'', in *The Magic Lantern at Work. Witnessing, Persuading, Experiencing and Connecting*, ed. by Martyn Jolly, and Elisa deCourcy (London: Routledge Studies in Cultural History, Taylor and Francis, 2020), p. 125.

65 Dellmann, Sarah, 'Visiting the European Neighbours', in *A Million Pictures: Magic Lantern Slides in the History of Learning*, ed. by Sarah Dellmann, and Frank Kessler (Herts: John Libbey & Co Ltd., 2020), p. 32.

66 Mackinder, *India*, p. 46.

In addition to the use of verbs related to the physical act of moving and of seeing, it is the use of the first-person plural pronoun *we* that fosters imperial consciousness. This language of belonging, as Yeandle argues, positions the reader as part of a community and, as he reflects, since the pronoun *we* encloses all the other personal pronouns, 'it can [...] be used to refer both to immediately localised settings and vague, abstract, and all-encompassing historical moments'.[67] In COVIC's lectures the text had to support the visual, hence the linguistic approach had to contribute significantly to sustain the narrative. The pronoun *we* emphasised that children, and the teacher who read the book, are part of a community, the whole classroom forms a unity. Rudolf De Cillia, Martin Reisigl and Ruth Wodak, in their work on discursive strategies and linguistic devices to construct national identities, have discerned the importance of the pronoun *we* to convey national collectiveness.[68] Its use to express unity fosters the sense of belonging to a group of people that shares certain beliefs, in our case study a classroom of pupils whose common feature is their Britishness. Through the employment of these linguistic forms the narration hopes to produce an emotional identification with the nation which would extend, in COVIC's case study, to the Empire, contributing to strengthen national and imperial identity. Throughout the textbooks, authors frequently make references and comparisons with British cities and landscapes. For example, in describing the Esk River in Tasmania, Sargent evokes English counties: 'The Esk here flows through a rocky and wooded gorge; and we might easily imagine ourselves in Devon or Cornwall'.[69] This served to familiarize children with the unknown lands and also to foster the sense of national belonging by reminding them what they have in common. And again, as De Cillia, Reisigl and Wodak argue, this sense of belonging can go back in national time as readers share national history, what they refer to as the 'historically expanded we'.[70] Indeed, the narrating identity pervades COVIC's textbooks, the creation of an imperial vision is strengthened through the past, a history which goes back in time and is shared by the classroom because it is the history of the nation. *We* is used to include the forebears in the discourse, linking their work with the present, as in the following passage by Sargent of the description of New Zealand territory:

> To open up the country for farming we had to reduce the hunting of the Maori, and even to drain, in spite of his opposition, some of his favourite marshes. As a result of the settlement we find in the south-east, south and south-west, butter, wool, meat and timber produced in large quantities; while in the centre we have nothing but the scenery and some forest as yet untouched.[71]

The actions described cultivated a sense of inclusion and of exploration, fostering pupils' perception of themselves as part of a glorious past and a prosperous present. Additionally, the lantern projection had the power to encourage the feeling of classroom collectiveness, as children experienced the same journey together. As Heathorn wrote, 'the building of

---

67 Yeandle, p. 177.

68 Rudolf De Cillia, Martin Reisigl and Ruth Wodak, 'The Discursive Construction of National Identities', *Discourse and Society*, 10.2 (April 1999), 149–73.

69 Sargent, *Australasia*, p. 60.

70 De Cillia, Reisigl, and Wodak, p. 164.

71 Sargent, *Australasia*, p. 117.

communal identity requires, of course, both an "us" and a "them". The insider/outsider dichotomy is integral to the construction of a sense of community and belonging'.[72] By emphasizing differences the discourse perpetuates a colonial mindset sustained by racial stereotypes. This linguistic practice is used to propagate notions of identity to maintain culturally and socially a subordinate status of the colonized and to distance the 'Others' from home. Throughout the lectures the authors express and draw attention to the physical and cultural differences between the native populations and the white British or Europeans. The narration included judgements establishing behaviours, somatic characteristics and values, to be attributed to the 'Others', simplifying different characteristics into one. 'Because "subject peoples" were under the control of the writer, they could be summoned to support more than one aspect of the imperial ethos'.[73] In the following passage Mackinder provides a physical analysis of the Burmese people:

> The Burmese are a short, sturdy people, merry and happy, and akin rather to the Japanese in temperament than to the people of the Indian Peninsula. The features of their faces are obviously Mongolian. They have the oblique eyes of the Chinaman. Here is a typical Burman with a rose-coloured wrap round his head. The Burmese women, whose praises have been sung through the world, are dainty and, according to a more or less Chinese standard, not infrequently beautiful.[74]

Mackinder's account accompanies Fisher's painting on canvas *Jack Burman, A Typical Burman* (Fig. 9.5) which corresponds to the lantern slide *A Typical Burman* (*India*, Lecture II, Image 13).[75] Mackinder's description confirms the colonial discourse in representing the native population through a collective character over individuality. The painting is a close-up portrait of Jack Burman: the warm colours create and transmit the sitter's human character, yet Mackinder's detailed explanation does not focus on the man; he tells the reader what to see. When looking at the slide projected on the wall then, we are asked to imagine the stature of the man portrayed. He is short, as all the Burmese, robust, and happy; we are also asked to relate his features with the Mongolian's peoples and the Chinese. Through the powerful relation between text and image Mackinder provides the reader with a clear and fixed idea of the peoples of Burma.

In COVIC's textbooks, the interaction between the visual and the spoken word is supported by the structure of travelogue which made use of actions that were happening at the same moment of the lecture. Sargent accompanies the lantern slide *Cutting Grapes, Grantham* (*Canada*, Lecture V, Image 7) showing farmers in Ontario, with a sentence that enhances the feeling of experiencing the scene as in real life: 'Here, again, is a magnificent vineyard where they are gathering grapes such as grow only in hothouses in our own Islands'.[76] The audience is invited to experience the harvest of the grapes along with the pickers. Passages like this recur frequently in all the lectures to strike the pupils' attention. In the

---

72 Heathorn, p. 115.

73 Castle, p. 180.

74 Mackinder, *India*, p. 26.

75 Jack Burman (a typical Burman), 1907–12–1908–01, GBR/0115/RCS/RCMS 10/5/359. Cambridge University Library.

76 Sargent, *Canada and Newfoundland*, p. 68.

Fig. 9.5. Alfred Hugh Fisher, Jack Burman (a typical Burman), 1907–12–1908–01, Myanmar, oil on board. GBR/0115/RCS/RCMS 10/5/359. Reproduced by kind permission of the Syndics of Cambridge University Library.

Canadian albums there are two photographs titled *Cutting "Black Concord": R. Thompson's Farm, Grantham, Ontario* (Fig. 9.6; Fig. 9.7).[77] Their composition is highly studied and Fisher organized the images in order to emphasize the human figure and the work. In the first photograph the plants along both sides accompanies our eye along the strip of soil, accentuating its length; in the foreground are visible a few baskets brimful with grapes and a man bent down. The other three figures are captured in the process of cutting the fruit from the plants. Fisher positioned himself at a human level giving the viewer the sensation that by moving forward the following step would be in the countryside, in Grantham. It is likely that the image used as lantern slide is this one, as it provides the audience with more information and a broader view of the scene. Nonetheless, the second photograph also gives the perception of being in the scene. This time Fisher decided to focus the attention

77 Cutting Black Concord Grapes, 1908–09, GBR/0115/RCS/Fisher 11/2421 and no. 2422. Cambridge University Library. Since it is not possible to compare the slides' list with the physical objects, we are unable to tell which one of the two was used in the lecture. The two photographs though are very similar in their structure and subject depicted, thus they can be both used for our analysis.

Fig. 9.6 (left). Alfred Hugh Fisher, Cutting Black Concord Grapes [Cutting "Black Concord" Grapes: R. Thompson's Farm, Grantham, Ontario], 1908–09, Canada, photographic print. GBR/0115/RCS/Fisher 11/2421. Reproduced by kind permission of the Syndics of Cambridge University Library;

and Fig. 9.7 (right). Alfred Hugh Fisher, Cutting Black Concord Grapes, 1908–09, Canada, photographic print. GBR/0115/RCS/Fisher 11/2422. Reproduced by kind permission of the Syndics of Cambridge University Library.

on the plant and the fruit, and if we take a closer look we can see a hand coming out of the leaves showing a grape. The woman kneeling on her basket is looking up towards the fruit forcing our gaze to see it and to desire to take it from the person's hand. In the foreground, is visible a hat that seems to be placed alone on the plant, waiting for the viewer to put it on and join the pickers. By looking more carefully we can see the head of the person wearing that hat, suggesting that Fisher instructed the man to hide among the leaves. We have to imagine that the enlargement of the photograph and its projection on to a screen would provide the sensation of being in the image and words were used to enhance this sensation. COVIC's teaching methods preclude pupils' interaction with the material objects. Students did not have the possibility to touch and scrutinize the physical photographs as they were printed in the textbook which only teachers would use. The lanternist would operate the projector and handle the slides, thus what remained for students to do was just to immerse themselves into the image.

COVIC structured the lectures in such a way that the narration would follow the slides which would have been shown in a particular order. The lantern is an integral part of the course: expressions like *Here we see*, *Notice here*, etc., reflect the impossibility of using the textbooks without the slides. The printed photographs are very few compared to the slides and the fact that the narrators refer largely to the images definitely bonds the visual with the text. To facilitate the synchronization of the text with the images, marginal references to the numbers of the slides were printed next to the correspondent text. This system was employed after the publication of the Indian lectures which, in the preface, provided a note to the readers reminding them that the 'lectures are illustrated with lantern slides … and that reference is made in the text to scenes which have not been reproduced here'.[78] This system of synchronization forced teachers to follow strictly the text without the opportunity of changing the narration or the order. In a review of *The Sea Road to The East* written for *The Geographical Teacher* journal, the writer is very positive, stating that 'all teachers should read the book' but he also criticizes the 'constant use of the words "here we see"' which 'somewhat spoil the book for use as a reader'.[79] The rigid structure that characterizes the textbooks does not allow the teachers or children to have individual reactions to the images, as the synchronization with the text immediately instructs them how to read the slide. Moreover, the lecture's structure prevented teachers from inserting their own slides and observations.[80] In COVIC's textbooks, the meaning is constructed through assumptions supported by the spoken word affecting the interpretation of the visual information.

Although there are no reports on how teachers performed COVIC's lectures, it is probable that they actively participated in order to deliver a more convincing experience.

78 Mackinder, *India*, p. x.

79 J. W. P., 'Review, The Sea Road to The East', *The Geographical Teacher*, 7.2 (Summer 1913), 130.

80 During this period, in some lantern slide lectures teachers would also prepare their own sets of slides, giving them the opportunity to develop and personalize the presentation. In an article published in *The Geographical Teacher*, C. C. Carter suggests teachers to supplement the lantern slides at their disposal by photographing subjects themselves. Carter also invites the instructors to ask friends and colleagues to lend their negatives to enrich the material that could be used in the classroom. Carter, C. C., 'Amateur Photography as an aid in Teaching Geography', *The Geographical Teacher*, 4.2 (October 1901), 27–31.

Let us take, for example, an extract in which the author, in describing the city of Invercargill in New Zealand, asks the audience to listen to the citizens' talk:

> Let us look around us: here are Dee Street and Don Street, whose very names, like that of the town itself, carry us back to Scotland. Or let us look at the people and listen to their talk: we shall find Scots everywhere, not in this town only, but all over the south of the island.[81]

By reading this passage, that was written to be read aloud, one could imagine the teacher reciting it with a distinctive Scottish accent to help students immerse themselves in the atmosphere; by performing the representation of the photograph the teacher could draw attention to the sounds and look of the streets. Through the text the images come to life and the classroom's journey all-round the Empire is made possible.

Britain's imperial role is at the core of the six textbooks. Felix Driver has discerned how geographical knowledge represented a tool of Empire for the benefit of colonizers' profits as it enabled the acquisition and exploitation of the conquered territories.[82] Geography textbooks could function as tools through which to begin the process of imperial education in the formative years. Despite the fact that COVIC's lectures were rich in historical accounts, their primary focus was on geography and the textbooks were intended to show the economic importance and commercial opportunities within the Empire. Teresa Ploszajska, in her work on geographical education and empire, has noted how imperial topics characterized geography textbooks, persisting throughout the entire Age of Empire.[83] At the turn of the twentieth century geography textbooks relating to the British Empire flourished. Additionally, as Ryan has observed, contemporary textbooks included photographs and other visual material illustrating the overseas territories.[84] COVIC wanted to produce colonial culture through the use of lantern slides. Ryan has noted how COVIC's ideas of promoting imperial world views were not unique, and conformed to traditional practices, but that the difference with other educational programmes lay in the scheme's dimension.[85] COVIC's project shared the attitudes of contemporary organizations, and I would add that the uniqueness of its scheme was in its methods, in the ways of producing the material as well as of imparting notions.

Conclusion

The Committee invested photography with the authority and power to impart colonial knowledge to children. COVIC's photographic practice allows us to consider the pedagogical approach to the lantern and the interaction of the device with textbooks to

81 Sargent, *Australasia*, p. 86.

82 Driver, Felix, 'Geography's Empire: Histories of Geographical Knowledge', *Environment and Planning D: Society and Space*, 10.1 (1992), 27.

83 Ploszajska, Teresa, *Geographical Education, Empire and Citizenship: Geographical Teaching and Learning in English Schools, 1870–1944*, Historical Geography Research Series, 35 (London: Historical Geography Research Group, 1999).

84 Ryan, pp. 183–213.

85 Ryan, pp. 183–213.

create a specific teaching environment. Through the analyses of the textbooks, which guided children's understanding of the images, this chapter has provided an important analytical window into understanding how the pictures corroborated the written word. By providing a reading of the lantern slides that goes beyond mere illustration and takes into consideration the context of the technology, the narration of the textbooks and the objectives of the institution which produced them, this analysis has demonstrated the importance of exploring the impact expected from the visual media. The sole projection of images onto a wall, although aimed at engaging the whole room at once, could stimulate individual reactions, yet it is the textual aid, which supported the visual, that was expected to shape the audience's interpretations.

The Committee identified the lantern device as the strategic teaching tool to construct and deliver a structured visual framework. This chapter has argued for the centrality of the classroom experience; COVIC's intention to employ the lantern device in its educational project allows us to explore the role the lantern technology played in constructing an environment capable of creating a sense of belonging in the classroom. This case study based on the Fisher Photograph Collection deepens our understanding of the pedagogical potential of the lantern performance to educate schoolchildren by playing upon specific emotions.

KERRIN VON ENGELHARDT

Complex Associations: On the Emotional Impact of Educational Film in the German Democratic Republic (1950–1990)

Introduction

'It also and primarily means, to enable you, the youth of our country, to master modern technology and to develop it further, to know you as partners at our side.'[1] This message was given in an educational film in 1980 with approval from the German Democratic Republic (GDR) (East Germany), indicating that educational film not only provided knowledge but also aimed to form a specific attitude towards the country and motivate the youth to learn.

When I was browsing the archival resources on the usage of educational films in the former eastern part of Germany, I was suddenly confronted with uneasiness, a vague fear of a warlike threat that I sometimes used to feel when I was a child. In 1990, I was twelve years old when the German Democratic Republic came to an end and was reunited with the western part of Germany. Most likely, I experienced indoctrination in school, but these days, I have rarely felt its afterlife as strongly as I did when reading lesson descriptions drawn up by East-German teachers in the 1980s. This experience led me to question to what extent educational media and teaching aids were designed to cause specific emotions. Since the school film movement of the Weimar Republic, film as a teaching aid has been attributed with a particularly emotionalising effect.[2] In this movement individual teachers were engaged for the use of films for teaching purposes.

* This article is part of the project The Myth of Scientific Neutrality. The educational film in schools during the Cold War. It is part of the research network *Bildungs-Mythen — eine Diktatur und ihr Nachleben. Bilder(welten) über Praktiken und Wirkungen in Bildung, Erziehung und Schule der DDR*, funded by the German Federal Ministry of Education and Research (BMBF) (https://bildungsmythen-ddr.de/ddr-forschung/).

1 The text reads in German: '*Das heißt auch und an erster Stelle, euch, die Jugend unseres Landes zu befähigen, diese moderne Technik zu beherrschen und weiterzuentwickeln, euch als Partner an unserer Seite zu wissen.*'

2 Annegarn-Gläß, Michael, *Neue Bildmedien revisited. Zur Einführung des Lehrfilms in der Zwischenkriegszeit* (Bad Heilbrunn: Klinkhardt, 2020), p. 11.

Kerrin von Engelhardt • Humboldt University Berlin, kerrin.engelhardt@hu-berlin.de

Learning with Light and Shadows: Educational Lantern and Film Projection, 1860-1990, ed. by Nelleke Teughels and Kaat Wils, TECHNE-MPH, 8 (Turnhout, 2022), pp. 245-267

10.1484/M.TECHNE-MPH-EB.5.131502

The German Democratic Republic's (GDR, 1949–1990) pedagogical agents spoke of attitudes towards the state, the collective, and in relation to the whole world. Accordingly, the goal of socialist education went beyond the transmission of knowledge to evoking the *right* attitude within the individual. Of course, ideologisation was central to socialist educational efforts.[3] It is important to point out, though, that 'ideology' was not a term used in critical analysis in the GDR but instead was used positively to describe specific contents and attitudes that should be transferred within educational processes. Indeed, every state-approved educational medium can be in general considered inherently ideological.[4] In particular, educational films functioned similarly to textbooks as 'politicum, informatorium, and paedagogicum'.[5] As *informatorium*, they provided an order of knowledge in which the processes of knowledge transmission could be objectified, thus enabling the constitution and circulation of knowledge. As *pedagogicum*, they reflected the pedagogical demands of their time. Finally, as *politicum*, these teaching media were authorised by state approval procedures and disseminated through public schools — institutions that serve as a means of exercising power. There is also a fourth category that can typify educational films and textbooks as well. Both media tools do not simply pass on information; rather, they act as a *constructorium*, resulting from social processes of negotiation about what should be passed on and how. Therefore, sociocultural knowledge can be 'observed' in textbooks and educational films.[6] Although educational films do not share the same 'leading medium'[7] status as textbooks, the aspects mentioned above also play a role in their production and use in the classroom. It is thus erroneous to assume that there are ideology-free educational films. For the national socialist educational film, for example, ideological embedding took place not only on the level of the commentary but also visually.[8] It, therefore, follows that an educational film, even a silent one, can hardly be free from ideological colouration.

The examination of German educational films in the study of historical education begins with the school film movement and focuses especially on the Weimar Republic.[9] For the period of National Socialism (1933–45), as mentioned above, media analyses of the visual language and its ideologisation have already been undertaken from the perspective of the history of education.[10] Likewise, a study from 2001 on the educational film of the GDR has focused on the history film investigating the representation of National Socialism in

3 Tenorth, Heinz-Elmar, 'Die "Erziehung gebildeter Kommunisten" als politische Aufgabe und theoretisches Problem. Erziehungsforschung in der DDR zwischen Theorie und Politik', *Zeitschrift für Pädagogik*, 63 (2017), 207–75.

4 Tenorth, Heinz-Elmar, 'Grenzen der Indoktrination', in *Ambivalenzen der Pädagogik — Zur Bildungsgeschichte der Aufklärung und des 20. Jh.*, ed. by Peter Drewek and others (Weinheim: Beltz-Verlag, 1995), pp. 335–50.

5 Stein, Gerd, 'Schulbücher in berufsfeldbezogener Lehrerbildung und pädagogischer Praxis', in *Pädagogik. Handbuch für Studium und Praxis*, ed. by Leo Roth (2., überarb. u. erw. Aufl., München: Oldenbourg, 2001) pp. 839–47.

6 Höhne, Thomas, 'Schulbuchwissen. Umrisse einer Wissens- und Medientheorie des Schulbuches' (published doctoral thesis, Johann-Wolfgang-Goethe-Universität Frankfurt a. M., 2003).

7 Stöber, Georg, *Schulbuchzulassung in Deutschland. Grundlagen, Verfahrensweisen und Diskussionen* (Braunschweig: Georg-Eckert-Institut für internationale Schulbuchforschung, 2010), p. 6.

8 Imai, Yasuo, 'Ding und Medium in der Filmpädagogik unter dem Nationalsozialismus', *Zeitschrift für Erziehungswissenschaft*, 25 (2015), 229–51.

9 See Annegarn-Gläß (2020).

10 See Imai (2015); and Niethammer, Verena, 'Indoktrination oder Innovation? Der Unterrichtsfilm als neues Lehrmedium im Nationalsozialismus', *Journal of Educational Media, Memory & Society*, 8.1 (2016), 30–60.

schools.[11] In addition, the collection of health films at the Hygiene Museum Dresden has been described and evaluated with regard to state strategies for health care in the GDR.[12] Anja Laukötter, meanwhile, elucidated how cinematic attempts were made to emotionalise health and medicine films to evoke certain reactions and convince viewers.[13] Konstantin Mitgutsch has similarly dealt with the aspect of indoctrination by media in educational contexts.[14] Additionally, the pioneers of scientific filmmaking in the German-speaking regions and the developments in the western part of Germany after 1945 have also been investigated.[15] In contrast, the GDR educational films in sciences have been largely unexplored. If one combines these two focal points of interest, with scientific film on the one hand and film as an instrument of influencing on the other, it is all the more startling that the GDR films in science education have hardly received any attention to date. This is all the more astonishing because technology and sciences were the main subjects in the self-concept of GDR socialist education.

The first part of this paper introduces the institutions of East German educational film and details the pedagogical research on the effects of audio-visual teaching aids. At first, the educational film was seen as a substitute for actual experience, thereby conceptualising a realistic audio-visual depiction as being able to put the viewer in the middle of the action. The power of persuasion was linked to what we would today call 'authenticity' but which the agents likened to reality and objectivity. However, as the pressure of political and economic planning on the educational sector began to increase, the efficient use of resources became an important issue for pedagogues too. As a result, the government officials initiated a number of empirical studies on the usage of educational aids like films and their effects and impacts. The aim was to systematically find out how teaching aids were currently used and how this could be improved. While socialist slogans are clearly recognisable in pedagogical texts about the value of educational films in the GDR, the question posed in this paper will be how the pedagogical agents referred to emotions. This focus was chosen to investigate to what possible extent emotionalising effects were used intentionally in the production of educational films in order to have an indoctrinating effect on pupils.

In the second part, the GDR educational film *Lernen für die Zukunft* ('Learning for the Future') will be examined. Produced in 1980, it thus stands at the end of the development

---

11 Kneile-Klenk, Karin, *Der Nationalsozialismus in Unterrichtsfilmen und Schulfernsehsendungen der DDR* (Weinheim: Beltz, 2001).

12 Schwarz, Uta, 'Vom Jahrmarktspektakel zum Aufklärungsinstrument. Gesundheitsfilme in Deutschland und der historische Filmbestand des Deutschen Hygiene-Museums Dresden', in *Kamera! Licht! Aktion! Filme über Körper und Gesundheit 1915 bis 1990*, ed. by Susanne Rößiger (Dresden: Sandstein, 2011) pp. 12–49.

13 Laukötter, Anja, 'How Films Entered the Classroom. Emotional Education of Youth through Health Education Films in the US and Germany, 1910–1930', *Osiris*, 31.1 (2016), 181–200; Laukötter, Anja, '"One feels so much in these times": Visual History in the GDR', in *Body, Capital & Screens. Moving Images and Individual's Health in Economy-based 20th Century Societies*, ed. by Anja Laukötter, and Christian Bonah (Amsterdam: Amsterdam University Press, 2020) pp. 205–30.

14 Mitgutsch, Konstatin, 'Indoktrination als Phantom. Über die Intentionalität des Medieneinsatzes im Lehr-Lernprozess', in *Indoktrination und Erziehung*, ed. by Henning Schluß (Wiesbaden: VS Verlag für Sozialwissenschaften, 2007), pp. 93–112.

15 See for example: Sattelmacher, Anja and others, 'Introduction: Reusing Research Film and the Institute for Scientific Film', in: *Isis*, 112 (2021), 291–98.

presented here. The paper will investigate the cinematic strategies employed in this film using the motivational strategies elaborated by Eef Masson as its framework. Masson's systematic analysis, *Watch and Learn. Rhetorical Devices in Classroom Films after 1940*, is a fundamental study of Dutch educational films and begins from a pedagogical dispositive, which she understands as 'framing' in the sense used by media studies.[16] Through this dispositive, the reception of certain cinematic means can thus be activated. So, the film is recognized as learning resource and the content is understood as relevant school knowledge. This is supported, for example, by an authorial commentary or certain illustrative practices such as the use of arrows. Consequently, the students are conditioned to know what to look for when they are shown a film in the classroom. In her study, Masson identifies different strategies that are associated with the specific camera angles, motifs, and dramaturgies that characterise historical educational films. She explicitly emphasises the potential of cinematic realisation (in contrast to the editing of school knowledge in printed media, for example), which also includes the dramaturgical combination of certain visual attractions and auditory stimulations like music pieces, in producing persuasive educational films.

Institutions of Educational Film

Although the educational system of the GDR was organised centrally and hierarchically,[17] from today's point of view, the field of educational film is surprisingly confusing. The names and functions of the responsible institutions and departments, for example, changed several times. Therefore, to keeping pace with both the different names and the GDR's typical and sometimes confusing abbreviations, it is helpful to have an overview of the institutional developments.

In 1950, a *Zentralinstitut für Film und Bild in Unterricht, Erziehung und Wissenschaft* (ZFB) (Central Institute for Film and Image in Education and Science) was established. This Central Institute was under the authority of the *Ministerium für Volksbildung* (Ministry of National Education) and was accountable to the *Zentralkomitee der Sozialistischen Einheitspartei Deutschlands der DDR* (State Council of the East German Communist Party).[18] The Central Institute's responsibilities included planning, producing, and testing films, photographs, LPs, and audiotapes for use in schools, vocational schools, and universities.[19] In addition, the Central Institute issued research commissions,[20] developed guidelines for the use of media, and conducted professional training sessions for educators.[21] In 1954, it was renamed the *Deutsches Zentralinstitut für Lehrmittel* (DZL) (German Central Institute

16 Masson, Eef, *Watch and Learn: Rhetorical Devices in Classroom Films after 1940* (Amsterdam: Amsterdam University Press, 2012), pp. 117–24.

17 See Waterkamp, Dietmar, *Handbuch zum Bildungswesen der DDR* (Berlin: Spitz, 1987); and Geißler, Gert, *Schule und Erziehung in der DDR* (2nd ed., Erfurt: Landeszentrale für Politische Bildung Thüringen, 2015).

18 Kneile-Klenk, Karin, *Der Nationalsozialismus in Unterrichtsfilmen und Schulfernsehsendungen der DDR* (Weinheim: Beltz, 2001), pp. 51–52.

19 Ibid., p. 51.

20 See on pedagogical research in the GDR: Tenorth, pp. 207–75.

21 Kneile-Klenk, p. 53.

for Teaching Aids).[22] Five years later, this newly named institute underwent internal restructuring; the previous organisation of departments by media type was abolished in favour of a subject-based structure to pursue an *Unterrichtsmittelsystem* (system of teaching aids) in line with the Soviet model.[23] In the GDR, educational films were understood as elements of a system in which all teaching and learning materials were coordinated with each other and with the curriculum. In 1962, the institute was incorporated into the *Deutsche Pädagogische Zentralinstitut* (DPZI) (German Central Pedagogical Institute) to provide a more scientific basis for the development of teaching aids.[24] Then, in 1970, the Pedagogical Institute was replaced by the *Akademie der Pädagogischen Wissenschaften der DDR* (APW) (Academy of Pedagogical Sciences of the GDR). The commissioning body for educational films was the *Institut für Lehrmittel* (Institute for Teaching Aids) at the Academy. This brought the production of teaching materials under the umbrella of a research institution. By 1989, fifty-three scientific and eighteen technical employees worked in this Institute for Teaching Aids.[25] The Academy of Pedagogical Sciences was a state research institution that encouraged reflection and set the theoretical foundation of everyday teaching practices; it also governed the educational system through curricula and educational media. The theory, authorisation, and distribution of teaching aids for schools were thus united under the umbrella of this Academy.

The timeline above refers to authority at the highest levels; however, the GDR educational films for schools were also centrally regulated at the lower levels. Since the 1970s, the highest level of responsibility was held by the main administration for teaching materials and school supplies of the *Ministerium für Volksbildung* (Ministry of National Education), followed by the Institute for Teaching Aids at the Academy of Pedagogical Sciences, then the *Bezirksstellen* (District Offices) and the *Kreisstellen* (Local Offices) for teaching materials and finally the individual schools and their staff.[26] The *DEFA-Studio für Dokumentarfilm* (DEFA-Studio for Documentary Film) and the *DEFA-Studio für Populärwissenschaftlichen Film* (DEFA-Studio for Popular Science Film) were responsible for the cinematic realisation.[27] Teams from the DEFA studios produced the respective films in constant consultations with officials in the Academy of Pedagogical Sciences and experts in the disciplines. The DEFA (Deutsche Film-Aktiengesellschaft), founded in 1946, was the state-owned Production Company of the German Democratic Republic,

---

22 Ibid., p. 56.

23 See for schools Kneile-Klenk, p. 57; for higher education, see Fuchs, Rolf, and Klaus Kroll, *Audiovisuelle Lehrmittel* (2nd ed., Leipzig: Fotokinoverlag, 1982), pp. 65–66.

24 Kneile-Klenk, p. 57; and Zabel, Nicole, *Zur Geschichte des Deutschen Pädagogischen Zentralinstituts der DDR* (Chemnitz: Dissertation, 2009), p. 124.

25 Malycha, Andreas, *Die Akademie der Pädagogischen Wissenschaften der DDR 1970–1990* (Leipzig: Akademische Verlagsanstalt, 2009), p. 147.

26 Today, the inventory of GDR educational films is kept in the film archive of the 'Bundesarchiv' (federal archive) but has not yet been made accessible to the public. Individual media sources can, however, be viewed via the holdings of the Oldenburg University Library. In addition, there is a collection of GDR educational films even in the USA. In California, 'The Wende Museum' has also undertaken digitisation efforts and, thus, individual DDR educational films are accessible via its homepage.

27 Knopfe, Gerhard, 'Der populärwissenschaftliche Film der DEFA', in *Schwarzweiß und Farbe. DEFA-Dokumentarfilme 1946–92*, ed. by Günter Jordan, and Ralf Schenk (Berlin: Jovis, 1996), pp. 294–341.

and its departments and studios were responsible for film productions in all fields from utility to entertainment.

Furthermore, there was an institutional junction that also had effects in the film sector. In 1950, the *Ministerium für Volksbildung* (Ministry of National Education) also established a *Hauptabteilung Hochschulen und Wissenschaft bzw. wissenschaftlicher Einrichtungen* (Central Department of Higher Education and Science or Scientific Institutions).[28] This department was then transformed a year later into a separate *Staatssekretariat* (State Secretariat).[29] This State Secretariat was independent of the Ministry of National Education and had its own department for educational media, the *Abteilung wissenschaftliche Bibliotheken, Museen und Hochschulfilm* (Department of Scientific Libraries, Museums, and University Film).[30] It was not until 1964, however, that the *Institut für Film, Bild und Ton* (IFBT) (Institute for Film, Image, and Sound) was established and could take on the film holdings relevant to higher and technical education from both the former pedagogical institutes for teaching aids. Furthermore, audio-visual teaching aids for universities were not standardised and institutionalised in the same way as teaching aids for schools. Only in some study subjects, such as Marxist-Leninist basic studies, foreign language education, mathematics, physics, chemistry, and biology, and medicine,[31] did the IFBT centrally store and loan audio-visual media for teaching purposes.[32] In some of these films, footage of research projects was used, meaning that one might address the IFBT films as science films as well.

Positions on the Impact of Audio-Visual Media

These institutional structures also laid the fundaments for theoretical reflections and empirical research on the effects of media usage in schools and universities. The results were presented in dissertations and research reports and published in magazines, newsletters, or handbooks and were often edited by the various state institutions. In the introductions to their studies, the authors justified their respective research concerns with the obligatory references to recent party congress resolutions and the valid socialist state doctrine.[33] All positions exhibited a strict Marxist-Leninist framing and were rhetorically integrated into the specific and complex research landscape of the German Democratic Republic.

---

28 Sperlich, Cordula, *Die Umwandlung des Staatssekretariats für das Hoch- und Fachschulwesen in ein Ministerium für Hoch- und Fachschulwesen und die sich daraus ergebende Organisation und Arbeitsweise* (Potsdam: Diplomarbeit, 2009), p. 25.

29 Ibid., p. 27.

30 Ibid., p. 30.

31 Heun, Hans-Georg, 'Die Organisation und Leitung der Arbeit mit audiovisuellen Lehr- und Lernmitteln an der Humboldt-Universität zu Berlin', *Wissenschaftliche Zeitschrift der Humboldt-Universität zu Berlin — Gesellschaftswissenschaftliche Reihe*, 35.7 (1986), 620–25 (p. 621).

32 Today, the University of Applied Sciences (HTW) in Berlin holds the IFBT media collection for teaching purposes in the field of preservation and restoration, audiovisual and photographic cultural heritage.

33 *Medizin, Wissenschaft und Technik in der SBZ und DDR: Organisationsformen, Inhalte, Realitäten*, ed. by Sabine Schleiermacher (Husum: Matthiesen-Verl., 2009); Geißler, Gert, and Ulrich Wiegmann, *Pädagogik und Herrschaft in der DDR. Die parteilichen, geheimdienstlichen und vormilitärischen Erziehungsverhältnisse* (Frankfurt a.M./Bern: Lang, 1996); Tenorth, Heinz-Elmar, 'Die "Erziehung gebildeter Kommunisten" als politische Aufgabe und theoretisches Problem. Erziehungsforschung in der DDR zwischen Theorie und Politik', *Zeitschrift für Pädagogik*, 63 (2017), 207–75.

Living Observation

Werner Hortschansky was the first director of the Central Institute for Film and Image in Education and Science and later of the German Central Institute for Teaching Aids. He was a protagonist in the development of educational film in the GDR until the 1960s and emphasised not only the 'increasing efficiency in schoolwork' stemming from projected films and photographs but also the way that audio-visual teaching aids helped connect school and life and stimulated the child's interests.[34] As Hortschansky argued, this would be especially useful in science education, a subject area that carried great significance. In his understanding, which was as influenced by Comenius and Pestalozzi as well as by the Russian pedagogy, a dialectical cognition connected *lebendige Anschauung und abstraktes Denken* ('living observation and abstract thinking'). As such, film and slide projection were particularly valuable teaching aids 'because they are derived from the grasp of natural conditions themselves, because they are, in a certain sense, *Naturkunden*' ('Natural histories'). Hortschansky emphasised that film should provide an image that was true-to-nature[35] and thereby capture 'only what is actually moving'. Furthermore, Hortschansky underlined how important it was for teachers to 'work through' the accompanying booklets that were issued with each film since this was the only way to prepare for the use of the film in the classroom.[36]

Conditioning

In addition to the mainly programmatic statements on the benefits of educational film by the spokesmen of the responsible institutions, reflections on the psychological effects of film can be found in the 1950s issues of the magazine *Bild und Ton*. At the time, Pavlov's reflex theory was not only being adopted in the field of pedagogy but also in the field of film. Even if this was only a short episode overall — at the end of the 1950s, the pedagogical researcher of the GDR would distance itself from the theory — it is an important one. Here, a precise computation of psychological effects was, for the first time, explicitly formulated for the educational film. In 1952, Albert Wilkening advocated for the adaptation of Pavlov's science of "conditional reflexes" and "temporary connections"' for the creation of films. Wilkening was not only editor-in-chief of the magazine *Bild und Ton* but also active in the DEFA, the state film studio. He was initially responsible for production and technology as co-director, and, from 1969 onwards, he became the main director of DEFA and was also made professor at the *Deutsche Hochschule für Filmkunst* (East-German University of Cinema). He stated that the 'film montage, the aperture, and the vision' were cinematic devices with which a knowledgeable viewer could establish 'temporary connections' and understand that which was shown within its contexts:

> After all, film is based to an extraordinarily strong degree, perhaps even exclusively, upon the formation of temporary connections. It is not 'illusion,' as it is mystically

34 Hortschansky, Werner, *Unterrichtsfilme und Lichtbilder in der Schularbeit* (Berlin: Volk u. Wissen Verl., 1951), p. 5.

35 Ibid., p. 14.

36 Ibid., pp. 9–18.

> called by the dream makers but a quite natural physiological process. To master it and to be able to apply it masterfully is the great task.[37]

The term 'dream makers' (or *Traumfabrikanten* in German) refers to Hollywood's so-called 'dream factory' and its movies. Wilkening primarily pursued a physiological approach, an orientation that, in the 1960s, was also compatible with the information-theoretical concepts like cybernetics that were fashionable at the time. However, the most enduring aspect for the further developments in the sector of teaching aids was probably the consideration that films could have controllable reflex effects on the viewer.

Emotionalizing

In his 1962 dissertation on educational film for chemistry instructions, Helmut Boeck discussed how students' emotions could be generated and influenced. He referred to his own experience as a teacher in the classroom and analysed different educational films in detail. Boeck argued that through the professional use and design of educational films, it was possible to make students associate positive feelings with the subject of chemistry, to motivate them to actively participate in class, and even to positively influence their physical condition. As he stated:

> In chemistry instruction, it is therefore necessary to awaken in students, for example, the joy of learning, of performing experiments, of dealing with the science of chemistry. By influencing emotions, pupils can be educated to love their homeland, to despise the application of chemistry's scientific knowledge for war, to respect the achievements of workers in factories and physical labour.

Boeck further described how educational films could trigger an 'emotionally enhanced learning process', thereby securing a more long-term acquisition of knowledge. As he argued, films could arouse excitement, respect, love, and interest and could thus 'effectively support certain tasks of moral education'.[38]

To grasp the possible emotional impact of educational films, the authors' collective,[39] under the supervision of Prof. Dr Günther Schwarze, demanded in 1969 that teachers should intensively study the written supplements to the audio-visual teaching aids. Like Hortschansky before them, the authors' collective assumed that the teachers had to be guided by the film producers to use the educational film correctly, because:

---

37 Wilkening, Albert, 'Pawlows bedingte Reflexe und der Film', *Bild und Ton*, 5.3 (1952), 160. Tanja Munz has elaborated on this principle for the behavioural scientist Konrad Lorenz, who conditioned the viewer to perceive certain phenomena — such as colour in black-and-white film — with the dramaturgy of his films: Munz, Tania, 'Die Ethologie des wissenschaftlichen Cineasten. Karl von Frisch, Konrad Lorenz und das Verhalten der Tiere im Film', *montage AV. Zeitschrift für Theorie und Geschichte audiovisueller Kommunikation*, 14.2 (2005), 52–68.

38 Boeck, Helmut, 'Über die Eigenarten, die Aufgaben und die Gestaltung von Unterrichtsfilmen für den Chemieunterricht der zehnklassigen allgemeinbildenden polytechnischen Oberschule' (unpublished doctoral thesis, Martin-Luther-Universität Halle/Saale, 1962), pp. 156–65.

39 The term collective, here used by the authors themselves, was fundamental to GDR society and referred to a group of socialist labourer.

> Only in this way is it possible for him [the teacher] to recognize the emotional effects on the students, be aware of impulses or necessary interruptions, etc., and to adjust his approach in the lesson to them.[40]

This statement in the handbook for the usage of audio-visual teaching aids for teachers hints that there was a gap at times between the expectations of those officially responsible for teaching aids for the effectiveness of educational media in the classroom and the actual teaching practices. Therefore, it was considered necessary for teachers at schools as well as lecturers at universities to be trained to use educational films properly.

In addition to the best practice examples given in the film institutions' newsletters, handbooks on educational film usage were published. For instance, Rolf Fuchs and Klaus Kroll edited a compilation of theoretical reflections and specific application scenarios of educational films in lessons that were specifically targeted at university lecturers. The authors considered 'a largely trustworthy reproduction of objective reality' and, consequently, vividness and practicality, to be characteristics of audio-visual teaching and learning aids — thus staying true to Hortschansky's argumentation. Audio-visual teaching aids made it possible to make otherwise inaccessible things visible, to make them 'relevant to ideology', and to react to current research developments. But beyond that, they argued, the 'unity of rational cognition and emotional experience' could be implemented specifically for learning processes, thereby leading to more intense studying and greater engagement.[41] The compendium on educational film in the university, published by Fuchs and Kroll in 1969, classified these positive emotional and thus stimulating effects, within the contemporary educational theory, under effect motivational function. They further emphasised that audio-visual information sources could be designed in such a way that they address the learner not only rationally but also emotionally.[42] This emotional impact was particularly considered in connection with the communication of ideological content.

In 1977, the teacher Walter Kraus exemplified how emotions could play a role in the teaching of political attitudes in physics lessons.[43] In his *Pädagogische Lesung* (Pedagogical Reading) on a teaching unit on 'nuclear weapons', which simultaneously served the purpose of military education, he coupled different teaching aids to achieve the strongest possible emotional impact. He stated that the central facts and numbers concerning the use of the atomic bomb as well as a blackboard sketch on its mechanisms were all comprehensible for the students, so that 'for the emotional conclusion of the lesson there was enough time left for the use of the following taped texts'.[44] According to the transcription, the narrated texts described the atomic test, the dropping of the atomic bomb on Hiroshima, and the victims. Kraus underscored that regarding

> [t]he effect on the students I would like to describe in a few words. Thoughtful, moved faces — silence, even after the bell rang — unquestioning, disciplined departure from

40 Schwarze and others, p. 19.

41 Fuchs and Kroll, p. 12.

42 Ibid., p. 19.

43 Klaus, Walter, *Möglichkeiten der sozialistischen Wehrerziehung im Physikunterricht: dargestellt am Einsatz audiovisueller Lehrmittel* (manuscript, Archiv Bibliothek für Bildungsgeschichtliche Forschung des DIPF, PL 4419a, 1977).

44 Ibid., p. 15.

> the classroom. My realization, resulting from many lessons held in a similar manner on this subject: Few lessons are better suited to illustrate the abominations of imperialism to our students.[45]

Here, the emotional impact of audio-visual resources was specifically used to create specific attitudes to emphasise the socialist worldview.

Programming

Fuchs and Kroll already aimed to use audio-visual media as tools to enable students to learn on their own. The authors referred to a 'control function' of audio-visual teaching and learning aids,[46] wherein the 'sequence of individual acts of thought [can be] firmly pre-planned for the learner' and 'the control of thought processes [can also be] expressed in the directing, provoking or stimulation of certain methods of thought'[47] through the sequence of images and camera angles. To some extent, this longing for control referred to the earlier reflex theory proposed by Pavlov but also correlated with the then-modern information science concepts.

Ewald Topp combined 1973 an economics approach with an information science approach with the intention to design the most efficient and systematic arrangement of technical equipment for schools that would last well beyond the 1980s and well into the 2000s.[48] From 1964 to 1968, Topp headed the section for teaching aids at the German Central Institute for Teaching Aids and was appointed deputy director of the Institute for Teaching Aids at the Academy of Pedagogical Sciences in 1971. It was his goal to support '[t]he independent learning of pupils, apprentices, and students [...] above all by introducing them to techniques and methods of intellectual work, by using modern information and educational aids.'[49]

In his remarks on the function of teaching aids for the 'formation of attitudes, ideological convictions, and character traits',[50] Topp also discussed emotionality, concluding that one must 'arrange the method and organization of the students' cognitive activity in such a way that rational process and emotional experience are properly combined'.[51] Topp thereby directed his analysis entirely towards the intensification and rationalisation of teaching and learning processes, which he saw realised through programmed, i.e., pre-planned, instruction.[52] However, Topp stated that how teaching aids could take on the individual functions of 'information transmission', 'presentation of the subject matter', 'control of the

45 Ibid., pp. 15–16.
46 Ibid., p. 21.
47 Ibid., p. 21.
48 Topp, Ewald, *Zur Funktion, Nutzung und Weiterentwicklung der technischen Grundausstattung der Oberschulen der DDR* (Berlin: Volk und Wissen Verlag, 1973), p. 63. The text was taken from his dissertation conceived with educational politics in mind.
49 Ibid., p. 8. In the context of information science, electronic data processing and cybernetics were important frameworks (Ibid., p. 63).
50 Ibid., p. 20.
51 Ibid.
52 Ibid., pp. 25–27.

learning process', and 'routine work' would 'still [need] to be investigated experimentally'.[53] This desideratum was met by various dissertations that investigated, through pedagogical experiments, the role of educational films in teaching and learning processes. For example, in his 1976 dissertation on the effectiveness of audio-visual teaching aids, Jürgen Küster focused on the didactics of physics and — referencing both Pavlov and cybernetics — likewise adopted a computational perspective. He specifically mentioned the 'unity of the rational and the emotional',[54] one of the seven 'basic positions of Marxist-Leninist pedagogy', thus making it clear that emotions were not simply understood as the flip side of factual knowledge but were thought of as an inseparable part of a dialectical unity.

Therefore, after the infrastructure for the educational film had been built up in the GDR in the 1950s and had become established as a teaching tool, its pedagogical effectiveness began to be considered in the 1960s. In the 1970s, the evaluation of pedagogical experiments increased in the didactic discourse on educational film in the GDR. Moreover, learning and teaching aids were understood as part of a system that was to connect teacher, teaching unit, and material-technical instrumentation, wherein the learner should be involved as an active agent. An important aspect in relation to all teaching aids was the activation of the learner. Indeed, controlling thought and action processes by means of audio-visual teaching and learning aids was framed by the theories of cybernetics and programmed learning.[55] The calculation of learning effects was increasingly rationalised by recourse to economic principles based on information science. Films were now discussed in terms of their control function and understood as elements of programmed instruction — in contrast to the simpler understanding of films as a substitute for otherwise impossible experiences. In general, GDR pedagogy adopted influences from abroad — explicitly from the USSR but also from the USA — both with regards to empirical research on media effects and the technification of the classroom. In essence, a rather positive attitude towards technical innovations is noticeable in the writings — perhaps in contrast to the more media-critical research in the Federal Republic of Germany (FRG) (West Germany). At the same time, contemporary studies also reveal sobering results when it came to the actual use of media in schools.[56] Topp, therefore, appeals to the 'socialist educator', stating: 'It is important that the educator recognizes how necessary this technology is in order to make the pedagogical process more rational and effective, and that the teacher's work can no longer be 'technology-free' in the future'.[57]

---

53 Ibid., p. 25.

54 Küster, Jürgen, 'Untersuchungen zur Wirksamkeit von audio-visuellen Hochschulunterrichtsmitteln im Prozeß der Festlegung des Wissens, durchgeführt im Forschungsbereich der Methodik des Physikunterrichts' (unpublished doctoral thesis, Pädagogische Hochschule 'Theodor Neubauer' Erfurt/Mühlhausen, 1976), p. 44.

55 Segal, Jérôme, 'Kybernetik in der DDR: Begegnung mit der marxistischen Ideologie' (published doctoral thesis, Technische Universität Dresden, 2001).

56 Heichel, Günter, 'Analyse des Ausstattungsstandes, des Ausnutzungsgrades, des methodischen Einsatzes von Unterrichtsmitteln und der erzieherischen Einflußnahme bei der Arbeit mit Unterrichtsmitteln im Biologieunterricht der allgemeinbildenden polytechnischen Oberschulen der DDR' (unpublished doctoral thesis, Martin-Luther-Universität Halle-Wittenberg, 1975). Heichel found that the usage level of audio-visual teaching aids was only between six and twenty-three per cent in biology class at general polytechnic secondary schools in the GDR (Heichel, p. 55).

57 Topp, p. 89.

Implementing

The educational process was to be pre-designed, pre-programmed, and stored in films so that the learners could learn independently. The GDR pedagogues distinguished *Unterrichtsmittel*, teaching aids for classes, from *Lehr- und Lernmittel*, aids for lessons and self-study at universities; as such, educational films were not only intended for use in school contexts but also in academic contexts as well. This may also be related to the special position of universities in the GDR, which were regarded by the state primarily as educational institutions rather than research institutions. In the 1970s, research on film effects intensified, as did efforts to train teachers and lecturers to work with audio-visual teaching aids. Not only was it determined how the teaching aids were to be used at schools and universities but also whether educational films were effective to derive conclusions regarding their design. Often, the research on educational media was connected to university departments for teacher training and pedagogy. The institutions involved in the production of teaching materials published the research results and guides in their own information journals. For the university sector, the newsletter was called, simply, *Informationen* and was edited by the IFBT (Institute for Film, Image, and Sound); the newsletter for the teacher-training sector was called *Erfahrungsberichte* (field reports) and edited by the *Zentralstelle für Rationalisierungsmittel der Lehreraus- und Weiterbildung* (ZRL, Central Office for Rationalisation Resources for Teacher Training). For teachers at school, the *Pädagogische Lesungen* (pedagogical readings) also gave examples of lessons structured by audio-visual media.

How this could be done cinematically was explained in issue nine of the *Informationen* in 1979, in which the discussion of the auditory component of the university teaching film the 'generation of emotions'[58] was seen as realised primarily through the use of music.[59] Then, in 1981, issue 15 of *Information* dealt with image design in photo, film, and television for the didactics of higher education, similarly describing 'emotional involvement'[60] and a so-called 'effect of presence'.[61] Here, the author described educational films' different camera angles and lighting situations that could, for example, trigger the feeling of fear.[62] In this way, the empirical studies and best practice examples of educational film usage also led to cinematic advice for film production.

In 1986, Bernd Denecke submitted his dissertation on the pedagogical and psychological foundations of the design of audio-visual moving teaching aids.[63] Denecke was

---

58 Artzt, Wolfgang, and Ingeborg Fischer, *Zu Fragen der auditiven Komponente des Hochschullehrfilms, ihrer Formen und ihrer Anwendung zur Erhöhung der didaktisch-methodischen Wirksamkeit der Filme*, IFBT-Informationen 9 (Berlin: IFBT, 1979), p. 7.

59 Ibid., p. 12; 16.

60 Mühlstedt, Gerhard, *Bildgestaltung. 1. Teil (Formlehre für Foto, Film, Fernsehen) nach Lehrbriefen von Dr Heinz Wolf zusammengestellt und auf die besonderen Belange zugeschnitten und ergänzt von Gerhard Mühlstedt*, IFBT-Informationen 15 (Berlin: IFBT, 1981), p. 7.

61 Ibid., p. 14.

62 Ibid., p. 33–42.

63 Denecke submitted a dissertation entitled 'Pädagogisch-psychologische Grundlagen der Bild/Ton-Gestaltung audiovisueller Lehr- und Studiermittel' (Dresden 1986). Central text segments had already appeared in 1984 in: Denecke, Bernd, *Zur Gestaltung und Wirksamkeit von Bild und Ton dynamischer audiovisueller Lehr- und Lernmittel*, IFBT-Informationen 22 (Berlin: IFBT, 1984).

part of various research teams that dealt with questions of perceptual psychology in extensive experiments with large study groups. In his dissertation, he discussed the role of associations in the success of teaching and learning processes. Denecke described how audio-visual effects evoked certain associations that could lead to physical reactions like 'the hormone release of the adrenal gland (adrenaline and noradrenaline)',[64] thereby hinting at Wilkening's ideas and thus of Pavlov's conditioned reflexes. Denecke wrote, with reference to the Russian director Sergei Eisenstein:

> The content of the image should be selected in such a way that, in addition to the visual illustration, an 'associative complex' is also created that brings forth the emotional-cognitive contents of the image. Camera and light subordinate themselves to the intention of not only depicting the object but revealing its innermost cognitive and emotional aspects.[65]

However, Denecke particularly attributed special effectiveness to music:

> It is able to promote the intended didactic unity of thinking and feeling in the learner by reinforcing the visual interpretation, for example, through musical punctuation (comma, dash, exclamation, and question mark). Music has a predominantly emotional effect, triggers associations, and can promote the acquisition of knowledge.[66]

During that same year (1986), Rolf Fuchs and his colleges at the Berlin Humboldt University were also more specific in describing the 'use of audio-visual teaching and learning aids in the methodical procedure for the formation of convictions' in the *Informationen*. They understood beliefs as 'complex mental phenomena' that included 'cognition, certainty, valuations, emotions, and attitudes, including conscious readiness to act, in mutual interpenetration and with changing proportions'.[67] In doing so, Fuchs emphasised, '[t]he fundamental beliefs that shape a person's overall behaviour most strongly are the political, ideological, and moral beliefs; consequently, audio-visual teaching and learning materials, with the evidential power available to them, should be used to a greater extent precisely with this objective in mind'.[68] Fuchs and his colleagues argued for the integration of educational audio-visual media in academic areas, such as law, pedagogy, and language education. They not only gave practical examples but also made suggestions for the qualification of university lecturers in the usage of educational media like films.

At the end of the 1980s, Steuding replaced the concept of *effect* with that of '*Bewährung*' (proof, in the sense of 'proven as useful') in his second dissertation. In his first dissertation in 1972, Steuding had addressed the psychophysical peculiarities of educational films in chemistry instruction, focusing more on such issues as the 'information quantity and

64 Denecke, Bernd, *Zur Gestaltung und Wirksamkeit von Bild und Ton dynamischer audiovisueller Lehr- und Lernmittel*, IFBT-Informationen 22 (Berlin: IFBT, 1984), p. 10.

65 Denecke, pp. 10–11.

66 Ibid., p. 16.

67 *Einführung in die Hochschulpädagogik. Teil 1: Grundlagen und Aufgaben der kommunistischen Erziehung an den Hochschulen der DDR*, ed. by Zentralinst. für Hochschulbildung (Berlin: Dt. Verl. der Wiss., 1984), p. 83 as cited in Fuchs, Rolf and others, *Zum wirksamen Einsatz audiovisueller Lehr- und Lernmittel*, IFBT-Informationen 30 (Berlin: IFBT, 1986), p. 24.

68 Fuchs and others, p. 25.

information density of films'.[69] Here, similar basic assumptions guided his research as in Ewald Topp's analysis. However, in 1988, the practice of learning effect calculations declined with the confidence in the effectiveness of educational films and was replaced by a more cautious view that carefully examined actual applications and did not take the film's usefulness for granted. Even after the dissolution of the GDR, some of these research initiatives were continued. Holger Donle's 1990 analysis of work with audio-visuals in teacher training, for instance, can be considered one of the last dissertations directly descending from the GDR research tradition on film didactics. It, however, mentions only in passing the 'high emotional impact'[70] and the resulting motivation and stimulation of learning.

In writings on film didactics and research into its educational effects, there was repeated talk of the emotional effect of film, which was seen as particularly relevant for ideological education and what was called 'attitude' and 'conviction'. In some texts on film effectiveness, emotions were seen as connected with physical reactions, and so there are also nods to Pavlovian reflex theory in these writings. However, after such experiments with cybernetic *learning machines*, the ensemble of audio-visual teaching aids again took centre stage, with television (or video recording) now being treated as the new hope to reach and influence youth through media.[71] In addition, studies on young people's media use by the *Zentralistitut für Jugendforschung der DDR* (Central Institute for Youth Research of the GDR) foregrounded the realisation that if one wanted to reach young people, one had to address them through their media and do so in a targeted manner.[72] Thus, it was concluded that only aesthetically pleasing media would be convincing.[73]

Complex and Fragment Films

For the most part, emotionality or emotional impact were only mentioned briefly in the studies on educational film; sometimes they were little more than phrases which were repeated as a matter of course to ensure that everything relevant had been said. Even the in the writings occurring slogan 'unity of the rational and the emotional' does not make matters more concrete. Furthermore, the emotional was somehow coupled with the aesthetic — after all, aesthetic design is capable of evoking emotions. Therefore, to direct this process towards generating the 'right' attitude, film didacticists were convinced that the

---

69 Steuding, Karl, 'Untersuchung zur Bestimmung der pädagogisch-methodischen Wirksamkeit von Unterrichtsfilmenim Unterrichtsprozeß des Faches Chemie der Zehnklassigen Allgemeinbildenden Polytechnischen Oberschule' (unpublished doctoral thesis, Pädagogische Hochschule 'Theodor Neubauer' Erfurt/Mühlhausen, 1972), p. 24.

70 Donle, Holger, 'Die koordinierte pädagogische Führungstätigkeit zur Ausbildung der Lehrerstudenten im Lehrgebiet "Technik der Arbeit mit audiovisuellen Unterrichtsmitteln" am der kombinierten Lehrerbildungseinrichtung unter Berücksichtigung der Erfahrungen an der Pädagogischen Hochschule Neubrandenburg' (unpublished doctoral thesis, Pädagogische Hochschule 'Theodor Neubauer' Erfurt/Mühlhausen, 1990), p. 26.

71 Zabel, Nicole, 'Die Lehrmaschine und der Programmierte Unterricht — Chancen und Grenzen im Bildungswesen der DDR in den 1960er und 1970er Jahren', *Jahrbuch für Historische Bildungsforschung*, 20 (2014), 123–52.

72 Wiedemann, Dieter, 'DDR-Jugend als Gegenstand empirischer Sozialforschung', in *Handbuch der deutschen Bildungsgeschichte, Bd. VI: 1945 bis zur Gegenwart, Zweiter Teilband: Deutsche Demokratische Republik und neue Bundesländer* (München: Beck, 1998), pp. 117–36.

73 See for example: Heun, Hans-Georg and others, *Erfahrungen beim Einsatz audiovisueller Lehr- und Lernmittel an Hochschulen'*, IFBT-Informationen 21 (Berlin: IFBT, 1983) p. 14.

teaching medium had to have a certain aesthetic quality. What was meant by aesthetic quality mostly concerned the auditory elements, from the spoken commentaries, noises, music, but also the silence and lighting moods. The use of these cinematic means was intended to enhance the effect, to illustrate statements, or to contrast them for greater impact. In summary, the didactics of audio-visual teaching aids explored questions of emotionality mostly in connection with learning motivation and the shaping of convictions, both of which were to be controlled by the conscious design of the educational films.

In the random samples of GDR educational films on technology and sciences — the entire stock has not yet been indexed — two types of educational films stand out. On the one hand, there are rather classical educational films with rigid framing focused on experimental setups and schematic illustrations, either mute or with a solemn male voice-over, and without other plot elements. On the other hand, there are rather cinematic and narrative compositions of footage with diverse origins. In this second film type, different camera framings, music, and characters are more common.

In the GDR research on educational film, attempts have also been made to distinguish between these different types of film. In particular, Fuchs and Kroll differentiated '*Fragmentfilme*' and '*Komplexfilme*'. The fragment films gave specific facts more technically and were often silent. Steuding's study of the educational film for chemistry classes suggested that fragment films were preferred by teachers as more subject-specific teaching aids in the sense of a 'working tool'.[74] In addition, habituation effects (in other words the films were dull) limited the effectiveness of the films — this was probably especially true for the 'ideological content' presented therein. As the GDR's educational films were usually no longer than 15 minutes, it was necessary to integrate them into the 45-minute lessons. In this context, however, a teaching practice was never entirely determined by the teaching material. Thus, the extent to which the ideological impetus of a teaching aid came to bear depended not least on the pedagogical work of the teacher.[75]

The 'complex film', in contrast, could provide 'a comprehensive presentation that revealed the connections',[76] for example, through the fact that references to the everyday were staged according to socialist understandings.[77] The complex films were intended for the consolidation, repetition, and linking of knowledge after the information had been delivered in class. Some scientific complex films, such as *Das Geiger-Müller-Zählrohr*[78] or *Das Wachstum der Bohne*[79] and *Lernen für die Zukunft*, exhibited a special musicality in sound and image. In the latter film, which will be discussed in more detail in the following passages, dance scenes were also interwoven that actually had nothing to do with the

74 Steuding, p. 29.

75 *Politisierung im Schulalltag der DDR: Durchsetzung und Scheitern einer Erziehungsambition*, ed. by Heinz-Elmar Tenorth and others (Weinheim: Deutscher Studien-Verlag, 1996).

76 Fuchs and Kroll, p. 46.

77 A further investigation regarding a specific visual culture would be interesting here. See in this regard: *Ästhetiken des Sozialismus. Populäre Bildmedien im späten Sozialismus*, ed. by Alexandra Köhring and Monica Rüthers (Wien, Köln, Weimar: Böhlau, 2018).

78 Discussed in: Klinger, Kerrin, 'Functional Bodies On Scientific Educational Film During the Cold War', *Medienimpulse*, 54.4 (2016), 1–18.

79 Discussed in: Degler, Wiebke and others, 'Staging nature in twentieth-century teacher education in classrooms', *Paedagogica Historica*, 56 (2020), 121–49.

Fig. 10.1 (left). Students in class; Fig. 10.2 (middle). Young adults dancing; Fig. 10.3 (right). At the production site. Film stills from *Lernen für die Zukunft* (GDR, 1980).

subject matter. However, even the fragmentary film that would concentrate only 'on the presentation of a specific learning step or a concise sequence of learning steps'[80] would be integrated into the respective socialist structures.

Film Analysis: Microelectronics and the Future of the Youth

The film *Lernen für die Zukunft* ('Learning for the Future') was produced in 1980 at the DEFA-Studio for Documentary Film on behalf of the Academy of Pedagogical Sciences. The 11:30 min long film was produced for the subject of electronics/*BMSR* technology, which was taught at vocational schools in the GDR. BMSR technology is the acronym for *Betriebsmess-, Steuerungs- und Regelungstechnik* (operational measurement, monitoring and control technology). If one were to ask whether this film is supposed to impart knowledge, however, it becomes more difficult to say, as a fair amount of prior knowledge is assumed. In other words, this film is clearly not an explanatory film. The thematic range — from the work of technologists to cargo transport, automatic fabrication, energy transfer, mining, and young people in school and in their spare time — alone makes the film quite complex. Appropriately, the film could thus be addressed as a 'complex film'.

Settings and Protagonists

The film has three major settings.: First, there are training situations in classrooms and workbenches (see Figure 10.1). The protagonists here are young adults, with the middle-aged teacher only seen once in the background. Secondly, leisure situations, including a youth club, sports facilities, a park, and an unspecified landscape, are shown (see Figure 10.2). Again, young adults are the protagonists. Thirdly, the film features industrial production sites, such as cargo stations, opencast lignite mines, and electricity substations or factories (see Figure 10.3). Here, the actors are young to middle-aged adults, indicating a biographical progression through working life. Portraying subjects as youthful in films, according to Masson, is a rhetorical means of persuasion — because it is demonstrated in this way that

80 Fuchs and Kroll, p. 46.

Fig. 10.4 (left). Teacher and students in class; Fig. 10.5 (middle). Young adults at leisure; Fig. 10.6 (right). Labourer programming a machine. Film stills from "Lernen für die Zukunft" (GDR, 1980).

the film and the knowledge it conveys are relevant to the youth target audience.[81] Hence, in the film *Lernen für die Zukunft*, vocational training is presented as desirable.

Visually, the motif structure alternates between people that are controlling and machines that are working. The film thus works with what Masson has called 'recognizable structural patterns' of certain motives and sounds.[82] In addition, scenes of young people dancing in different locations are cyclically interspersed. Furthermore, scenes from a classroom recur, using three different camera views: the class viewed from the teacher's desk, a student at the blackboard, and a shot of the back row of benches, where the teacher is seen standing behind two female students (see Figure 10.4). The groups of people depicted are identifiable by their clothing and surroundings: the teacher in a white coat and the students in casual clothing in the classroom, young adults in casual clothing at a club or outdoors (see Figure 10.5), vocational trainees at workbenches, and workers in smock aprons and overalls in industrial plants. Workers are generally depicted individually, implying that every single person counts (see Figure 10.6). According to Anton Makarenko, a Soviet pedagogue who also left his mark on education in the GDR, individuals are united in work with the common goal of improving life for one and all. Accordingly, the film depicts workers as clearly identifiable individuals, thereby referring to collective work as a higher principle.

Leisure time, on the other hand, takes place in pairs or groups. Dancing, though presented as a group activity, is also interspersed with close-ups of individuals (usually young women), thus singling them out. In general, these young women are shown relatively often in the film, smiling as they dance but also working hard with serious expressions on their faces. Overall, though, the gender ratio seems balanced, represented by the roughly equal frequency of appearances, while the average age corresponded to the target group — teenage vocational students. In the eastern part of Germany, gender equality as a state policy focused primarily on women's employment.[83]

The characters' clothes, too, seem carefully chosen and to be of high quality — there are never any dirty work clothes to be seen. Furthermore, the characters are all neatly or even fashionably coiffed, lending an overall believable and authentic appearance to them. Notably, they do not look at the camera, though the hand movements do seem staged. This can be explained, however, with reference to the rather complicated shots in terms

81 Masson, p. 146.
82 Ibid., p. 164.
83 See Trappe, Heike, *Emanzipation oder Zwang? Frauen in der DDR zwischen Familie, Beruf und Sozialpolitik* (Berlin: Akademie Verlag, 1995).

Fig. 10.7 (left). In the control centre; Fig. 10.8 (middle). Automated production; Fig. 10.9. (right). Cargo transport wagon. Film stills from *Lernen für die Zukunft* (GDR, 1980).

of the angles of movement and camera work. Masson has suggested that zoology films were supposed to evoke an "impression of procedural logic and transparency — a goal which often seems to justify a certain amount of staging".[84] This can be said to apply to the documentary-like film sequences of industrial work in this particular GDR film as well. Here, the film strives to present the workflows as natural, which not only refers to the documentation of real-life but also states an internal logic between the control and implementation of work actions.

Industrial Complexes and Attention

The industrial complexes on display, wherein the computer-controlled machines are used, are mainly for vehicle construction. In addition, several technical facilities are presented for cargo transport, power line grids, the chemical industry, opencast mining, and shipbuilding. Here, the montage of images alternates between small, neat control centres and large automatically operating machines. There are detailed shots of instructions on paper being matched with the displays of the control panels (see Fig. 10.7), and, sometimes, there is a view out of the window to the railroad tracks or the mining area. Here, the viewer's gaze is meant to follow the effect of pressing a button. The exterior shots all show recognisable landscapes. Parallel to Masson's descriptions, this educational film, then, is also a film 'that depict[s] industrial production processes'[85] in the style of 'what happens behind the scenes'.[86] In this film, too, viewers are audio-visually taken to places that are not open to the public. They look over the shoulders of the experts and are able to see their view on switchboards or electronic devises. Masson names this the strategy of 'Providing Access', one of several 'that focus on the matter's unfamiliarity: the fact that the subject dealt with, or certain aspects thereof, are fundamentally unknown to the audience addressed'.[87] Although it can be assumed that the vocational students already knew individual industrial settings (see Fig. 10.8), the filmmakers, according to Masson, hoped for 'a certain measure of admiration: an appreciation of the intricacy or rigid organization of the phenomena or procedures shown'.[88]

84 Masson, p. 164.
85 Ibid., p. 153.
86 Ibid.
87 Ibid., p. 149.
88 Ibid., p. 154.

As for specific cinematic techniques, this educational film also uses dramatic shots, such as a wagon (see Fig. 10.9) or a machine's arm moving head-on towards the viewer. With Masson, this is a means of holding the viewer's attention. Many views from different angles and distances are combined in the film; there is a constant change between shots in detail, long shots, and those shot in the middle distance. As pictorial motifs, working hands and the faces of relaxed dancers or concentrated workers and learners recur — so that the audience feels addressed and can identify with those portrayed. The film has many cuts, creating an elaborate montage of images that play with movement patterns and the repetition of motifs. In this way, the educational film also has a rhythmic and almost musical effect on the image level.

Voiceover

If the sound is included in the analysis, this rhythmic impression is further enhanced. Music and original sounds accompanying the commentary are interlaced with the images in a dynamic way. Sometimes, the commentary — provided by a middle-aged, male, and presumably trained speaker — runs across image-sound breaks, or it ends or begins in the middle of an image sequence. Technical terms in the commentary (the characters have no spoken lines) are also used frequently throughout, making the language appear somewhat artificial and not at all like everyday language. This effect is exacerbated by unusual sentence order, nominal style, and the omission of verbs:

> Modern technology, BMSR technology, data processing, electronics, microelectronics, industrial robots. They will determine the face of our operations. They will be inseparable from the future. In this development, we are intentionally relying on you, on young people who are not afraid to take on the risk of the unknown.

As the film was made for vocational students, the producers could assume that the viewers knew the technical devices shown and were already familiar with the terms used in the commentary. Thus, terms like 'BMSR-Technik' are not explained.

Overall, language is used as a political instrument, and *ideologems*[89] pervade the film, such as 'economic superiority', '(mechanised) future', 'control instead of routine and dangerous, physical work', and not least the slogan, 'We need you!' Who is meant by 'you' is made unquestionably clear: the dancing young people and thus the audience that should be addressed. But who is the 'we' of the authorial commentary? It can be assumed that it was older people, the functionaries, but also working people in general; in other words, it was the state community. The film seeks to promote a vision of a technological future that not only frees humankind from physical labour but also requires commitment and dedication in the name of socialism. This distinctive type of rhetoric has been described by

---

89 Demke, Elena, 'Indoktrination als Code in der SED-Diktatur', in *Indoktrination und Erziehung. Aspekte der Rückseite der Erziehung*, ed. by Henning Schluß (Wiesbaden: VS Verlag für Sozialwissenschaften, 2007), pp. 35–44 (p. 37).

Sheila Jasanoff in her article 'Future Imperfect: Science, Technology, and the Imaginations of Modernity',[90] in which she defines

> sociotechnical imaginaries [...] as collectively held, institutionally stabilized, and publicly performed visions of desirable futures, animated by shared understandings of forms of social life and social order attainable through, and supportive of, advances in science and technology.[91]

Pattern

Ideologems, such as those found in the commentary, also characterise the pictorial level. The motif of the well-groomed worker is pervasive, as is the skilled worker in a coat. The film shows attractive though still relatable young adults, who could also have been depicted in a GDR fashion magazine, with fashionable hairstyles, wristwatches, and freshly ironed clothes. Exceptions, here, are the older teacher and the older railroad worker — they could be representatives of the authority of experienced elders. The faces of the younger ones are sometimes seen close-up, centring attention upon these individuals (see Figure 10.10 and 10.11). However, even in such shots, the emphasis is on the community rather than the individual, focusing on people in their role as workers, for example, who are relevant to society. In this educational film, too, we find the motif of 'creating hands',[92] one that Verena Niethammer has characterised as a visual stereotype for the national socialist educational film (see Figure 10.12). The 'creating hands' symbolise good German craftsmanship and educational values like diligence and efficiency. However, I would prefer to speak of patterns in the sense of cultural patterns since the term 'stereotype' already imports critical connotations.

This film also plays on the spectacularism of automated manufacturing with robots moving on command. The film is meant to trigger the viewer's ambitions; it is meant to evoke 'a desire — and ideally, a determination — to perform the actions shown as well as they are done on screen',[93] speaking with Masson. Rather, it is the kind of film that large organisations use today, for example, to legitimise their actions or acquire investors or applicants. The future of everyday professional life — mostly familiar and possibly unattractive — was thus to be given a modern *image* that combined being a 'key worker' (and, therefore, relevant to the system) with fun.

In this film, well-dressed working people — some in neat work clothes — stand out. One could define them as *Facharbeiter*, a type of specialist worker. This 'skilled worker' is not only a central figure in the GDR ideology but was 'the dominant occupational group' in the 1970s and thus the 'core of the workers' society of the GDR'.[94] For comparison: In

---

90 Jasanoff, Sheila, 'Future Imperfect: Science, Technology, and the Imaginations of Modernity', in *Dreamscapes of Modernity. Sociotechnical Imaginaries and the Fabrication of Power*, ed. by Sheila Jasanoff and Sang-Hyun Kim (Chicago/London: The University of Chicago Press, 2015), pp. 1–33.

91 Jasanoff, p. 4.

92 Niethammer, p. 45.

93 Masson, p. 169.

94 Kreutzer, Florian, *Die gesellschaftliche Konstitution des Berufs. Zur Divergenz von formaler und reflexiver Modernisierung in der DDR Kreutzer* (Frankfurt am Main: Campus-Verlag, 2001), p. 149.

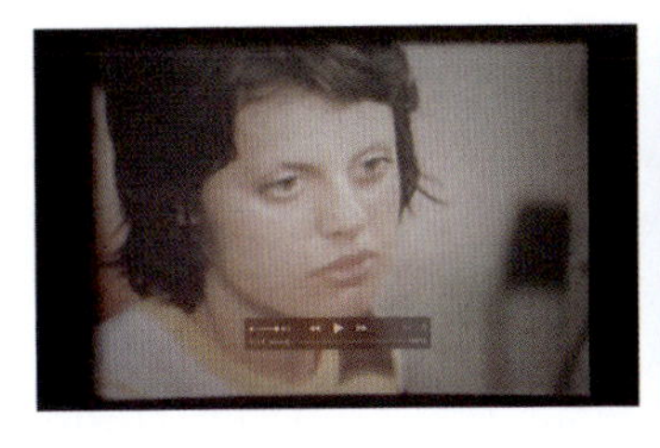

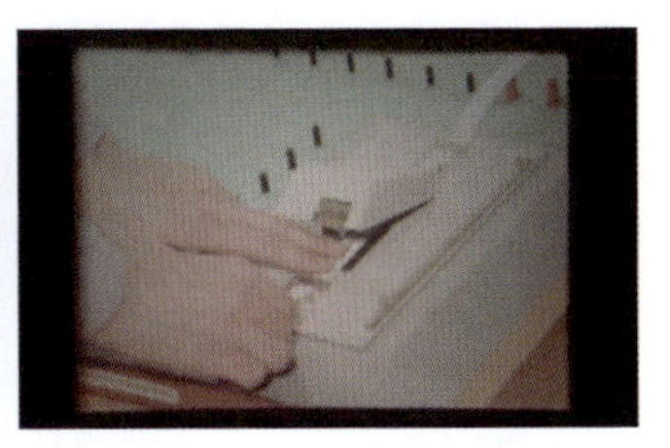

Fig. 10.10 (left). Young woman dancing; Fig. 10.11 (middle). Young woman soldering; Fig. 10.12 (right). Hands at punched paper tape (Film stills from *Lernen für die Zukunft* (GDR, 1980).

1945, twenty-one per cent of the working population were qualified workers and master craftsmen, and seventy-six per cent were unskilled workers and farmers. In 1985, sixty-four per cent of the working population consisted of formally qualified workers and master craftsmen, and only fifteen per cent were semi-skilled and unskilled workers.[95] Accordingly, in the mid-1980s, about two-thirds of the working population (both male and female) had a skilled worker or master craftsman degree. It is noteworthy, however, that only about two per cent of the working population was trainees.[96] It thus follows that the GDR economy was dependent, especially in the rapidly developing technological sector, on an increasing number of vocational students entering the labour market.[97]

Based on the sequence analysis of this film, the GDR youth should learn for professional positions in control centres. The monotony of such supervisory activities is evident to today's viewers, even when captured in dynamic shots. But why are young women brought into focus especially? These visualisations could be understood as an attempt to attract women, in particular, to this form of vocational training or work and to participate in the workforce, in general. However, it could also be a classic advertising strategy to make the job more attractive by presenting beautiful women to entice the male gaze.

Emotional charging

The way in which setting 1, the classroom, and setting 3, the production sites, are connected is given in the title and the commentary: current training prepares for future occupation. But, what role does the second setting of the youth club play? The music that goes with the dancing adds to the commentary and the images of the other settings. Here, an atmospheric association is forged. This can be seen as a link to the state's efforts to emotionally charge ideological content and to use the medium of film specifically for this purpose. The cyclical principle in the music and image montage connects all of the subject areas with each other. The educational film, analysed here, emphasised enjoyment rather than information transfer — the music and the rhythmisation of the

95 Kreutzer, p. 149.

96 Dietrich, Rainer, *Das System beruflicher Erwachsenenbildung in derehemaligen DDR mit Ausblick auf künftige Strukturprobleme in den neuen Bundesländern* (Stuttgart: Kohlhammer, 1991), p. 433.

97 For the development of microelectronics in the GDR specifically, see: Salomon, Peter, *Die Geschichte der Mikroelektronik-Halbleiterindustrie der DDR* (Dessau: Funk Verl. Hein, 2003).

image sequences generate 'a visual pleasure'[98] and 'a cinematic attraction'[99] and equated club and workplace in an audio-visual spectacularisation — suggesting that technology is just as modern and enjoyable as music since they follow the same rhythm. In the sense of Masson, sequences with dance and leisure activities were mounted to appeal to the viewers.[100] Music is persuasively associated with a non-professional context. Thus, the joy of leisure time with like-minded people was filmic and brought into the workplace. Conversely, work was seen as a prerequisite for recreational pleasure, making private life relevant not only collectively but also for society at large.

Conclusion

In the example studied here, motivational strategies are observable, such as the use of rhythmic music, a dynamic montage, close-ups of the target group's peers, spectacular camera angles and scenery, and the appeal to a socialist vision of the future. Together, this creates an aesthetic presence that virtually places the viewer into the action. In this way, the guidelines formulated by pedagogues for an emotionally effective educational film are nearly perfectly implemented. It is a complex film that conveys a general mood rather than concrete technical knowledge. Visually and auditorily, a specific image of how the theme is to be perceived is thus created. The viewers are supposed to identify with technological progress, to understand it as a core part of their identities. As my example shows, the educational film can employ an auditory and visual rhetoric that aims to persuade its viewers. The message of the commentary is dynamic and embedded emotionally through images and sounds. Then again, the fact that Masson's analytical framework was able to provide insight here also shows that educational films, at least in the Western European context, employed a similar cinematic rhetoric.

This film example is certainly an extreme one. The majority of GDR educational films have been more 'classical', which means that the entertaining and more spectacular cinematic elements were used more sparingly. Nevertheless, especially in the final decades of the GDR, greater emphasis was placed on a carefully arranged picture-sound ratio in educational films. The educational system of the GDR sought instruments to permanently implement not only educational knowledge but also politically approved convictions and attitudes in students. This not only applied to school subjects like German language and literature, history, and geography (as Kneile-Klenk has proven for the educational film in history) but also to the sciences as well. This can be seen, for instance, in the fact that GDR pedagogues referred to the indivisible dialectic of *Wissenschaftlichkeit und Parteilichkeit* (scientificity and partisanship) in teaching. Indeed, pedagogical-psychological research was conducted on the effect of educational films to implement this aim as efficiently as possible. In essence, it was recognised that the aesthetic quality of the film's realisation has an influence on the learners' reception. Guidelines for the production of educational films were formulated on the basis of this research, which the film production, in turn,

98 Ibid., p. 156.
99 Ibid.
100 Ibid., p. 145.

attempted to implement. For the GDR educational films, there are indications that specific cinematic patterns were used more consciously with the increasing theorisation of the educational film production. To support the intended effect, there are noticeable references to practice, for example, through the portrayal of labourers and production sites, but also the use of original sounds and of specially composed film music. This also closes the circle to Imai's studies. He identified a specifically politically connoted visual language in German educational films from the National Socialist era, which draws attention to the fact that the intention of indoctrination was not implemented on a verbal level alone, but on the visual level as well. The educational films of the GDR also reveal that a specific political view of the world can be conveyed even without the use of linguistic ideologems. Moreover, it becomes apparent that emotionalising effects were not only used in medical or health films (as demonstrated by Laukötter and Schwartz) but also in educational films on science. However, it should be noted that most likely, these aspects also apply to the educational film production in the Soviet Union and the former Eastern Bloc as well, as there was professional exchange and exchange of film material.

Finally, I would like to suggest a more complex view on the pairing of science and educational films because science and research were relevant to films in educational contexts on several levels. The following questions can be derived from the present study as impulses for further reflections: How was scientific content represented in educational films for schools and universities? How did educational films contribute as teaching aids to the pedagogical-didactic appropriateness of teaching? How was the research on the use and effect of educational film conducted, and with what consequences? What do educational films as cultural media say about the socio-political embedding of science teaching?

TECHNE

Media Performance Histories

This series focuses on the intersections between media developments and performative culture since the early nineteenth century. The modern era witnessed a proliferation of media performances and exhibitions, encouraged by the burgeoning rise of science and technology, and supported by changes in transportation, communication, education, and social mobility. These popular events were part of nascent culture industries that took root in learning environments and lecture halls but also in theatre and opera houses, spilling out into public space, the boulevards, and the fairgrounds. Academics and science enthusiasts but also illusionists, artists, and amateur savants, all shared a knack for understanding what would entice different audiences, coupled with a delicate balance between scientific demonstration and sensational entertainment. While relying on international networks, media performances contributed to the circulation of knowledge, technologies, and visual culture between European cities and across the Atlantic.

Media Performance Histories explores the ways in which cultural change, new forms of knowledge, science, and technology were turned into modern spectacles that addressed different audiences and produced different modes of reception. It provides readers with a unique guide to how transnational performance created a culturally shared repertoire of signs and shaped modern Western culture. The books in this series offer accounts that cut across disciplinary and geographical boundaries, while being sensitive to how specific historical contexts and institutional circumstances constituted media and performance cultures. By also considering the interplay between present-day media performances and the archaeological traces that they carry, the series moreover aims to unearth previously overlooked but resurgent prehistories of so-called "new" media.

The series is situated at the intersection of performance studies, media studies, and the history of science. It welcomes edited collections and monographs on issues including (but not limited to) the interaction between media (archaeology) and performance; the role of theatre and performance in the circulation of knowledge; the way (early) media and technologies are staged; the agency of human observers as part of intermedial interactions or as part of viewing strategies.

Edited by Frank Kessler, Sabine Lenk, Kurt Vanhoutte, and Nele Wynants